ENZYMES

SECOND EDITION

ENZYMES
A Practical Introduction to Structure, Mechanism, and Data Analysis

SECOND EDITION

Robert A. Copeland

WILEY-VCH

A JOHN WILEY & SONS, INC., PUBLICATION

New York / Chichester / Weinheim / Brisbane / Singapore / Toronto

For ordering and customer service, call 1-800-CALL-WILEY.

Library of Congress Cataloging-in-Publication Data:

Copeland, Robert Allen.
   Enzymes: a practical introduction to structure, mechanism, and data analysis / Robert A. Copeland.--2nd ed.
      p. cm.
   "Published simultaneously in Canada."
   Includes bibliographical references and index.
   ISBN 0-471-35929-7 (acid-free paper)
   1. Enzymes. 2. Enzymology. I. Title.
   QP601.C753 2000
   572.7--dc21   99-050087

Printed in the United States of America

10 9 8 7 6 5 4 3 2 1

10019924446

To Clyde Worthen
for teaching me all the important lessons:
arigato sensei.

And to Theodore (Doc) Janner
for stoking the fire.

CONTENTS

PREFACE

In the four years since the first edition of *Enzymes* was published, I have been delighted to learn of the wide acceptance of the book throughout the biochemical community, and particularly in the pharmaceutical community. During this time a number of colleagues have contacted me to express their views on the value of the text, and importantly to make suggestions for improvements to the content and presentation of some concepts. I have used the first edition as a teaching supplement for a course in which I lecture at the University of Pennsylvania School of Medicine. From my lecture experiences and from conversations with students, I have developed some new ideas for how to better explain some of the concepts in the text and have identified areas that deserve expanded coverage. Finally, while the first edition has become popular with students and industrial scientists, some of my academic colleagues have suggested a need for a more in-depth treatment of chemical mechanisms in enzymology.

In this second edition I have refined and expanded the coverage of many of the concepts in the text. To help the reader better understand some of the interactions between enzymes and their substrates and inhibitors, a new chapter on protein–ligand binding equilibria has been added (Chapter 4). The chapters on chemical mechanisms in enzyme catalysis (Chapter 6) and on experimental measures of enzyme activity (Chapter 7) have been expanded significantly. The discussions of enzyme inhibitors and multiple substrate reactions (Chapters 8 through 11) have been refined, and in some cases alternative treatments have been presented. In all of this, however, I have tried to maintain the introductory nature of the book. There are many excellent advanced texts on catalysis, enzyme mechanisms, and enzyme kinetics, but the level at which these are generally written is often intimidating to the beginner. Hence, as stated in the preface to the first edition, this book is intended to serve as a mechanism for those new to the field of enzymology to develop a reasonable understanding of the science and experimental methods, allowing them to competently begin laboratory studies with enzymes. I have continued to rely on extensive citations to more advanced texts and primary literature as a means for the interested reader to go beyond the treatments offered here and delve more deeply into specific areas of enzymology.

In developing this second edition I have had fruitful conversations and advice from a number of colleagues. In particular, I wish to thank Andy Stern, Ross Stein, Trevor Penning, Bill Pitts, John Blanchard, Dennis Murphy, and the members of the Chemical Enzymology Department at the DuPont Pharmaceuticals Company. As always, the love and support of my family has been most important in making this work possible.

Robert A. Copeland
Wilmington, Delaware

ACKNOWLEDGMENTS

It is a great pleasure for me to thank the many friends and coworkers who have helped me in the preparation of this work. Many of the original lecture notes from which this text has developed were generated while I was teaching a course on biochemistry for first-year medical students at the University of Chicago, along with the late Howard S. Tager. Howard contributed greatly to my development as a teacher and writer. His untimely death was a great loss to many of us in the biomedical community; I dearly miss his guidance and friendship.

As described in the Preface, the notes on which this text is based were significantly expanded and reorganized to develop a course of enzymology for employees and students at the DuPont Merck Pharmaceutical Company. I am grateful for the many discussions with students during this course, which helped to refine the final presentation. I especially thank Diana Blessington for the original suggestion of a course of this nature. That a graduate-level course of this type could be presented within the structure of a for-profit pharmaceutical company speaks volumes for the insight and progressiveness of the management of DuPont Merck. I particularly thank James M. Trzaskos, Robert C. Newton, Ronald L. Magolda, and Pieter B. Timmermans for not only tolerating, but embracing this endeavor.

Many colleagues and coworkers contributed suggestions and artwork for this text. I thank June Davis, Petra Marchand, Diane Lombardo, Robert Lombardo, John Giannaras, Jean Williams, Randi Dowling, Drew Van Dyk, Rob Bruckner, Bill Pitts, Carl Decicco, Pieter Stouten, Jim Meek, Bill De-Grado, Steve Betz, Hank George, Jim Wells, and Charles Craik for their contributions.

Finally, and most importantly, I wish to thank my wife, Nancy, and our children, Lindsey and Amanda, for their constant love, support, and encouragement, without which this work could not have been completed.

PREFACE
TO THE FIRST EDITION

The latter half of this century has seen an unprecedented expansion in our knowledge and use of enzymes in a broad range of basic research and industrial applications. Enzymes are the catalytic cornerstones of metabolism, and as such are the focus of intense research within the biomedical community. Indeed enzymes remain the most common targets for therapeutic intervention within the pharmaceutical industry. Since ancient times enzymes also have played central roles in many manufacturing processes, such as in the production of wine, cheese, and breads. During the 1970s and 1980s much of the focus of the biochemical community shifted to the cloning and expression of proteins through the methods of molecular biology. Recently, some attention has shifted back to physicochemical characterization of these proteins, and their interactions with other macromolecules and small molecular weight ligands (e.g., substrates, activators, and inhibitors). Hence, there has been a resurgence of interest in the study of enzyme structures, kinetics, and mechanisms of catalysis.

The availability of up-to-date, introductory-level textbooks, however, has not kept up with the growing demand. I first became aware of this void while teaching introductory courses at the medical and graduate student level at the University of Chicago. I found that there were a number of excellent advanced texts that covered different aspects of enzymology with heavy emphasis on the theoretical basis for much of the science. The more introductory texts that I found were often quite dated and did not offer the blend of theoretical and practical information that I felt was most appropriate for a broad audience of students. I thus developed my own set of lecture notes for these courses, drawing material from a wide range of textbooks and primary literature.

In 1993, I left Chicago to focus my research on the utilization of basic enzymology and protein science for the development of therapeutic agents to combat human diseases. To pursue this goal I joined the scientific staff of the DuPont Merck Pharmaceutical Company. During my first year with this company, a group of associate scientists expressed to me their frustration at being unable to find a textbook on enzymology that met their needs for guidance in laboratory protocols and data analysis at an appropriate level and at the same time provide them with some relevant background on the scientific basis of their experiments. These dedicated individuals asked if I would prepare and present a course on enzymology at this introductory level.

Using my lecture notes from Chicago as a foundation, I prepared an extensive set of notes and intended to present a year-long course to a small group of associate scientists in an informal, over-brown-bag-lunch fashion. After the lectures had been announced, however, I was shocked and delighted to find that more than 200 people were registered for this course! The makeup of the student body ranged from individuals with associate degrees in medical technology to chemists and molecular biologists who had doctorates. This convinced me that there was indeed a growing interest and need for a new introductory enzymology text that would attempt to balance the theoretical and practical aspects of enzymology in such a way as to fill the needs of graduate and medical students, as well as research scientists and technicians who are actively involved in enzyme studies.

The text that follows is based on the lecture notes for the enzymology course just described. It attempts to fill the practical needs I have articulated, while also giving a reasonable introduction to the theoretical basis for the laboratory methods and data analyses that are covered. I hope that this text will be of use to a broad range of scientists interested in enzymes. The material covered should be of direct use to those actively involved in enzyme research in academic, industrial, and government laboratories. It also should be useful as a primary text for senior undergraduate or first-year graduate course, in introductory enzymology. However, in teaching a subject as broad and dynamic as enzymology, I have never found a single text that would cover all of my students' needs; I doubt that the present text will be an exception. Thus, while I believe this text can serve as a useful foundation, I encourage faculty and students to supplement the material with additional readings from the literature cited at the end of each chapter, and the primary literature that is continuously expanding our view of enzymes and catalysis.

In attempting to provide a balanced introduction to enzymes in a single, readable volume I have had to present some of the material in a rather cursory fashion; it is simply not possible, in a text of this format, to be comprehensive in such an expansive field as enzymology. I hope that the literature citations will at least pave the way for readers who wish to delve more deeply into particular areas. Overall, the intent of this book is to get people *started* in the laboratory and in their thinking about enzymes. It provides sufficient experimental and data handling methodologies to permit one to begin to design and perform experiments with enzymes, while at the same time providing a theoretical framework in which to understand the basis of the experimental work. Beyond this, if the book functions as a stepping-stone for the reader to move on to more comprehensive and in-depth treatments of enzymology, it will have served its purpose.

ROBERT A. COPELAND
Wilmington, Delaware

"All the mathematics in the world is no substitute for a reasonable amount of common sense."

W. W. Cleland

1

A BRIEF HISTORY
OF ENZYMOLOGY

Life depends on a well-orchestrated series of chemical reactions. Many of these reactions, however, proceed too slowly on their own to sustain life. Hence nature has designed catalysts, which we now refer to as *enzymes*, to greatly accelerate the rates of these chemical reactions. The catalytic power of enzymes facilitates life processes in essentially all life-forms from viruses to man. Many enzymes retain their catalytic potential after extraction from the living organism, and it did not take long for mankind to recognize and exploit the catalytic power of enzyme for commercial purposes. In fact, the earliest known references to enzymes are from ancient texts dealing with the manufacture of cheeses, breads, and alcoholic beverages, and for the tenderizing of meats. Today enzymes continue to play key roles in many food and beverage manufacturing processes and are ingredients in numerous consumer products, such as laundry detergents (which dissolve protein-based stains with the help of proteolytic enzymes). Enzymes are also of fundamental interest in the health sciences, since many disease processes can be linked to the aberrant activities of one or a few enzymes. Hence, much of modern pharmaceutical research is based on the search for potent and specific inhibitors of these enzymes. The study of enzymes and the action of enzymes has thus fascinated scientists since the dawn of history, not only to satisfy erudite interest but also because of the utility of such knowledge for many practical needs of society. This brief chapter sets the stage for our studies of these remarkable catalysts by providing a historic background of the development of enzymology as a science. We shall see that while enzymes are today the focus of basic academic research, much of the early history of enzymology is linked to the practical application of enzyme activity in industry.

1.1 ENZYMES IN ANTIQUITY

The oldest known reference to the commercial use of enzymes comes from a description of wine making in the Codex of Hammurabi (ancient Babylon, circa 2100 B.C.). The use of microorganisms as enzyme sources for fermentation was widespread among ancient people. References to these processes can be found in writings not only from Babylon but also from the early civilizations of Rome, Greece, Egypt, China, India. Ancient texts also contain a number of references to the related process of vinegar production, which is based on the enzymatic conversion of alcohol to acetic acid. Vinegar, it appears, was a common staple of ancient life, being used not only for food storage and preparation but also for medicinal purposes.

Dairy products were another important food source in ancient societies. Because in those days fresh milk could not be stored for any reasonable length of time, the conversion of milk to cheese became a vital part of food production, making it possible for the farmer to bring his product to distant markets in an acceptable form. Cheese is prepared by curdling milk via the action of any of a number of enzymes. The substances most commonly used for this purpose in ancient times were ficin, obtained as an extract from fig trees, and rennin, as rennet, an extract of the lining of the fourth stomach of a multiple-stomach animal, such as a cow. A reference to the enzymatic activity of ficin can, in fact, be found in Homer's classic, the Iliad:

> As the juice of the fig tree curdles milk, and thickens it in a moment though it be liquid, even so instantly did Paeëon cure fierce Mars.

The philosopher Aristotle likewise wrote several times about the process of milk curdling and offered the following hypothesis for the action of rennet:

> Rennet is a sort of milk; it is formed in the stomach of young animals while still being suckled. Rennet is thus milk which contains fire, which comes from the heat of the animal while the milk is undergoing concoction.

Another food staple throughout the ages is bread. The leavening of bread by yeast, which results from the enzymatic production of carbon dioxide, was well known and widely used in ancient times. The importance of this process to ancient society can hardly be overstated.

Meat tenderizing is another enzyme-based process that has been used since antiquity. Inhabitants of many Pacific islands have known for centuries that the juice of the papaya fruit will soften even the toughest meats. The active enzyme in this plant extract is a protease known as papain, which is used even today in commercial meat tenderizers. When the British Navy began exploring the Pacific islands in the 1700s, they encountered the use of the papaya fruit as a meat tenderizer and as a treatment for ringworm. Reports of these native uses of the papaya sparked a great deal of interest in eighteenth-century

Europe, and may, in part, have led to some of the more systematic studies of digestive enzymes that ensued soon after.

1.2 EARLY ENZYMOLOGY

While the ancients made much practical use of enzymatic activity, these early applications were based purely on empirical observations and folklore, rather than any systematic studies or appreciation for the chemical basis of the processes being utilized. In the eighteenth and nineteenth centuries scientists began to study the actions of enzymes in a more systematic fashion. The process of digestion seems to have been a popular subject of investigation during the years of the enlightenment. Wondering how predatory birds manage to digest meat without a gizzard, the famous French scientist Réaumur (1683–1757) performed some of the earliest studies on the digestion of buzzards. Réaumur designed a metal tube with a wire mesh at one end that would hold a small piece of meat immobilized, to protect it from the physical action of the stomach tissue. He found that when a tube containing meat was inserted into the stomach of a buzzard, the meat was digested within 24 hours. Thus he concluded that digestion must be a chemical rather than a merely physical process, since the meat in the tube had been digested by contact with the gastric juices (or, as he referred to them, "a solvent"). He tried the same experiment with a piece of bone and with a piece of a plant. He found that while meat was digested, and the bone was greatly softened by the action of the gastric juices, the plant material was impervious to the "solvent"; this was probably the first experimental demonstration of enzyme specificity.

Réaumur's work was expanded by Spallanzani (1729–1799), who showed that the digestion of meat encased in a metal tube took place in the stomachs of a wide variety of animals, including humans. Using his own gastric juices, Spallanzani was able to perform digestion experiments on pieces of meat in vitro (in the laboratory). These experiments illustrated some critical features of the active ingredient of gastric juices: by means of a control experiment in which meat treated with an equal volume of water did not undergo digestion Spallanzani demonstrated the presence of a specific active ingredient in gastric juices. He also showed that the process of digestion is temperature dependent, and that the time required for digestion is related to the amount of gastric juices applied to the meat. Finally, he demonstrated that the active ingredient in gastric juices is unstable outside the body; that is, its ability to digest meat wanes with storage time.

Today we recognize all the foregoing properties as common features of enzymatic reactions, but in Spallanzani's day these were novel and exciting findings. The same time period saw the discovery of enzyme activities in a large number of other biological systems. For example, a peroxidase from the horseradish was described, and the action of α-amylase in grain was observed. These early observations all pertained to materials — crude extract from plants or animals — that contained enzymatic activity.

During the latter part of the nineteenth century scientists began to attempt fractionations of these extracts to obtain the active ingredients in pure form. For example, in 1897 Bertrand partially purified the enzyme laccase from tree sap, and Buchner, using the "pressed juice" from rehydrated dried yeast, demonstrated that alcoholic fermentation could be performed in the absence of living yeast cells. Buchner's report contained the interesting observation that the activity of the pressed juice diminished within 5 days of storage at ice temperatures. However, if the juice was supplemented with cane sugar, the activity remained intact for up to 2 weeks in the ice box. This is probably the first report of a now well-known phenomenon — the stabilization of enzymes by substrate. It was also during this period that Kühne, studying catalysis in yeast extracts, first coined the term "enzyme" (the word derives from the medieval Greek word *enzymos*, which relates to the process of leavening bread).

1.3 THE DEVELOPMENT OF MECHANISTIC ENZYMOLOGY

As enzymes became available in pure, or partially pure forms, scientists' attention turned to obtaining a better understanding of the details of the reaction mechanisms catalyzed by enzymes. The concept that enzymes form complexes with their substrate molecules was first articulated in the late nineteenth century. It is during this time period that Emil Fischer proposed the "lock and key" model for the stereochemical relationship between enzymes and their substrates; this model emerged as a result of a large body of experimental data on the stereospecificity of enzyme reactions. In the early twentieth century, experimental evidence for the formation of an enzyme–substrate complex as a reaction intermediate was reported. One of the earliest of these studies, reported by Brown in 1902, focused on the velocity of enzyme-catalyzed reactions. Brown made the insightful observation that unlike simple diffusion-limited chemical reactions, in enzyme-catalyzed reactions "it is quite conceivable...that the time elapsing during molecular union and transformation may be sufficiently prolonged to influence the general course of the action." Brown then went on to summarize the available data that supported the concept of formation of an enzyme–substrate complex:

> There is reason to believe that during inversion of cane sugar by invertase the sugar combines with the enzyme previous to inversion. C. O'Sullivan and Tompson...have shown that the activity of invertase in the presence of cane sugar survives a temperature which completely destroys it if cane sugar is not present, and regard this as indicating the existence of a combination of the enzyme and sugar molecules. Wurtz [1880] has shown that papain appears to form an insoluble compound with fibrin previous to hydrolysis. Moreover, the more recent conception of E. Fischer with regard to enzyme configuration and action, also implies some form of combination of enzyme and reacting substrate.

Observations like these set the stage for the derivation of enzyme rate equations, by mathematically modeling enzyme kinetics with the explicit involvement of an intermediate enzyme–substrate complex. In 1903 Victor Henri published the first successful mathematical model for describing enzyme kinetics. In 1913, in a much more widely read paper, Michaelis and Menten expanded on the earlier work of Henri and rederived the enzyme rate equation that today bears their names. The Michaelis–Menten equation, or more correctly the Henri–Michaelis–Menten equation, is a cornerstone of much of the modern analysis of enzyme reaction mechanisms.

The question of how enzymes accelerate the rates of chemical reactions puzzled scientists until the development of transition state theory in the first half of the twentieth century. In 1948 the famous physical chemist Linus Pauling suggested that enzymatic rate enhancement was achieved by stabilization of the transition state of the chemical reaction by interaction with the enzyme active site. This hypothesis, which was widely accepted, is supported by the experimental observation that enzymes bind very tightly to molecules designed to mimic the structure of the transition state of the catalyzed reaction.

In the 1950s and 1960s scientists reexamined the question of how enzymes achieve substrate specificity in light of the need for transition state stabilization by the enzyme active site. New hypotheses, such as the "induced fit" model of Koshland emerged at this time to help rationalize the competing needs of substrate binding affinity and reaction rate enhancement by enzymes. During this time period, scientists struggled to understand the observation that metabolic enzyme activities can be regulated by small molecules other than the substrates or direct products of an enzyme. Studies showed that indirect interactions between distinct binding sites within an enzyme molecule could occur, even though these binding sites were quite distant from one another. In 1965 Monod, Wyman, and Changeux developed the theory of allosteric transitions to explain these observations. Thanks in large part to this landmark paper, we now know that many enzymes, and nonenzymatic ligand binding proteins, display allosteric regulation

1.4   STUDIES OF ENZYME STRUCTURE

One of the tenets of modern enzymology is that catalysis is intimately related to the molecular interactions that take place between a substrate molecule and components of the enzyme molecule, the exact nature and sequence of these interactions defining per se the catalytic mechanism. Hence, the application of physical methods to elucidate the structures of enzymes has had a rich history and continues to be of paramount importance today. Spectroscopic methods, x-ray crystallography, and more recently, multidimensional NMR methods have all provided a wealth of structural insights on which theories of enzyme mechanisms have been built. In the early part of the twentieth century, x-ray crystallography became the premier method for solving the structures of small molecules. In 1926 James Sumner published the first crystallization of an enzyme, urease (Figure 1.1). Sumner's paper was a landmark contribution, not

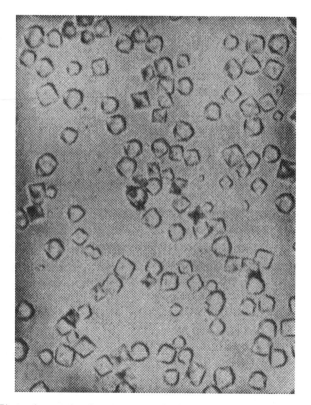

Figure 1.1 Photomicrograph of urease crystals (728× magnification), the first reported crystals of an enzyme. [From J. B. Sumner, *J. Biol. Chem.* **69**, 435–441 (1926), with permission.]

only because it portended the successful application of x-ray diffraction for solving enzyme structures, but also because a detailed analysis allowed Sumner to show unequivocally that the crystals were composed of protein and that their dissolution in solvent led to enzymatic activity. These observations were very important to the development of the science of enzymology because they firmly established the protein composition of enzymes, a view that had not been widely accepted by Sumner's contemporaries.

Sumner's crystallization of urease opened a floodgate and was quickly followed by reports of numerous other enzyme crystals. Within 20 years of Sumner's first paper more than 130 enzyme crystals had been documented. It was not, however, until the late 1950s that protein structures began to be solved through x-ray crystallography. In 1957 Kendrew became the first to deduce from x-ray diffraction the entire three-dimensional structure of a protein, myoglobin. Soon after, the crystal structures of many proteins, including enzymes, were solved by these methods. Today, the structural insights gained from x-ray crystallography and multidimensional NMR studies are commonly used to elucidate the mechanistic details of enzyme catalysis, and to design new ligands (substrate and inhibitor molecules) to bind at specific sites within the enzyme molecule.

The deduction of three-dimensional structures from x-ray diffraction or NMR methods depends on knowledge of the arrangement of amino acids along the polypeptide chain of the protein; this arrangement is known as the amino acid sequence. To determine the amino acid sequence of a protein, the component amino acids must be hydrolyzed in a sequential fashion from the polypeptide chain and identified by chemical or chromatographic analysis. Edman and coworkers developed a method for the sequential hydrolysis of amino acids from the N-terminus of a polypeptide chain. In 1957 Sanger reported the first complete amino acid sequence of a protein, the hormone insulin, utilizing the chemistry developed by Edman. In 1963 the first amino acid sequence of an enzyme, ribonuclease, was reported.

1.5 ENZYMOLOGY TODAY

Fundamental questions still remain regarding the detailed mechanisms of enzyme activity and its relationship to enzyme structure. The two most powerful tools that have been brought to bear on these questions in modern times are the continued development and use of biophysical probes of protein structure, and the application of molecular biological methods to enzymology. X-ray crystallography continues to be used routinely to solve the structures of enzymes and of enzyme–ligand complexes. In addition, new NMR methods and magnetization transfer methods make possible the assessment of the three-dimensional structures of small enzymes in solution, and the structure of ligands bound to enzymes, respectively.

The application of Laue diffraction with synchrotron radiation sources holds the promise of allowing scientists to determine the structures of reaction intermediates during enzyme turnover, hence to develop detailed pictures of the individual steps in enzyme catalysis. Other biophysical methods, such as optical (e.g., circular dichroism, UV–visible, fluorescence) and vibrational (e.g., infrared, Raman) spectroscopies, have likewise been applied to questions of enzyme structure and reactivity in solution. Technical advances in many of these spectroscopic methods have made them extremely powerful and accessible tools for the enzymologist. Furthermore, the tools of molecular biology have allowed scientists to clone and express enzymes in foreign host organisms with great efficiency. Enzymes that had never before been isolated have been identified and characterized by molecular cloning. Overexpression of enzymes in prokaryotic hosts has allowed the purification and characterization of enzymes that are available only in minute amounts from their natural sources. This has been a tremendous advance for protein science in general.

The tools of molecular biology also allow investigators to manipulate the amino acid sequence of an enzyme at will. The use of site-directed mutagenesis (in which one amino acid residue is substituted for another) and deletional mutagenesis (in which sections of the polypeptide chain of a protein are eliminated) have allowed enzymologists to pinpoint the chemical groups that participate in ligand binding and in specific chemical steps during enzyme catalysis.

The study of enzymes remains of great importance to the scientific community and to society in general. We continue to utilize enzymes in many industrial applications. Moreover enzymes are still in use in their traditional roles in food and beverage manufacturing. In modern times, the role of enzymes in consumer products and in chemical manufacturing has expanded greatly. Enzymes are used today in such varied applications as stereospecific chemical synthesis, laundry detergents, and cleaning kits for contact lenses.

Perhaps one of the most exciting fields of modern enzymology is the application of enzyme inhibitors as drugs in human and veterinary medicine. Many of the drugs that are commonly used today function by inhibiting specific enzymes that are associated with the disease process. Aspirin, for example, one of the most widely used drugs in the world, elicits its anti-inflammatory efficacy by acting as an inhibitor of the enzyme prostaglandin synthase. As illustrated in Table 1.1, enzymes take part in a wide range of human pathophysiologies, and many specific enzyme inhibitors have been developed to combat their activities, thus acting as therapeutic agents. Several of the inhibitors listed in Table 1.1 are the result of the combined use of biophysical methods for assessing enzyme structure and classical pharmacology in what is commonly referred to as rational or structure-based drug design. This approach uses the structural information obtained from x-ray crystallography or NMR spectroscopy to determine the topology of the enzyme active site. Next, model building is performed to design molecules that would fit well into this active site pocket. These molecules are then synthesized and tested as inhibitors. Several iterations of this procedure often lead to extremely potent inhibitors of the target enzyme.

The list in Table 1.1 will continue to grow as our understanding of disease state physiology increases. There remain thousands of enzymes involved in human physiology that have yet to be isolated or characterized. As more and more disease-related enzymes are discovered and characterized, new inhibitors will need to be designed to arrest the actions of these catalysts, in the continuing effort to fulfill unmet human medical needs.

1.6 SUMMARY

We have seen in this chapter that the science of enzymology has a long and rich history. From phenomenological observations, enzymology has grown to a quantitative molecular science. For the rest of this book we shall view enzymes from a chemical prospective, attempting to understand the actions of

Table 1.1   Examples of enzyme inhibitors as potential drugs

Inhibitor/Drug	Disease/Condition	Enzyme Target
Acetazolamide	Glaucoma	Carbonic anhydrase
Acyclovir	Herpes	Viral DNA polymerase
Allopurinol	Gout	Xanthine oxidase
Argatroban	Coagulation	Thrombin
Aspirin, ibuprofen, DuP697	Inflammation, pain, fever	Prostaglandin synthase
β-Lactam antibiotics	Bacterial infections	D-Ala-D-Ala transpeptidase
Brequinar	Organ transplantation	Dihydroorotate dehydrogenase
Candoxatril	Hypertension, congestive heart failure	Atriopeptidase
Captopril	Hypertension	Angiotensin-converting enzyme
Clavulanate	Bacterial resistance	β-Lactamase
Cyclosporin	Organ transplantation	Cyclophilin/calcineurin
DuP450	AIDS	HIV protease
Enoximone	Congestive heart failure ischemia	cAMP phosphodiesterase
Finazteride	Benign prostate hyperplasia	Testosterone-5-α-reductase
FK-506	Organ transplantation, autoimmune disease	FK-506 binding protein
Fluorouracyl	Cancer	Thymidilate synthase
3-Fluorovinylglycine	Bacterial infection	Alanine racemase
(2-Furyl)-acryloyl-Gly-Phe-Phe	Lung elastin degradation in cystic fibrosis	*Pseudomonas* elastase
ICI-200,808	Emphysema	Neutrophil elastase
Lovastatin	High cholesterol	HMG CoA reductase
Ly-256548	Inflammation	Phospholipase A_2
Methotrexate	Cancer	Dihydrofolate reductase
Nitecapone	Parkinson's disease	Catechol-*O*-methyltransferase
Norfloxacin	Urinary tract infections	DNA gyrase
Omeprazole	Peptic ulcers	H^+, K^+-ATPase
PALA	Cancer	Aspartate transcarbamoylase
PD-116124	Metabolism of antineoplastic drugs	Purine nucleoside phosphorylase
Phenelzine	Depression	Brain monoamine oxidase
Ro 42-5892	Hypertension	Renin
Sorbinil	Diabetic retinopathy	Aldose reductase
SQ-29072	Hypertension, congestive heart failure, analgesia	Enkephalinase
Sulfamethoxazole	Bacterial infection, malaria	Dihydropteorate synthase
Testolactone	Hormone-dependent tumors	Aromatase
Threo-5-fluoro-L-dihydroorotate	Cancer	Dihydroorotase
Trimethoprim	Bacterial infection	Dihydrofolate reductase
WIN 51711	Common cold	Rhinovirus coat protein
Zidovudine	AIDS	HIV reverse transcriptase
Zileuton	Allergy	5-Lipoxygenase

Source: Adapted and expanded from M. A. Navia and M. A. Murcko, *Curr. Opin. Struct. Biol.* **2**, 202–210 (1992).

these proteins in the common language of chemical and physical forces. While the vital importance of enzymes in biology cannot be overstated, the understanding of their structures and functions remains a problem of chemistry.

REFERENCES AND FURTHER READING

Rather than providing an exhaustive list of primary references for this historical chapter, I refer the reader to a few modern texts that have done an excellent job of presenting a more detailed and comprehensive treatment of the history of enzymology. Not only do these books provide good descriptions of the history of science and the men and women who made that history, but they are also quite entertaining and inspiring reading—enjoy them!

Friedmann, H. C., Ed. (1981) *Enzymes*, Hutchinson Ross, Stroudsburg, PA. [This book is part of the series "Benchmark Papers in Biochemistry." In it, Friedmann has compiled reprints of many of the most influential publications in enzymology from the eighteenth through twentieth centuries, along with insightful commentaries on these papers and their importance in the development of the science.]

Judson, H. F. (1980) *The Eighth Day of Creation*, Simon & Schuster, New York. [This extremely entertaining book chronicles the history of molecular biology, including protein science and enzymology, in the twentieth century.]

Kornberg, A. (1989) *For the Love of Enzymes. The Odyssey of a Biochemist*, Harvard University Press, Cambridge, MA. [An autobiographical look at the career of a Nobel Prize–winning biochemist.]

Werth, B. (1994) *The Billion Dollar Molecule*, Simon & Schuster, New York. [An interesting, if biased, look at the modern science of structure-based drug design.]

2

CHEMICAL BONDS
AND REACTIONS
IN BIOCHEMISTRY

The hallmark of enzymes is their remarkable ability to catalyze very specific chemical reactions of biological importance. Some enzymes are so well designed for this purpose that they can accelerate the rate of a chemical reaction by as much as 10^{12}-fold over the spontaneous rate of the uncatalyzed reaction! This incredible rate enhancement results from the juxtaposition of chemically reactive groups within the binding pocket of the enzyme (the enzyme active site) and other groups from the target molecule (substrate), in a way that facilitates the reaction steps required to convert the substrate into the reaction product. In subsequent chapters we shall explore the structural details of these reactive groups and describe how their interactions with the substrate result in the enhanced reaction rates typical of enzymatic catalysis. First, however, we must understand the chemical bonding and chemical reactions that take place both in enzymes and in the simpler molecules on which enzymes act. This chapter is meant as a review of material covered in introductory chemistry courses (basic chemical bonds, some of the reactions associated with these bonds); however, a thorough understanding of the concepts covered here will be essential to understanding the material in Chapters 3–12.

2.1 ATOMIC AND MOLECULAR ORBITALS

2.1.1 Atomic Orbitals

Chemical reactions, whether enzyme-catalyzed or not, proceed mainly through the formation and cleavage of chemical bonds. The bonding patterns seen in molecules result from the interactions between electronic orbitals of individual atoms to form molecular orbitals. Here we shall review these orbitals and some properties of the chemical bonds they form.

Recall from your introductory chemical courses that electrons occupy discrete atomic orbitals surrounding the atomic nucleus. The first model of electronic orbitals, proposed by Niels Bohr, viewed these orbitals as a collection of simple concentric circular paths of electron motion orbiting the atomic nucleus. While this was a great intellectual leap in thinking about atomic structure, the Bohr model failed to explain many of the properties of atoms that were known at the time. For instance, the simple Bohr model does not explain many of the spectroscopic features of atoms. In 1926 Erwin Schrödinger applied a quantum mechanical treatment to the problem of describing the energy of a simple atomic system. This resulted in the now-famous Schrödinger wave equation, which can be solved exactly for a simple one-proton, one-electron system (the hydrogen atom).

Without going into great mathematical detail, we can say that the application of the Schrödinger equation to the hydrogen atom indicates that atomic orbitals are quantized; that is, only certain orbitals are possible, and these have well-defined, discrete energies associated with them. Any atomic orbital can be uniquely described by a set of three values associated with the orbital, known as *quantum numbers*. The first or principal quantum number describes the effective volume of the orbital and is given the symbol n. The second quantum number, l, is referred to as the orbital shape quantum number, because this value describes the general probability density over space of electrons occupying that orbital. Together the first two quantum numbers provide a description of the *spatial probability distribution* of electrons within the orbital. These descriptions lead to the familiar pictorial representations of atomic orbitals, as shown in Figure 2.1 for the 1s and 2p orbitals.

The third quantum number, m_l, describes the orbital angular momentum associated with the electronic orbital and can be thought of as describing the orientation of that orbital in space, relative to some arbitrary fixed axis. With these three quantum numbers, one can specify each particular electronic orbital of an atom. Since each of these orbitals is capable of accommodating two electrons, however, we require a fourth quantum number to uniquely identify each individual electron in the atom.

The fourth quantum number, m_s, is referred to as the electron spin quantum number. It describes the direction in which the electron is imagined to spin with respect to an arbitrary fixed axis in a magnetic field (Figure 2.2). Since no two electrons can have the same values for all four quantum numbers, it follows that two electrons within the same atomic orbital must be spin-paired; that is, if one is spinning clockwise ($m_s = +\frac{1}{2}$), the other must be spinning counterclockwise ($m_s = -\frac{1}{2}$). This concept, known as the Pauli exclusion principle, is often depicted graphically by representing the spinning electron as an arrow pointing either up or down, within an atomic orbital.

Thus we see that associated with each atomic orbital is a discrete amount of potential energy; that is, the orbitals are quantized. Electrons fill these

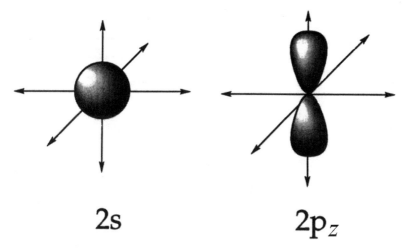

2s 2p$_z$

Figure 2.1 Spatial representations of the electron distribution in s and p orbitals.

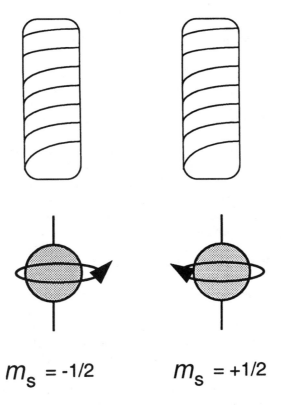

m_s = -1/2 m_s = +1/2

Figure 2.2 Electron spin represented as rotation of a particle in a magnetic field. The two spin "directions" of the electron are represented as clockwise $(m_s = +\frac{1}{2})$ and counterclockwise $(m_s = -\frac{1}{2})$ rotations. The coil-bearing rectangles schematically represent the magnetic fields.

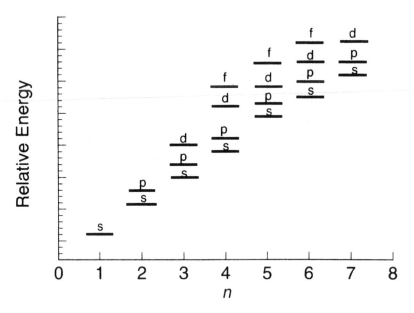

Figure 2.3   The aufbau principle for the order of filling of atomic orbitals, s, p, d, and f.

orbitals according to the potential energy associated with them; low energy orbitals fill first, followed by higher energy orbitals in ascending energetic order (the aufbau principle). By schematizing the energetic order of atomic orbitals, as illustrated in Figure 2.3, we can inventory the electrons in the orbitals of an atom. For example, each atom of the element helium contains two electrons; since both electrons occupy the 1s orbital, we designate this by the shorthand notation $1s^2$. Lithium contains 3 electrons and, according to Figure 2.3, has the configuration $1s^2 2s^1$. When dealing with nonspherical orbitals, such as the p, d, and f orbitals, we must keep in mind that more than one atomic orbital is associated with each orbital set that is designated by a combination of n and l quantum numbers. For example, the 2p orbital set consists of three atomic orbitals: $2p_x$, $2p_y$, and $2p_z$. Hence, the 2p orbital set can accommodate 6 electrons. Likewise, a d orbital set can accommodate a total of 10 electrons (5 orbitals, with 2 electrons per orbital), and an f orbital set can accommodate 14 electrons.

A survey of the biological tissues in which enzymes naturally occur indicates that the elements listed in Table 2.1 are present in highest abundance. Because of their abundance in biological tissue, these are the elements we most often encounter as components of enzyme molecules. For each of these atoms, the highest energy s and p orbital electrons are those that are capable of participating in chemical reactions, and these are referred to as *valence electrons* (the electrons in the lower energy orbitals are chemically inert and are referred to as closed-shell electrons). In the carbon atom, for example, the two

Table 2.1 Electronic configurations of the elements most commonly found in biological tissues

Element	Number of Electrons	Orbital Configuration
Hydrogen	1	$1s^1$
Carbon	6	$1s^2 2s^2 2p^2$
Nitrogen	7	$1s^2 2s^2 2p^3$
Oxygen	8	$1s^2 2s^2 2p^4$
Phosphorus	15	$1s^2 2s^2 2p^6 3s^2 3p^3$
Sulfur	16	$1s^2 2s^2 2p^6 3s^2 3p^4$

1s electrons are the closed-shell type, while the four electrons in the 2s and 2p orbitals are valence electrons, and are thus available for bond formation.

2.1.2 Molecular Orbitals

If two atoms can approach each other at close enough range, and if their valence orbitals are of appropriate energy and symmetry, the two valence atomic orbitals (one from each atom) can combine to form two molecular orbitals: a *bonding* and an *antibonding* molecular orbital. The bonding orbital occurs at a lower potential energy than the original two atomic orbitals; hence electron occupancy in this orbital promotes bonding between the atoms because of a net stabilization of the system (molecule). The antibonding orbital, in contrast, occurs at a higher energy than the original atomic orbitals; electron occupancy in this molecular orbital would thus be destabilizing to the molecule.

Let us consider the molecule H_2. The two 1s orbitals from each hydrogen atom, each containing a single electron, approach each other until they overlap to the point that the two electrons are shared by both nuclei (i.e., a bond is formed). At this point the individual atomic orbital character is lost and the two electrons are said to occupy a *molecular orbital*, resulting from the mixing of the original two atomic orbitals. Since there were originally two atomic orbitals that mixed, there must result two molecular orbitals. As illustrated in Figure 2.4, one of these molecular orbitals occurs at a lower potential energy than the original atomic orbital, hence stabilizes the molecular bond; this orbital is referred to as a *bonding orbital* (in this case a σ-bonding orbital, as discussed shortly). The other molecular orbital occurs at higher potential energy (displaced by the same amount as the bonding orbital). Because the higher energy of this orbital makes it destabilizing relative to the atomic orbitals, it is referred to as an *antibonding orbital* (again, in this case a σ-antibonding orbital, σ*). The electrons fill the molecular orbitals in order of potential energy, each orbital being capable of accommodating two electrons. Thus for H_2 both electrons from the 1s orbitals of the atoms will occupy the σ-bonding molecular orbital in the molecule.

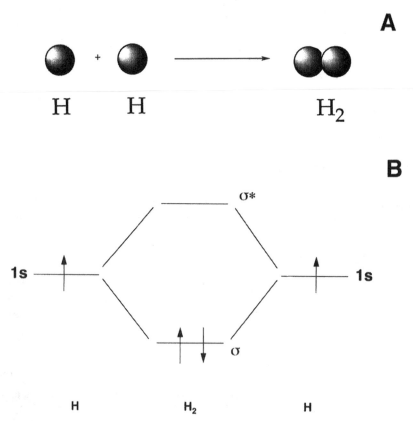

Figure 2.4 (A) Schematic representation of two s orbitals on separate hydrogen atoms combining to form a bonding σ molecular orbital. (B) Energy level diagram for the combination of two hydrogen s orbitals to form a bonding and antibonding molecular orbital in the H_2 molecule.

Now let us consider the diatomic molecule F_2. The orbital configuration of the fluorine atom is $1s^2 2s^2 2p^5$. The two s orbitals and two of the three p orbitals are filled and will form equal numbers of bonding and antibonding molecular orbitals, canceling any net stabilization of the molecule. The partially filled p orbitals, one on each atom of fluorine, can come together to form one bonding and one antibonding molecular orbital in the diatomic molecule F_2. As illustrated in Figure 2.5, the lobes of the two valence p_z orbitals overlap *end to end* in the bonding orbital; molecular orbitals that result from such end-to-end overlaps are referred to as sigma orbitals (σ). The bonding orbital is designated by the symbol σ, and the accompanying antibonding orbital is designated by the symbol σ^*. The bond formed between the two atoms in the F_2 molecules is therefore referred to as a *sigma bond*. Because the σ orbital is lower in energy than the σ^* orbital, both the electrons from the valence atomic

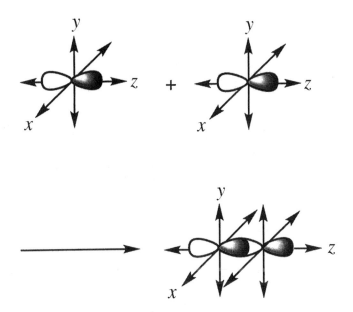

Figure 2.5 Combination of two p_z atomic orbitals by end-to-end overlap to form a σ-type molecular orbital.

orbit will reside in the σ molecular orbital when the molecule is at rest (i.e., when it is in its lowest energy form, referred to as the *ground state* of the molecule).

2.1.3 Hybrid Orbitals

For elements in the second row of the periodic table (Li, Be, C, N, O, F, and Ne), the 2s and 2p orbitals are so close in energy that they can interact to form orbitals with combined, or mixed, s and p orbital character. These *hybrid orbitals* provide a means of maximizing the number of bonds an atom can form, while retaining the greatest distance between bonds, to minimize repulsive forces. The hybrid orbitals formed by carbon are the most highly studied, and the most germane to our discussion of enzymes.

From the orbital configuration of carbon ($1s^2 2s^2 2p^2$), we can see that the similar energies of the 2s and 2p orbital sets in carbon provide four electrons that can act as valence electrons, giving carbon the ability to form four bonds to other atoms. Three types of hybrid orbital are possible, and they result in three different bonding patterns for carbon. The first type results from the combination of one 2s orbital with three 2p orbitals, yielding four hybrid orbitals referred to as *sp^3* orbitals (the exponent reflects the number of p orbitals that have combined with the one s orbital to produce the hybrids). The four sp^3 orbitals allow the carbon atom to form four σ bonds that lie along

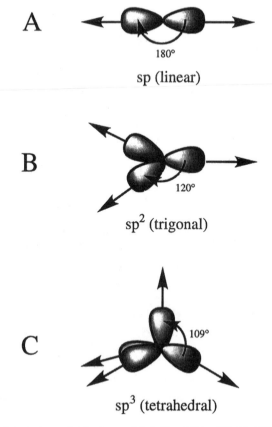

A

180°

sp (linear)

B

120°

sp^2 (trigonal)

C

109°

sp^3 (tetrahedral)

Figure 2.6 Spatial electron distributions of hybrid orbitals: (A) sp hybridization, (B) sp^2 hybridization, and (C) sp^3 hybridization.

the apices of a tetrahedron, as shown in Figure 2.6C. The second type of hybrid orbital, *sp^2*, results from the mixing of one 2s orbital and two 2p orbitals. These hybrid orbitals allow for three trigonal planar bonds to form (Figure 2.6B). When a single 2p orbital combines with a 2s orbital, the resulting single hybrid orbital is referred to as an *sp* orbital (Figure 2.6A).

Let us look at the sp^2 hybrid case in more detail. We have said that the 2p orbital set consists of three p orbitals that can accommodate a total of six electrons. With sp^2 hybridization, we have accounted for two of the three p orbitals available in forming three trigonal planar σ bonds, as in the case of ethylene (Figure 2.7A). On each carbon atom, this hybridization leaves one orbital, of pure p character, which is available for bond formation. These orbitals can interact with one another to form a bond by edge-to-edge orbital overlap, above and below the plane defined by the sp^2 σ bonds (Figure 2.7B). This type of edge-to-edge orbital overlap results in a different type of molecular orbital, referred to as a π orbital. As illustrated in Figure 2.7B, the overlap of

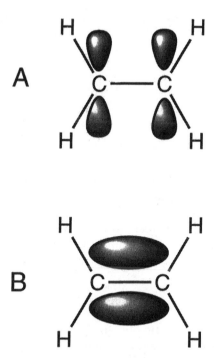

Figure 2.7 Hybrid bond formation in ethylene. (A) The bonds are illustrated as lines, and the remaining p orbitals lobes form edge-to-edge contacts. (B) The p orbitals combine to form a π bond with electron density above and below the interatomic bond axis defined by the σ bond between the carbon atoms.

the p orbitals provides for bonding electron density above and below the interatomic axis, resulting in a *pi bond* (π). Of course, as with σ bonds, for every π orbital formed, there must be an accompanying antibonding orbital at higher energy, which is denoted by the symbol π^*. Thus along the interatomic axis of ethylene we find two bonds: one σ bond, and one π bond. This combination is said to form a *double bond* between the carbon atoms. A shorthand notation for this bonding situation is to draw two parallel lines connecting the carbon atoms:

$$
\begin{array}{ccc}
H & & H \\
\diagdown & & \diagup \\
& C = C & \\
\diagup & & \diagdown \\
H & & H
\end{array}
$$

A similar situation arises when we consider sp hybridization. In this case we have two mutually perpendicular p orbitals on each carbon atom available for π orbital formation. Hence, we find one σ bond (from the sp hybrid orbitals) and two π bonds along the interatomic axis. This *triple bond* is denoted by drawing three parallel lines connecting the two carbon atoms, as in acetylene:

$$H-C \equiv C-H$$

Not all the valence electrons of the atoms in a molecule are shared in the form of covalent bonds. In many cases it is energetically advantageous to the molecule to have *unshared* electrons that are essentially localized to a single atom; these electrons are often referred to as nonbonding or *lone pair* electrons. Whereas electrons within bonding orbitals are denoted as lines drawn between atoms of the molecule, lone pair electrons are usually depicted as a pair of dots surrounding a particular atom. (Combinations of atoms and molecules represented by means of these conventions are referred to as Lewis structures.)

2.1.4   Resonance and Aromaticity

Let us consider the ionized form of acetic acid that occurs in aqueous solution at neutral pH (i.e., near physiological conditions). The carbon bound to the oxygen atoms uses sp^2 hybridization: it forms a σ bond to the other carbon, a σ bond to each oxygen atom, and one π bond to one of the oxygen atoms. Thus, one oxygen atom would have a double bond to the carbon atom, while the other has a single bond to the carbon and is negatively charged. Suppose that we could somehow identify the individual oxygen atoms in this molecule — by, for example, using an isotopically labeled oxygen (^{18}O rather than ^{16}O) at one site. Which of the two would form the double bond to carbon, and which would act as the anionic center?

Both of these are reasonable electronic forms, and there is no basis on which to choose one over the other. In fact, neither is truly correct, because in reality we find that the π bond (or more correctly, the π-electron density) is *delocalized* over both oxygen atoms. In some sense neither forms a single bond nor a double bond to the carbon atom, but rather both behave as if they shared the π bond between them. We refer to these two alternative electronic forms of the molecule as *resonance structures* and sometimes represent this arrangement by drawing a double-headed arrow between the two forms:

$$H_3C-C\underset{^{16}O}{\overset{^{18}O^-}{}} \longleftrightarrow H_3C-C\underset{^{16}O^-}{\overset{^{18}O}{}}$$

Alternatively, the resonance form is illustrated as follows, to emphasize the delocalization of the π-electron density:

$$H_3C-C\underset{^{16}O^{\delta^-}}{\overset{^{18}O^{\delta^-}}{}}$$

Now let us consider the organic molecule benzene (C_6H_6). The carbon atoms are arranged in a cyclic pattern, forming a planar hexagon. To account for this, we must assume that there are three double bonds among the carbon–carbon bonds of the molecule. Here are the two resonance structures:

Now a typical carbon–carbon single bond has a bond length of roughly 1.54 Å, while a carbon–carbon double bond is only about 1.35 Å long. When the crystal structure of benzene was determined, it was found that all the carbon–carbon bonds were the same length, 1.45 Å, which is intermediate between the expected lengths for single and double bonds. How can we rationalize this result? The answer is that the π orbitals are not localized to the p_z orbitals of two adjacent carbon atoms (Figure 2.8, left: here the plane defined by the carbon ring system is arbitrarily assigned as the x,y plane); rather, they are delocalized over all six carbon p_z orbitals. To emphasize this π system delocalization, many organic chemists choose to draw benzene as a hexagon enclosing a circle (Figure 2.8, right) rather than a hexagon of carbon with three discrete double bonds.

The delocalization of the π system in molecules like benzene tends to stabilize the molecule relative to what one would predict on the basis of three isolated double bonds. This difference in stability is referred to as the *resonance energy stabilization*. For example, consider the heats of hydrogenation (breaking the carbon–carbon double bond and adding two atoms of hydrogen), using

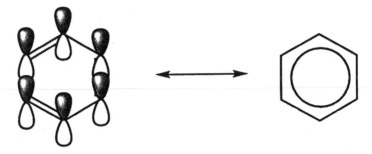

Figure 2.8 Two common representations for the benzene molecule. The representation on the right emphasizes the π-system delocalization in this molecule.

H_2 and platinum catalysis, for the series cyclohexene ($\Delta H = 28.6\,\text{kcal/mol}$), cyclohexadiene, benzene. If each double bond were energetically equivalent, one would expect the ΔH value for cyclohexadiene hydrogenation to be twice that of cyclohexene ($-57.2\,\text{kcal/mol}$), and that is approximately what is observed. Extending this argument further, one would expect the ΔH value for benzene (if it behaved energetically equivalent to cyclohexatriene) to be three times that of cyclohexene, $85.8\,\text{kcal/mol}$. Experimentally, however, the ΔH of hydrogenation of benzene is found to be only $-49.8\,\text{kcal/mol}$, a resonance energy stabilization of 36 kcal/mol! This stabilizing effect of π-orbital delocalization has an important influence over the structure and chemical reactivities of these molecules, as we shall see in later chapters.

2.1.5 Different Electronic Configurations Have Different Potential Energies

We have seen how electrons distribute themselves among molecular orbitals according to the potential energies of those molecular orbitals. The specific distribution of the electrons within a molecule among the different electronic molecular orbitals defines the electronic configuration or *electronic state* of that molecule. The electronic state that imparts the least potential energy to that molecule will be the most stable form of that molecule under normal conditions. This electronic configuration is referred to as the *ground state* of the molecule. Any alternative electronic configuration of higher potential energy than the ground state is referred to as an *excited state* of the molecule.

Let us consider the simple carbonyl formaldehyde (CH_2O):

$$\begin{array}{c} H \\ \diagdown \\ C{=}O \\ \diagup \\ H \end{array}$$

In the ground state electronic configuration of this molecule, the π-bonding orbital is the highest energy orbital that contains electrons. This orbital is referred to as the highest occupied molecular orbital (HOMO). The π^* molecular orbital is the next highest energy molecular orbital and, in the ground state, does not contain any electron density. This orbital is said to be the lowest unoccupied molecular orbital (LUMO). Suppose that somehow we were able to move an electron from the π to the π^* orbital. The molecule would now have a different electronic configuration that would impart to the overall molecule more potential energy; that is, the molecule would be in an excited electronic state. Now, since in this excited state we have moved an electron from a bonding (π) to an antibonding (π^*) orbital, the overall molecule has acquired more antibonding character. As a consequence, the nuclei will occur at a longer equilibrium interatomic distance, relative to the ground state of the molecule.

In other words, the potential energy minimum (also referred to as the zero-point energy) for the excited state occurs when the atoms are further apart from one another than they are for the potential energy minimum of the ground state. Since the π electrons are localized between the carbon and oxygen atoms in this molecule, it will be the carbon–oxygen bond length that is most affected by the change in electronic configuration; the carbon–hydrogen bond lengths are essentially invariant between the ground and excited states. The nuclei, however, are not fixed in space, but can vibrate in both the ground and excited electronic states of the molecule. Hence, each electronic state of a molecule has built upon it a manifold of vibrational substates.

The foregoing concepts are summarized in Figure 2.9, which shows a potential energy diagram for the ground and one excited state of the molecule. An important point to glean from this figure is that even though the potential minima of the ground and excited states occur at different equilibrium interatomic distances, vibrational excursions within either electronic state can bring the nuclei into register with their equilibrium positions at the potential minimum of the other electronic state. In other words, a molecule in the ground electronic state can, through vibrational motions, transiently sample the interatomic distances associated with the potential energy minimum of the excited electronic state, and vice versa.

2.2 THERMODYNAMICS OF CHEMICAL REACTIONS

In freshman chemistry we were introduced to the concept of free energy, ΔG, which combined the first and second laws of thermodynamics to yield the familiar formula:

$$\Delta G = \Delta H - T\Delta S \tag{2.1}$$

where ΔG is the change in free energy of the system during a reaction at

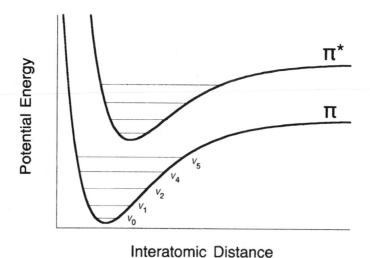

Figure 2.9 Potential energy diagram for the ground and one excited electronic state of a molecule. The potential wells labeled π and π^* represent the potential energy profiles of the ground and excited electronic states, respectively. The sublevels within each of these potential wells, labeled v_n, represent the vibrational substates of the electronic states.

constant temperature (T) and pressure, ΔH is the change in enthalpy (heat), and ΔS is the change in entropy (a measure of disorder or randomness) associated with the reaction. Some properties of ΔG should be kept in mind. First, ΔG is less than zero (negative) for a spontaneous reaction and greater than zero (positive) for a nonspontaneous reaction. That is, a reaction for which ΔG is negative will proceed spontaneously with the liberation of energy. A reaction for which ΔG is positive will proceed only if energy is supplied to drive the reaction. Second, ΔG is always zero at equilibrium. Third, ΔG is a *path-independent function*. That is, the value of ΔG is dependent on the starting and ending states of the system but not on the path used to go from the starting point to the end point. Finally, while the value of ΔG gives information on the spontaneity of a reaction, it does not tell us anything about the rate at which the reaction will proceed.

Consider the following reaction:

$$A + B \rightleftharpoons C + D$$

Recall that the ΔG for such a reaction is given by:

$$\Delta G = \Delta G^0 + RT \ln \left(\frac{[C][D]}{[A][B]} \right) \tag{2.2}$$

where ΔG^0 is the free energy for the reaction under standard conditions of all reactants and products at a concentration of 1.0 M (1.0 atm for gases). The terms in brackets, such as [C], are the molar concentrations of the reactants and products of the reaction, the symbol "ln" is shorthand for the natural, or base e, logarithm, and R and T refer to the ideal gas constant (1.98×10^{-3} kcal/mol·degree) and the temperature in degrees Kelvin (298 K for average room temperature, 25°C, and 310 K for physiological temperature, 37°C), respectively. Since, by definition, $\Delta G = 0$ at equilibrium, it follows that under equilibrium conditions:

$$\Delta G^0 = -RT \ln \left(\frac{[C][D]}{[A][B]} \right) \qquad (2.3)$$

For many reactions, including many enzyme-catalyzed reactions, the values of ΔG^0 have been tabulated. Thus knowing the value of ΔG^0 one can easily calculate the value of ΔG for the reaction at any displacement from equilibrium. Examples of these types of calculation can be found in any introductory chemistry or biochemistry text.

Because free energy of reaction is a path-independent quantity, it is possible to drive an unfavorable (nonspontaneous) reaction by *coupling* it to a favorable (spontaneous) one. Suppose, for example, that the product of an unfavorable reaction was also a reactant for a thermodynamically favorable reaction. As long as the absolute value of ΔG was greater for the second reaction, the overall reaction would proceed spontaneously. Suppose that the reaction $A \rightleftharpoons B$ had a ΔG^0 of $+5$ kcal/mol, and the reaction $B \rightleftharpoons C$ had a ΔG^0 of -8 kcal/mol. What would be the ΔG^0 value for the net reaction $A \rightleftharpoons C$?

$A \rightleftharpoons B$	$\Delta G^0 = +5$ kcal/mol
$B \rightleftharpoons C$	$\Delta G^0 = -8$ kcal/mol
$A \rightleftharpoons C$	$\Delta G^0 = -3$ kcal/mol

Thus, the overall reaction would proceed spontaneously. In our scheme, B would appear on both sides of the overall reaction and thus could be ignored. Such a species is referred to as a *common intermediate*. This mechanism of providing a thermodynamic driving force for unfavorable reactions is quite common in biological catalysis.

As we shall see in Chapter 3, many enzymes use nonprotein cofactors in the course of their catalytic reactions. In some cases these cofactors participate directly in the chemical transformations of the reactants (referred to as *substrates* by enzymologists) to products of the enzymatic reaction. In many other cases, however, the reactions of the cofactors are used to provide the thermodynamic driving force for catalysis. Oxidation and reduction reactions of metals, flavins, and reduced nicotinamide adenine dinucleotide (NADH) are commonly used for this purpose in enzymes. For example, the enzyme cytochrome c oxidase uses the energy derived from reduction of its metal cofactors to drive the transport of protons across the inner mitochondrial membrane, from a region of low proton concentration to an area of high proton concentration. This energetically unfavorable transport of protons could not proceed without coupling to the exothermic electrochemical reactions of the metal centers. Another very common coupling reaction is the hydrolysis of adenosine triphosphate (ATP) to adenosine diphosphate (ADP) and inorganic phosphate (P_i). Numerous enzymes drive their catalytic reactions by coupling to ATP hydrolysis, because of the high energy yield of this reaction.

2.2.1 The Transition State of Chemical Reactions

A chemical reaction proceeds spontaneously when the free energy of the product state is lower than that of the reactant state (i.e., $\Delta G < 0$). As we have stated, the path taken from reactant to product does not influence the free energies of these beginning and ending states, hence cannot affect the spontaneity of the reaction. The path can, however, greatly influence the rate at which a reaction will proceed, depending on the free energies associated with any intermediate state the molecule must access as it proceeds through the reaction. Most of the chemical transformations observed in enzyme-catalyzed reactions involve the breaking and formation of covalent bonds. If we consider a reaction in which an existing bond between two nuclei is replaced by an alternative bond with a new nucleus, we could envision that at some instant during the reaction a chemical entity would exist that had both the old and new bonds partially formed, that is, a state in which the old and new bonds are simultaneously broken and formed. This molecular form would be extremely unstable, hence would be associated with a very large amount of free energy. For the reactant to be transformed into the product of the chemical reaction, the molecule must transiently access this unstable form, known as the *transition state* of the reaction. Consider, for example, the formation of an alcohol by the nucleophilic attack of a primary alkyl halide by a hydroxide ion:

$$RCH_2Br + OH^- \rightleftharpoons RCH_2OH + Br^-$$

We can consider that the reaction proceeds through a transition state in which the carbon is simultaneously involved in partial bonds between the oxygen and the bromine:

$$RCH_2Br + OH^- \rightarrow [HO\text{-}\text{-}\text{-}CH_2R\text{-}\text{-}\text{-}Br] \rightarrow RCH_2OH + Br^-$$

where the species in brackets is the transition state of the reaction and partial bonds are indicated by dashes. Figure 2.10 illustrates this reaction scheme in terms of the free energies of the species involved. (Note that for simplicity, the various molecular states are represented as lines designating the position of the potential minimum of each state. Each of these states is more correctly

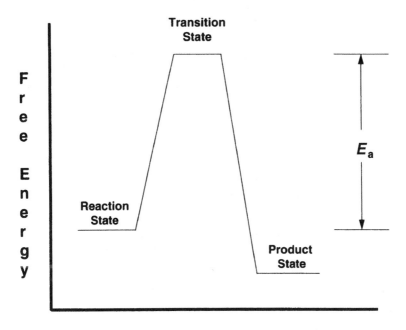

Reaction Coordinate

Figure 2.10 Free energy diagram for the reaction profile of a typical chemical reaction, a chemical reaction. The activation energy E_a is the energetic difference between the reactant state and the transition state of the reaction.

described by the potential wells shown in Figure 2.9, but diagrams constructed according to this convention are less easy to follow.)

In the free energy diagram of Figure 2.10, the x axis is referred to as the reaction coordinate and tracks the progressive steps in going from reactant to product. This figure makes it clear that the transition state represents an energy barrier that the reaction must overcome in order to proceed. The higher the energy of the transition state in relation to the reactant state, the more difficult it will be for the reaction to proceed. Once, however, the system has attained sufficient energy to reach the transition state, the reaction can proceed effortlessly downhill to the final product state (or, alternatively, collapse back to the reactant state). Most of us have experienced a macroscopic analogy of this situation in riding a bicycle. When we encounter a hill we must pedal hard, exerting energy to ascend the incline. Having reached the crest of the hill, however, we can take our feet off the pedals and coast downhill without further exertion.

The energy required to proceed from the reactant state to the transition state, which is known as the *activation energy* or energy barrier of the reaction, is the difference in free energy between these two states. The activation energy is given the symbol E_a or $\Delta G^{\ddagger}$. This energy barrier is an important concept for our subsequent discussions of enzyme catalysis. This is because the height of the activation energy barrier can be directly related to the rate of a chemical reaction. To illustrate, let us consider a unimolecular reaction in which the reactant A decomposes to B through the transition state $A^{\ddagger}$. The activation energy for this reaction is E_a. The equilibrium constant for A going to $A^{\ddagger}$ will be $[A^{\ddagger}]/[A]$. Using this, and rearranging Equation 2.3 with substitution of E_a for ΔG^0, we obtain:

$$[A^{\ddagger}] = [A] \exp\left(-\frac{E_a}{RT}\right) \tag{2.4}$$

The transition state will decay to product with the same frequency as that of the stretching vibration of the bond that is being ruptured to produce the product molecule. It can be shown that this vibrational frequency is given by:

$$v = \frac{k_B T}{h} \tag{2.5}$$

where v is the vibrational frequency, k_B is the Boltzmann constant, and h is Planck's constant. The rate of loss of $[A]$ is thus given by:

$$\frac{-d[A]}{dt} = v[A^{\ddagger}] = [A]\left(\frac{k_B T}{h}\right)\exp\left(-\frac{E_a}{RT}\right) \tag{2.6}$$

and the first-order rate constant for the reaction is thus given by the Arrhenius equation:

$$k = \left(\frac{k_B T}{h}\right)\exp\left(-\frac{E_a}{RT}\right) \tag{2.7}$$

From Equations 2.6 and 2.7 it is obvious that as the activation energy barrier increases (i.e., E_a becomes larger), the rate of reaction will decrease in an exponential fashion. We shall see in Chapter 6 that this concept relates directly to the mechanism by which enzymes achieve the acceleration of reaction rates characteristic of enzyme-catalyzed reactions.

It is important to recognize that the transition state of a chemical reaction is, under most conditions, an extremely unstable and short-lived species. Some chemical reactions go through intermediate states that are more long-lived and stable than the transition state. In some cases, these intermediate species exist long enough to be kinetically isolated and studied. When present, these intermediate states appear as local free energy minima (dips) in the free energy diagram of the reaction, as illustrated in Figure 2.11. Often these intermediate states structurally resemble the transition state of the reaction (Hammond,

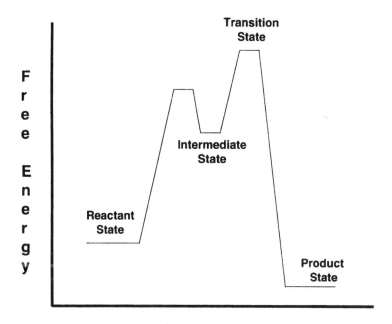

Reaction Coordinate

Figure 2.11 Free energy diagram for a chemical reaction that proceeds through the formation of a chemical intermediate.

1955). Therefore, when they can be trapped and studied, these intermediates provide a glimpse at what the true transition state may look like. Enzyme-catalyzed reactions go through intermediate states like this, mediated by the specific interactions of the protein and/or enzyme cofactors with the reactants and products of the chemical reaction being catalyzed. We shall have more to say about some of these intermediate species in Chapter 6.

2.3 ACID–BASE CHEMISTRY

In freshman chemistry we were introduced to the common Lewis definition of acids and bases: a *Lewis acid* is any substance that can act as an *electron pair acceptor*, and a *Lewis base* is any substance that can act as an *electron pair donor*. In many enzymatic reactions, protons are transferred from one chemical species to another, hence the alternative *Brønsted–Lowry* definition of acids and bases becomes very useful for dealing with these reactions. In the Brønsted–Lowry classification, an *acid* is any substance that can *donate a proton*, and a *base* is any substance that can *accept a proton* by reacting with a Brønsted–Lowry acid. After donating its proton, a Brønsted–Lowry acid is converted to its *conjugate base*.

Table 2.2 Examples of Brønsted–Lowry acids and their conjugate bases

Brønsted–Lowry Acid	Conjugate Base
H_2SO_4 (sulfuric acid)	HSO_4^- (bisulfate ion)
HCl (hydrochloric acid)	Cl^- (chloride ion)
H_3O^+ (hydronium ion)	H_2O (water)
NH_4^+ (ammonium ion)	NH_3 (ammonia)
CH_3COOH (acetic acid)	CH_3COO^- (acetate ion)
H_2O (water)	OH^- (hydroxide ion)

Table 2.2 gives some examples of Brønsted–Lowry acids and their conjugate bases. For all these pairs, we are dealing with the transfer of a hydrogen ion (proton) from the acid to some other species (often the solvent) to form the conjugate base. A convenient means of measuring the hydrogen ion concentration in aqueous solutions is the pH scale. The term "pH" is a shorthand notation for the negative base-10 logarithm of the hydrogen ion concentration:

$$pH = -\log[H^+] \tag{2.8}$$

Consider the dissociation of a weak Brønsted–Lowry acid (HA) into a proton (H^+) and its conjugate base (A^-) in aqueous solution.

$$HA \rightleftharpoons H^+ + A^-$$

The dissociation constant for the acid, K_a, is given by the ratio $[H^+][A^-]/[HA]$. Let us define the pK_a for this reaction as the negative base-10 logarithm of K_a:

$$pK_a = -\log\left(\frac{[H^+][A^-]}{[HA]}\right) \tag{2.9}$$

or, using our knowledge of logarithmic relationships, we can write:

$$pK_a = \log(HA) - \log(A^-) - \log(H^+) \tag{2.10}$$

Note that the last term in Equation 2.10 is identical to our definition of pH (Equation 2.8). Using this equality, and again using our knowledge of logarithmic relationships we obtain:

$$pK_a = \log\left(\frac{[HA]}{[A^-]}\right) + pH \tag{2.11}$$

or, rearranging (note the inversion of the logarithmic term):

$$pH = pK_a + \log\left(\frac{[A^-]}{[HA]}\right) \qquad (2.12)$$

Equation 2.12 is known as the *Henderson–Hasselbalch* equation, and it provides a convenient means of calculating the pH of a solution from the concentrations of a Brønsted–Lowry acid and its conjugate base. Note that when the concentrations of acid and conjugate base are equal, the value of $[A^-]/[HA]$ is 1.0, and thus the value of $\log([A^-]/[HA])$ is zero. At this point the pH will be exactly equal to the pK_a. This provides a useful working definition of pK_a:

> The pK_a is the pH value at which half the Brønsted–Lowry acid is dissociated to its conjugate base and a proton.

Let us consider a simple example of this concept. Suppose that we dissolve acetic acid into water and begin titrating the acid with hydroxide ion equivalents by addition of NaOH. If we measure the pH of the solution after each addition, we will obtain a titration curve similar to that shown in Figure 2.12. Two points should be drawn from this figure. First, such a titration curve provides a convenient means of graphically determining the pK_a value of the species being titrated. Second, we see that at pH values near the pK_a, it takes a great deal of NaOH to effect a change in the pH value. This resistance to pH change in the vicinity of the pK_a of the acid is referred to as *buffering capacity*, and it is an

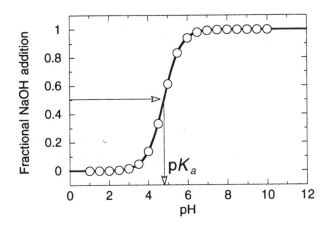

Figure 2.12 Hypothetical titration curve for a weak acid illustrating the graphical determination of the acid's pK_a.

important property to be considered in the preparation of solutions for enzyme studies. As we shall see in Chapter 7, the pH at which an enzyme reaction is performed can have a dramatic effect on the rate of reaction and on the overall stability of the protein. As a rule, therefore, specific buffering molecules, whose pK_a values match the pH for optimal enzyme activity, are added to enzyme solutions to maintain the solution pH near the pK_a of the buffer.

2.4 NONCOVALENT INTERACTIONS IN REVERSIBLE BINDING

All the properties of molecules we have discussed until now have led us to focus on the formation, stabilization, and breaking of covalent bonds between atoms of the molecule. These are important aspects of the chemical conversions that are catalyzed at the enzyme active site. Molecules can interact with one another by a number of noncovalent forces as well. These weaker attractive forces are very important in biochemical reactions because they are readily reversible. As we shall see in Chapters 4, 6, and 8, the reversible formation of binary complexes between enzymes and ligand molecules (i.e., substrates and inhibitors) is a critical aspect of both enzymatic catalysis and enzyme inhibition. Four types of noncovalent interaction are particularly important in protein structure (Chapter 3) and enzyme–ligand binding (Chapters 4, 6, and 8); these are electrostatic interactions, hydrogen bonding, hydrophobic interactions, and van der Waals forces. Here we describe these forces briefly. In subsequent chapters we shall see how each force can participate in stabilizing the protein structure of an enzyme and may also play an important role in the binding interactions between enzymes and their substrates and inhibitors.

2.4.1 Electrostatic Interactions

When two oppositely charged groups come into close proximity, they are attracted to one another through a Coulombic attractive force that is described by:

$$F = \frac{q_1 q_2}{r^2 D} \tag{2.13}$$

where q_1 and q_2 are the charges on the two atoms involved, r is the distance between them, and D is the dielectric constant of the medium in which the two atoms come together. Since D appears in the denominator, the attractive force is greatest in low dielectric solvents. Hence electrostatic forces are stronger in the hydrophobic interior of proteins than on the solvent-exposed surface. These attractive interactions are referred to as *ionic bonds*, *salt bridges*, and *ion pairs*.

Equation 2.13 describes the attractive force only. If two atoms, oppositely charged or not, approach each other too closely, a repulsive force between the outer shell electrons on each atom will come into play. Other factors being constant, it turns out that the balance between these attractive and repulsive forces is such that, on average, the optimal distance between atoms for salt bridge formation is about 2.8 Å (Stryer, 1989).

2.4.2 Hydrogen Bonding

A hydrogen bond (H bond) forms when a hydrogen atom is shared by two electronegative atoms. The atom to which the hydrogen is covalently bonded is referred to as the *hydrogen bond donor*, and the other atom is referred to as the *hydrogen bond acceptor*:

$$N^{\delta-}-H^{\delta+}\cdots O^{\delta-}$$

The donor and acceptors in H bonds are almost exclusively electronegative heteroatoms, and in proteins these are usually oxygen, nitrogen, or sometimes sulfur atoms. Hydrogen bonds are weaker than covalent bonds, varying in bond energy between 2.5 and 8 kcal/mol. The strength of a H bond depends on several factors, but mainly on the length of the bond between the hydrogen and acceptor heteroatom (Table 2.3). For example, NH—O hydrogen bonds between amides occur at bond lengths of about 3 Å and are estimated to have bond energies of about 5 kcal/mol. Networks of these bonds can occur in proteins, however, collectively adding great stability to certain structural motifs. We shall see examples of this in Chapter 3 when we discuss protein secondary structure. H-bonding also contributes to the binding energy of ligands to enzyme active sites and can play an important role in the catalytic mechanism of the enzyme.

2.4.3 Hydrophobic Interactions

When a nonpolar molecule is dissolved in a polar solvent, such as water, it disturbs the H-bonding network of the solvent without providing compensat-

Table 2.3 Hydrogen bond lengths for H bonds found in proteins

Bond Type	Typical Length (Å)
O—H---O	2.70
O—H---O^{-}	2.63
O—H---N	2.88
N—H---O	3.04
N^{+}—H---O	2.93
N—H---N	3.10

ing H-bonding opportunities. Hence there is an entropic cost to the presence of nonpolar molecules in aqueous solutions. Therefore, if such a solution is mixed with a more nonpolar solvent, such as *n*-octanol, there will be a thermodynamic advantage for the nonpolar molecule to partition into the more nonpolar solvent. The same hydrophobic effect is seen in proteins. For example, amino acids with nonpolar side chains are most commonly found in the core of the folded protein molecule, where they are shielded from the polar solvent. Conversely, amino acids with polar side chains are most commonly found on the exterior surface of the folded protein molecule (see Chapter 3 for further details). Likewise, in the active sites of enzymes hydrophobic regions of the protein tend to stabilize the binding of hydrophobic molecules. The partitioning of hydrophobic molecules from solution to the enzyme active site can be a strong component of the overall binding energy. We shall discuss this further in Chapters 4, 6 and 8 in our examination of enzyme–substrate interactions and reversible enzyme inhibitors.

2.4.4   Van der Waals Forces

The distribution of electrons around an atom is not fixed; rather, the character of the so-called electron cloud fluctuates with time. Through these fluctuations, a transient asymmetry of electron distribution, or *dipole moment*, can be established. When atoms are close enough together, this asymmetry on one atom can influence the electronic distribution of neighboring atoms. The result is a similar redistribution of electron density in the neighbors, hence an attractive force between the atoms is developed. This attractive force, referred to as a *van der Waals bond*, is much weaker than either salt bridges or H bonds. Typically a van der Waals bond is worth only about 1 kcal/mol in bond energy. When conditions permit large numbers of van der Waals bonds to simultaneously form, however, their collective attractive forces can provide a significant stabilizing energy to protein–protein and protein–ligand interactions.

As just described, the attractive force between electron clouds increases as the two atoms approach each other but is counterbalanced by a repulsive force at very short distances. The attractive force, being dipolar, depends on the interatomic distance, R, as $1/R^6$. The repulsive force is due to the overlapping of the electron clouds of the individual atoms that would occur at very close distances. This force wanes quickly with distance, showing a $1/R^{12}$ dependence. Hence, the overall potential energy of a van der Waals interaction depends on the distance between nuclei as the sum of these attractive and repulsive forces:

$$\text{PE} = \frac{A}{R^{12}} - \frac{B}{R^6} \tag{2.14}$$

where PE is the potential energy, and A and B can be considered to be characteristic constants for the pair of nuclei involved. From Equation 2.14 we see that the optimal attraction between atoms occurs when they are separated

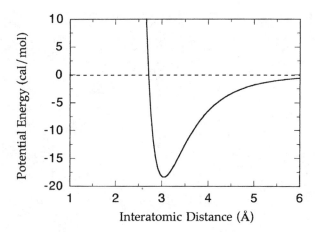

Figure 2.13 Potential energy diagram for the van der Waals attraction between two helium atoms. [Data adapted from Gray (1973) and fit to Equation 2.14.]

by a critical distance known as the *van der Waals contact distance* (Figure 2.13). The contact distance for a pair of atoms is determined by the individual van der Waals contact radius of each atom, which itself depends on the electronic configuration of the atom.

Table 2.4 provides the van der Waals radii for the most abundant atoms found in proteins. Imagine drawing a sphere around each atom with a radius defined by the van der Waals contact radius (Figure 2.14). These spheres, referred to as *van der Waals surfaces*, would define the closest contact that atoms in a molecule could make with one another, hence the possibilities for defining atom packing in a molecular structure. Because of the differences in radii, and the interplay between repulsive and attractive forces, van der Waals bonds and surfaces can play an important role in establishing the specificity of interactions between protein binding pockets and ligands. We shall have more to say about such specificity in Chapter 6, when we discuss enzyme active sites.

Table 2.4   Van der Waals radii for atoms in proteins

Atom	Radius (Å)
H	1.2
C	2.0
N	1.5
O	1.4
S	1.9
P	1.9

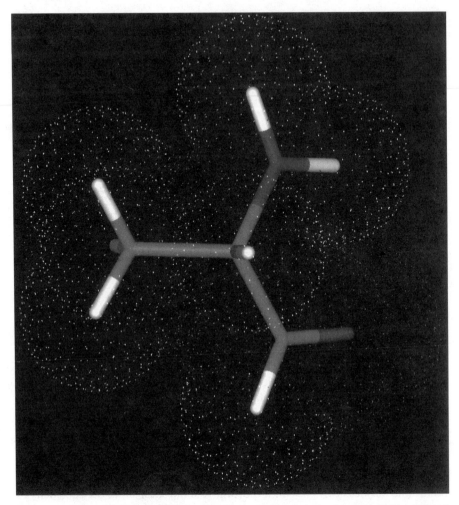

Figure 2.14 Van der Waals radii for the atoms of the amino acid alanine. The "tubes" represent the bonds between atoms. Oxygen is colored red, nitrogen is blue, carbon is green, and hydrogens are gray. The white dimpled spheres around each atom represent the van der Waals radii. (Diagram courtesy of Karen Rossi, Department of Computer Aided Drug Design, The DuPont Pharmaceuticals Company.) (See Color Plates.)

2.5 RATES OF CHEMICAL REACTIONS

The study of the rates at which chemical reactions occur is termed *kinetics*. We shall deal with the kinetics of enzyme-catalyzed reactions under steady state conditions in Chapter 5. Here we review basic kinetic principles for simple chemical reactions.

Let us consider a very simple chemical reaction in which a molecule S, decomposes irreversibly to a product P:

$$S \rightarrow P$$

The radioactive decay of tritium to helium is an example of such a chemical reactions:

$$^3H \rightarrow {_1}\beta^0 + {_2^3}He$$

At the start of the reaction we have some finite amount of S, symbolized by $[S]_0$. At any time later, the amount of S remaining will be less than $[S]_0$ and is symbolized by $[S]_t$. The amount of S will decline with time until there is no S left, at which point the reaction will stop. Hence, we expect that the reaction rate (also called reaction velocity) will be proportional to the amount of S present:

$$v = \frac{-d[S]}{dt} = k[S] \tag{2.15}$$

where v is the velocity and k is a constant of proportionality referred to at the *rate constant*. If we integrate this differential equation we obtain:

$$-\int d[S] = \int k[S]dt \tag{2.16}$$

Solving this integration we obtain:

$$[S]_t = [S]_0 \exp(-kt) \tag{2.17}$$

Equation 2.17 indicates that substrate concentration will decay exponentially from $[S]_t = [S]_0$ at $t = 0$ to $[S]_t = 0$ at infinite time. Over this same time period, the product concentration grows exponentially. At the start of the reaction ($t = 0$) there is no product; hence $[P]_0 = 0$. At infinite time, the maximum amount of product that can be produced is defined by the starting concentration of substrate, $[S]_0$; hence at infinite time $[P]_t = [S]_0$. At any time between 0 and infinity, we must have conservation of mass, so that:

$$[S]_t + [P]_t = [S]_0 \tag{2.18}$$

or

$$[P]_t = [S]_0 - [S]_t \tag{2.19}$$

Using Equation 2.17, we can recast Equation 2.19 as follows:

$$[P]_t = [S]_0 - [S]_0 e^{-kt} \tag{2.20}$$

or:

$$[P]_t = [S]_0(1 - e^{-kt}) \tag{2.21}$$

Hence, from Equations 2.17 and 2.21 we expect the concentrations of S and P to respectively decrease and increase exponentially, as illustrated in Figure 2.15.

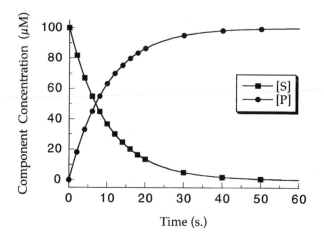

Figure 2.15 Progress curves of product development (circles) and substrate loss (squares) for a first-order reaction.

From Equation 2.17 we could ask the question, How much time is required to reduce the concentration of S to half its original value? To answer this we first rearrange Equation 2.17 as follows:

$$\frac{[S]_t}{[S]_0} = e^{-kt} \tag{2.22}$$

when $[S]_t$ is half of $[S]_0$ the ratio $[S]_t/[S]_0$ is obviously $\frac{1}{2}$. Using this equality and taking the natural logarithm of both sides and then dividing both sides by k, we obtain:

$$\frac{-\ln(\frac{1}{2})}{k} = \frac{0.6931}{k} = t_{1/2} \tag{2.23}$$

The value $t_{1/2}$ is referred to as the *half-life* of the reaction. This value is inversely related to the rate constant, but it provides a value in units of time that some people find easier to relate to. It is not uncommon, for example, for researchers to discuss radioactive decay in terms of isotope half-lives (Table 7.4 in Chapter 7 provides half-lives for four of the radioisotopes commonly used in enzyme studies).

2.5.1 Reaction Order

In the discussion above we considered the simplest of kinetic processes in which there was only one reactant and one product. From the rate equation, 2.15, we see that the velocity for this reaction depends linearly on initial reactant concentration. A reaction of this type is said to be a *first-order* reaction, and the rate constant, k, for the reaction is said to be a *first-order rate*

Table 2.5   Reaction order for a few simple chemical reactions

Order	Reaction	Rate Equation
1	$A \rightarrow P$	$v = k[A]$
2	$2A \rightarrow P$	$v = k[A]^2$
2	$A + B \rightarrow P$	$v = k[A][B]$

constant. Suppose that the form of our reaction was that two molecules of reactant A produced one molecule of product P:

$$2A \rightarrow P$$

If we now solve for the rate equation, we will find that it has the form:

$$v = k[A]^2 \qquad (2.24)$$

A reaction of this type would be said to be a *second-order* reaction. Generally, the order of a chemical reaction is the sum of the exponent terms to which reactant concentrations are raised in the velocity equation. A few examples of this are given in Table 2.5. A more comprehensive discussion of chemical reaction order and rate equations can be found in any good physical chemistry text (e.g., Atkins, 1978).

As we have just seen, reactions involving two reactants, such as $A + B \rightarrow P$, are strictly speaking always second order. Often, however, the reaction can be made to appear to be first order in one reactant when the second reactant is held at a constant, excess concentration. Under such conditions the reaction is said to be *pseudo–first order* with respect to the nonsaturating reactant. Such reactions appear kinetically to be first order and can be well described by a first-order rate equation. As we shall see in Chapters 4 and 5, under most experimental conditions the rate of ligand binding to receptors and the rates of enzyme-catalyzed reactions are most often pseudo–first order.

2.5.2   Reversible Chemical Reactions

Suppose that our simple chemical reaction $S \rightarrow P$ is reversible so that there is some rate of the forward reaction of S going to P, defined by rate constant k_f, and also some rate of the reverse reaction of P going to S defined by the rate constant k_r. Because of the reverse reaction, the reactant S is never completely converted to product. Instead, an equilibrium concentration of both S and P is established after sufficient time. The equilibrium constant for the forward reaction is given by:

$$K_{eq} = \frac{[P]}{[S]} = \frac{k_f}{k_r} \qquad (2.25)$$

The rate equation for the forward reaction must take into account the reverse reaction rate so that:

$$v = \frac{-d[S]}{dt} = k_f[S] - k_r[P] \tag{2.26}$$

Integrating this equation with the usual boundary conditions, we obtain:

$$[S]_t = [S]_0 \frac{k_r + k_f \exp[-(k_f + k_r)t]}{k_f + k_r} \tag{2.27}$$

Integration of the velocity equation with respect to product formation yields:

$$[P]_t = \frac{k_f[S]_0\{1 - \exp[-(k_f + k_r)t]\}}{k_f + k_r} \tag{2.28}$$

At infinite time (i.e., when equilibrium is reached), the final concentrations of S and P are given by the following equations:

$$[S]_{eq} = \frac{k_r[S]_0}{k_f + k_r} \tag{2.29}$$

$$[P]_{eq} = \frac{k_f[S]_0}{k_f + k_r} \tag{2.30}$$

Hence a reaction progress curve for a reversible reaction will follow the same exponential growth of product and loss of substrate seen in Figure 2.15, but now, instead of $[S] = 0$ and $[P] = [S]_0$ at infinite time, the curves will asymptotically approach the equilibrium values of the two species.

2.5.3   Measurement of Initial Velocity

The complexity of the rate equations presented in Section 2.5.2 made graphic determination of reaction rates a significant challenge prior to the widespread availability of personal computers with nonlinear curve-fitting software. For this reason researchers established the convention of measuring the initial rate or *initial velocity* of the reaction as a means of quantifying reaction kinetics. At the initiation of reaction no product is present, only substrate at concentration $[S]_0$. For a brief time after initiation, $[P] \ll [S]$, so that formation of P is quasi-linear with time. Hence, during this initial phase of the reaction one can define the velocity, $d[P]/dt = -d[S]/dt$, as the slope of a linear fit of [P] or [S] as a function of time. As a rule of thumb, this initial quasi-linear phase of the reaction usually extends over the time period between $[P] = 0$ and $[P] = 0.1[S]_0$ (i.e., until about 10% of substrate has been utilized). This rule, however, is merely a guideline, and the time window of the linear phase of any particular reaction must be determined empirically. Initial velocity measurements are used extensively in enzyme kinetics, as we shall see in Chapters 5 and 7.

2.6 SUMMARY

In this chapter we have briefly reviewed atomic and molecular orbitals and the types of bond formed within molecules as a result of these electronic configurations. We have seen that noncovalent forces also can stabilize interatomic interactions in molecules. Most notably, hydrogen bonds, salt bridges, hydrophobic interactions, and van der Waals forces can take on important roles in protein structure and function. We have also reviewed some basic kinetics and thermodynamics as well as acid–base theories that provide a framework for describing the reactivities of protein components in enzymology. In the chapters to come we shall see how these fundamental forces of chemistry come into play in defining the structures and reactivities of enzyme.

REFERENCES AND FURTHER READING

Atkins, P. W. (1978) *Physical Chemistry*, Freeman, New York.

Fersht, A. (1985) *Enzyme Structure and Mechanism*, 2nd ed., Freeman, New York.

Gray, H. B. (1973) *Chemical Bonds: An Introduction to Atomic and Molecular Structure*, Benjamin/Cummings, Menlo Park, CA.

Kemp, D. S., and Vellaccio, F. (1980) *Organic Chemistry*, Worth, New York.

Hammond, G. S. (1955) *J. Am. Chem. Soc.* 77, 334.

Lowry, T. H., and Richardson, K. S. (1981) *Mechanism and Theory in Organic Chemistry*, 2nd ed., Harper & Row, New York.

Palmer, T. (1985) *Understanding Enzymes*, Wiley, New York.

Pauling, L. (1960) *The Nature of the Chemical Bond*, 3rd ed., Cornell University Press, Ithaca, NY.

Stryer, L. (1989) *Molecular Design of Life*, Freeman, New York.

3

STRUCTURAL
COMPONENTS
OF ENZYMES

In Chapter 2 we reviewed the forces that come to play in chemical reactions, such as those catalyzed by enzymes. In this chapter we introduce the specific molecular components of enzymes that bring these forces to bear on the reactants and products of the catalyzed reaction. Like all proteins, enzymes are composed mainly of the 20 naturally occurring amino acids. We shall discuss how these amino acids link together to form the polypeptide backbone of proteins, and how these macromolecules fold to form the three-dimensional conformations of enzymes that facilitate catalysis. Individual amino acid side chains supply chemical reactivities of different types that are exploited by the enzyme in catalyzing specific chemical transformations. In addition to the amino acids, many enzymes utilize nonprotein *cofactors* to add additional chemical reactivities to their repertoire. We shall describe some of the more common cofactors found in enzymes, and discuss how they are utilized in catalysis.

3.1 THE AMINO ACIDS

An amino acid is any molecule that conforms, at neural pH, to the general formula:

$$H_3N^+—CH(R)—COO^-$$

The central carbon atom in this structure is referred to as the *alpha carbon* (C_α), and the substituent, R, is known as the *amino acid side chain*. Of all the possible chemical entities that could be classified as amino acids, nature has chosen to use 20 as the most common building blocks for constructing proteins and peptides. The structures of the side chains of the 20 naturally occurring amino acids are illustrated in Figure 3.1, and some of the physical properties of these molecules are summarized in Table 3.1. Since the alpha carbon is a chiral center, all the naturally occurring amino acids, except glycine, exist in two enantiomeric forms, L and D. All naturally occurring proteins are composed exclusively from the L enantiomers of the amino acids.

As we shall see later in this chapter, most of the amino acids in a protein or peptide have their charged amino and carboxylate groups neutralized through peptide bond formation (in this situation the amino acid structure that remains is referred to as an amino acid *residue* of the protein or peptide). Hence, what chemically and physically distinguishes one amino acid from another in a protein is the identity of the side chain of the amino acid. As seen in Figure 3.1, these side chains vary in chemical structure from simple substituents, like a proton in the case of glycine, to complex bicyclic ring systems in the case of tryptophan. These different chemical structures of the side chains impart vastly different chemical reactivities to the amino acids of a protein. Let us review some of the chemical properties of the amino acid side chains that can participate in the interaction of proteins with other molecular and macro-molecular species.

3.1.1 Properties of Amino Acid Side Chains

3.1.1.1 Hydrophobicity Scanning Figure 3.1, we note that several of the amino acids (valine, leucine, alanine, etc.) are composed entirely of hydrocar-bons. One would expect that solvation of such nonpolar amino acids in a polar solvent like water would be thermodynamically costly. In general, when hydrophobic molecules are dissolved in a polar solvent, they tend to cluster together to minimize the amount of surface areas exposed to the solvent; this phenomenon is known as *hydrophobic attraction*. The repulsion from water of amino acids in a protein provides a strong driving force for proteins to fold into three-dimensional forms that sequester the nonpolar amino acids within the interior, or *hydrophobic core*, of the protein. Hydrophobic amino acids also help to stabilize the binding of nonpolar substrate molecules in the binding pockets of enzymes.

The hydrophobicity of the amino acids is measured by their tendency to partition into a polar solvent in mixed solvent systems. For example, a molecule can be dissolved in a 50:50 mixture of water and octanol. After mixing, the polar and nonpolar solvents are allowed to separate, and the concentration of the test molecule in each phase is measured. The equilibrium constant for transfer of the molecule from octanol to water is then given by:

$$K_{transfer} = \frac{[C_{H_2O}]}{[C_{oct}]} \tag{3.1}$$

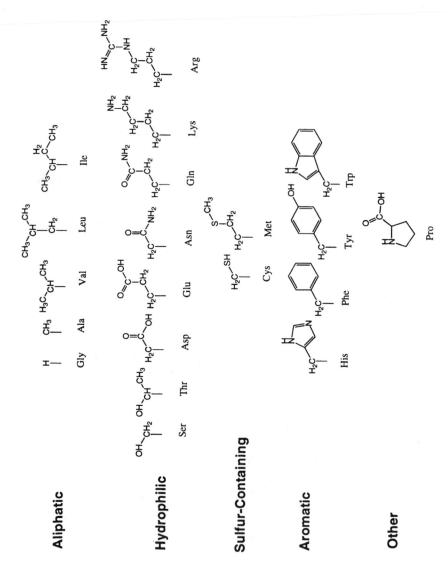

Figure 3.1 Side chain structures of the 20 natural amino acids. The entire proline molecule is shown.

Table 3.1 Physicochemical properties of the natural amino acids

Amino Acid	Three-Letter Code	One-Letter Code	Mass of Residue in Proteins[a]	Accessible Surface Area (Å²)[b]	Hydro-phobicity[c]	pK$_a$ of Ionizable Side Chain	Occurrence in Proteins (%)[d]	Relative Mutability[e]	Van der Waals Volume (Å³)
Alanine	Ala	A	71.08	115	+1.8		9.0	100	67
Arginine	Arg	R	156.20	225	-4.5	12.5	4.7	65	148
Asparagine	Asn	N	114.11	160	-3.5		4.4	134	96
Aspartate	Asp	D	115.09	150	-3.5	3.9	5.5	106	91
Cysteine	Cys	C	103.14	135	+2.5	8.4	2.8	20	86
Glutamate	Glu	E	128.14	180	-3.5	4.1	3.9	102	109
Glutamine	Gln	Q	129.12	190	-3.5		6.2	93	114
Glycine	Gly	G	57.06	75	-0.4		7.5	49	48
Histidine	His	H	137.15	195	-3.2	6.0	2.1	66	118
Isoleucine	Ile	I	113.17	175	+4.5		4.6	96	124
Leucine	Leu	L	113.17	170	+3.8		7.5	40	124
Lysine	Lys	K	128.18	200	-3.9	10.8	7.0	56	135
Methionine	Met	M	131.21	185	+1.9		1.7	94	124
Phenylalanine	Phe	F	147.18	210	+2.8		3.5	41	135
Proline	Pro	P	97.12	145	-1.6		4.6	56	90
Serine	Ser	S	87.08	115	-0.8		7.1	120	73
Threonine	Thr	T	101.11	140	-0.7		6.0	97	93
Tryptophan	Trp	W	186.21	255	-0.9		1.1	18	163
Tyrosine	Tyr	Y	163.18	230	-1.3	10.1	3.5	41	141
Valine	Val	V	99.14	155	+4.2		6.9	74	105

[a]Values reflect the molecular weights of the amino acids minus that of water.

[b]Accessible surface are for residues as part of a polypeptide chain. Data from Chothia (1975).

[c]Hydrophobicity indices from Kyte and Doolittle (1982).

[d]Based on the frequency of occurrence for each residue in the sequence of 207 unrelated proteins. Data from Klapper (1977).

[e]Likelihood that a residue will mutate within a specified time period during evolution. Data from Dayoff et al. (1978).

where $[C_{H_2O}]$ and $[C_{oct}]$ are the molar concentrations of the molecule in the aqueous and octanol phases, respectively. The free energy of transfer can then be calculated from the $K_{transfer}$ value using Equation 2.3. Such thermodynamic studies have been performed for the transfer of the naturally occurring amino acids from a number of nonpolar solvent to water. To make these measurements more representative of the hydrophobicities of amino acids within proteins, workers use analogues of the amino acids in which the amino and carboxylate charges are neutralized (e.g., using N-acetyl ethyl esters of the amino acids). Combining this type of thermodynamic information for the different solvent systems, one can develop a rank order of hydrophobicities for the 20 amino acids. A popular rank order used in this regard is that developed by Kyte and Doolittle (1982); the Kyte and Doolittle hydrophobicity indices for the amino acid are listed in Table 3.1. In general, hydrophobic amino acids are found on the interior of folded proteins, where they are shielded from the repulsive forces of the polar solvent, and polar amino acids tend to be found on the solvent-exposed surfaces of folded proteins.

3.1.1.2 Hydrogen Bonding

Associated with the heteroatoms of the side chains of several amino acids are exchangeable protons that can serve as hydrogen donors for H-bonding. Other amino acids can participate as H-bond acceptors through the lone pair electrons on heteroatoms of their side chains. Hydrogen bonding of amino acid side chains and polypeptide backbone groups can greatly stabilize protein structures, as we shall see later in this chapter. Additionally, hydrogen bonds can be formed between amino acid side chains and ligand (substrates, products, inhibitors, etc.) atoms and can contribute to the overall binding energy for the interactions of enzymes with such molecules. Side chains that are capable of acting as H-bond donors include tyrosine (—O—H), serine (—O—H), threonine (—O—H), tryptophan (—N—H), histidine (—N—H), and cysteine (—S—H). At low pH the side chains of glutamic and aspartic acid can also act as H-bond donors (—COO—H). Heteroatoms on the side chains of the following amino acids can serve as H-bond acceptors: tyrosine (—:O:—H), glutamic and aspartic acid (—COO⁻), serine and threonine (—:O:—H), histidine (N:), cysteine (—:S:—H), and methionine (:S:). Several of the amino acids can serve as both donors and acceptors, of H bonds, as illustrated for tyrosine in Figure 3.2.

3.1.1.3 Salt Bridge Formation

Noncovalent electrostatic interactions can occur between electonegative and electropositive species within proteins. Figure 3.3 illustrates the formation of such an electrostatic interaction between the side chains of a lysine residue on one polypeptide chain and a glutamic acid residue on another polypeptide. Because these interactions resemble the ionic interactions associated with small molecule salt formation, they are often referred to as *salt bridges*. Salt bridges can occur intramolecularly, between a charged amino acid side chain and other groups within the protein, or intermolecularly, between the amino acid side chain and charged groups on a

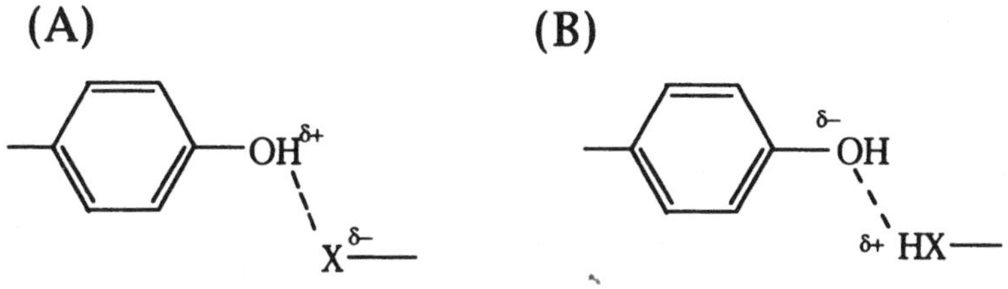

(A)   **(B)**

Figure 3.2 Tyrosine participation in hydrogen bonding as (A) a hydrogen donor and (B) a hydrogen acceptor.

ligand or other macromolecule. For example, many proteins that bind to nucleic acids derive a significant portion of their binding energy by forming electrostatic interactions between positively charged amino acid residues on their surfaces (usually lysine and arginine residues) and the negatively charged phosphate groups of the nucleic acid backbone.

Another example of the importance of these electrostatic interactions comes from the mitochondrial electron transfer cascade. Here electrons flow from the protein cytochrome c to the enzyme cytochrome oxidase, where they are used to reduce oxygen to water during cellular respiration. For the electron to jump from one protein to the other, the two must form a tight (dissociation constant $\approx 10^{-9}$ M) complex. When the crystal structure of cytochrome c was solved, it became obvious that the surface of this molecule contained an area with an unusually high density of positively charged lysine residues. The putative binding site for cytochrome c on the cytochrome oxidase molecule has a corresponding high density of aspartic and glutamic acid residues. It is thus believed that the tight complex formed between these two proteins is facilitated by forming a large number of salt bridges at this interface. This suggestion is supported by the ability of the complex to be dissociated by adding high concentrations of salt to the solution. As the ionic strength increases, the salt ions compete for the counterions from

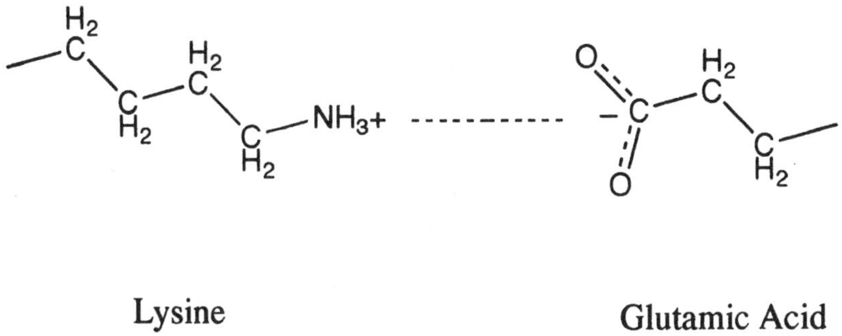

Lysine   Glutamic Acid

Figure 3.3 Salt bridge formation between a lysine and a glutamic acid residue at neutral pH.

the amino acid residues that would otherwise participate in salt bridge formation.

3.1.2 Amino Acids as Acids and Bases

Surveying Table 3.1, we see that there are seven amino acid side chains, with titratable protons that can act as Brønsted–Lowry acids and conjugate bases. These are tyrosine, histidine, cysteine, lysine, and arginine, and aspartic and glutamic acids. The ability of these side chains to participate in acid–base chemistry provides enzymes with a mechanism for proton transfer to and from reactant and product molecules. In addition to proton transfer, side chain Lewis acids and bases can participate in nucleophilic and electrophilic reactions with the reactant molecules, leading to bond cleavage and formation. The placement of acid and base groups from amino acid side chains, at critical positions within the active site, is a common mechanism exploited by enzymes to facilitate rapid chemical reactions with the molecules that bind in the active site. For example, hydrolysis of peptide and ester bonds can occur through nucleophilic attack of the peptide by water. This reaction goes through a transition state in which the carbonyl oxygen of the peptide has a partial negative charge, and the oxygen of water has a partial positive charge:

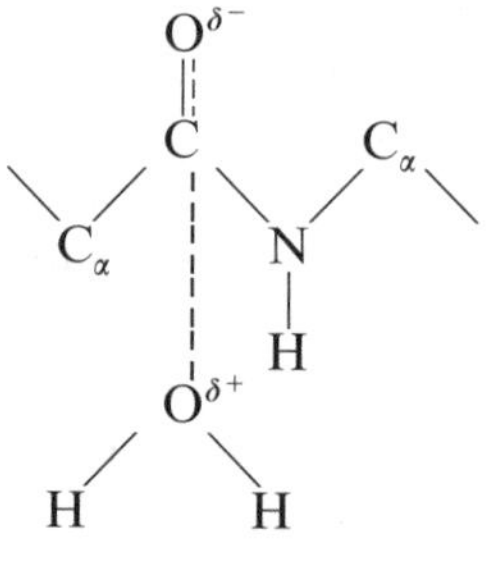

If one could place a basic group at a fixed position close to the water molecule, it would be possible to stabilize this transition state by partial transfer of one of the water protons to the base. This stabilization of the transition state would allow the reaction to proceed rapidly:

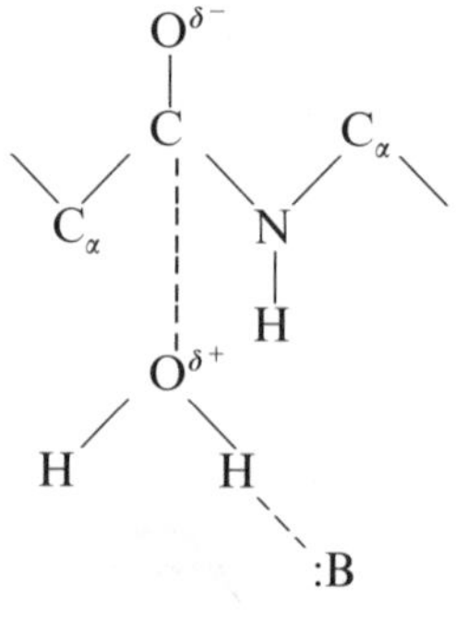

Alternatively, one could achieve the same stabilization by placing an acidic group at a fixed position in close proximity to the carbonyl oxygen, so that partial transfer of the proton from the acid to the carbonyl would stabilize the partial negative charge at this oxygen in the transition state. When one surveys the active sites of enzymes that catalyze peptide bond cleavage (a family of enzymes known as the proteases), one finds that there are usually acidic or basic amino acid side chains (or both) present at positions that are optimized for this type of transition state stabilization. We shall have more to say about stabilization of the transition states of enzyme reactions in Chapter 6.

In discussing the acid and base character of amino acid side chains, it is important to recognize that the pK_a values listed in Table 3.1 are for the side chain groups in aqueous solution. In proteins, however, these pK_a values can be greatly affected by the local environment that is experienced by the amino acid residue. For example, the pK_a of glutamic acid in aqueous solution is 4.1, but the pK_a of particular glutamic acid residues in some proteins can be as high as 6.5. Thus, while the pK_a values listed in Table 3.1 can provide some insights into the probable roles of certain side chains in chemical reactions, caution must be exercised to avoid making oversimplifications.

3.1.3 Cation and Metal Binding

Many enzymes incorporate divalent cations (Mg^{2+}, Ca^{2+}, Zn^{2+}) and transition metal ions (Fe, Cu, Ni, Co, etc.) within their structures to stabilize the folded conformation of the protein or to make possible direct participation in the chemical reactions catalyzed by the enzyme. Metals can provide a template for protein folding, as in the zinc finger domain of nucleic acid binding proteins, the calcium ions of calmodulin, and the zinc structural center of insulin. Metal ions can also serve as redox centers for catalysis; examples include heme–iron centers, copper ions, and nonheme irons. Other metal ions can serve as electrophilic reactants in catalysis, as in the case of the active site zinc ions of the metalloproteases. Most commonly metals are bound to the protein portion of the enzyme by formation of coordinate bonds with certain amino acid side chains: histidine, tyrosine, cysteine, and methionine, and aspartic and glutamic acids. Examples of metal coordination by each of these side chains can be found in the protein literature.

The side chain imidazole ring of histidine is a particularly common metal coordinator. Histidine residues are almost always found in association with transition metal binding sites on proteins and are very often associated with divalent metal ion binding as well. Figure 3.4, for example, illustrates the coordination sphere of the active site zinc of the enzyme carbonic anhydrase. Zinc typically forms four coordinate bonds in a tetrahedral arrangement about the metal ion. In carbonic anhydrase, three of the four bonds are formed by coordination to the side chains of histidine residues from the protein. The fourth coordination site is occupied by a water molecule that participates directly in catalysis. During the course of the enzyme-catalyzed reaction, the

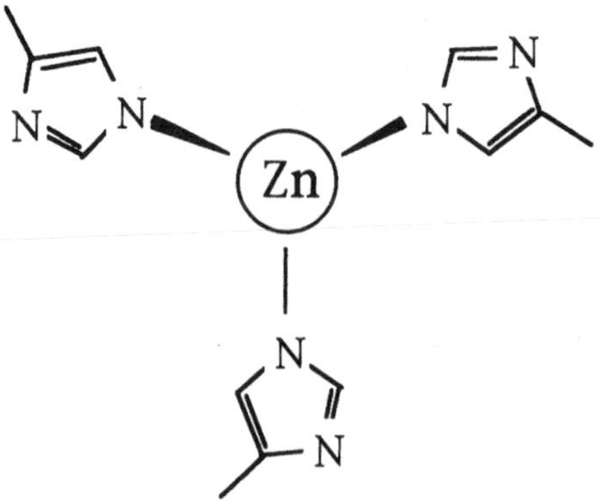

Figure 3.6 The coordination sphere of the active site zinc of carbonic anhydrase.

zinc–water bond is broken and replaced transiently by a bond between the metal and the carbon dioxide substrate of the reaction.

3.1.4 Anion and Polyanion Binding

The positively charged amino acids lysine and arginine can serve as counterions for anion and polyanion binding. Interactions of this type are important in binding of cofactors, reactants, and inhibitors to enzymes. Examples of anionic reactants and cofactors utilized by enzymes include phosphate groups, nucleotides and their analogues, nucleic acids, and heparin.

3.1.5 Covalent Bond Formation

We have pointed out that the chemical reactivities of amino acid residues within proteins are determined by the structures of their side chains. Several amino acids can undergo posttranslational modification (i.e., alterations that occur after the polypeptide chain has been synthesized at the ribosome) that alter their structure, hence reactivity, by means of covalent bond formation. In some cases, reversible modification of amino acid side chains is a critical step in the catalytic mechanism of the enzyme. Sections 3.1.5.1–3.1.5.3 give some examples of covalent bonds formed by amino acid side chains.

3.1.5.1 Disulfide Bonds   Two cysteine residues can cross-link, through an oxidative process, to form a sulfur–sulfur bond, referred to as a disulfide bond. These cross links can occur intramolecularly, between two cysteines within a single polypeptide, or intermolecularly, to join two polypeptides together. Such disulfide bond cross-linking can provide stabilizing energy to the folded conformation of the protein. Numerous examples exist of proteins that utilize both inter- and intramolecular disulfide bonds in their folded forms. Intermolecular disulfide bonds can also occur between a cysteine residue on a protein and a sulfhydryl group on a small molecule ligand or modifying reagent. For example, 4,4'-dithioldipyridine is a reagent used to quantify the number of free cysteines (those not involved in disulfide bonds) in proteins. The reagent reacts with the free sulfhydryls to form intermolecular disulfide bonds, with the liberation of a chromophoric by-product. The formation of the by-product is stoichiometric with reactive cysteines. Thus, one can quantify the number of cysteines that reacted from the absorbance of the by-product.

3.1.5.2 Phosphorylation

Certain amino acid side chains can be covalently modified by addition of a phosphate from inorganic phosphate (P_i). In nature, the phosphorylation of specific residues within proteins is facilitated by a class of enzymes known as the *kinases*. Another class of enzymes, the *phosphatases*, will selectively remove phosphate groups from these amino acids. This reversible phosphorylation/dephosphorylation can greatly affect the biological activity of enzymes, receptors, and proteins involved in protein–protein and protein–nucleic acid complex formation.

The most common sites for phosphorylation on proteins are the hydroxyl groups of threonine and serine residues; however, the side chains of tyrosine, histidine, and lysine can also be modified in this way (Figure 3.5). Tyrosine kinases, enzymes that specifically phosphorylate tyrosine residues within certain proteins, are of great current interest in biochemistry and cell biology. This is because it is recognized that tyrosine phosphorylation and dephosphorylation are critical in the transmission of chemical signals within cells (signal transduction).

Enzymes can also transiently form covalent bonds to phosphate groups during the course of catalytic turnover. In these cases, a *phosphoryl–enzyme intermediate* is formed by the transfer of an phosphate from substrate molecule or inorganic phosphate to specific amino acid side chains within the enzyme

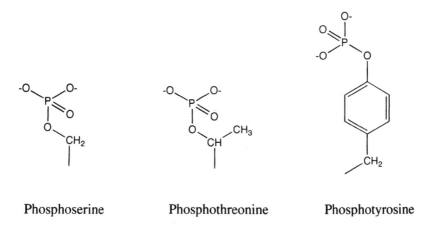

| Phosphoserine | Phosphothreonine | Phosphotyrosine |

Figure 3.5 The structures of the phosphorylated forms of serine, threonine, and tyrosine.

active site. Several examples of phosphoryl–enzyme intermediates are now known, which involve phosphoserine, phosphohistidine, and even phosphoaspartate formation. For example, the ATPases are enzymes that catalyze the hydrolysis of ATP (adenosine triphosphate) to form ADP (adenosine diphosphate) and P_i. In a subgroup of these enzymes, the Na^+, K^+, and the Ca^{2+} ATPases, the γ-phosphate of ATP is transferred to the β-carboxylate of an aspartic acid residue of the enzyme during the reaction. Since the phosphoaspartate is thermodynamically unstable, it very quickly dissociates to liberate inorganic phosphate.

3.1.5.3 *Glycosylation*

3.1.5.3 Glycosylation In eukaryotic cells, sugars can attach to proteins by covalent bond formation at the hydroxyl groups of serine and threonine residues (O-linked glycosylation) or at the nitrogen of asparagine side chains (N-linked glycosylation). The resulting protein–sugar complex is referred to as a *glycoprotein*. The sugars used for this purpose are composed of monomeric units of galactose, glucose, manose, *N*-acetylglucosamine, *N*-acetylgalactosamine, sialic acid, fructose, and xylose. The presence of these sugar moieties can significantly affect the solubility, folding, and biological reactivity of proteins.

3.1.6 Steric Bulk

Aside from the chemical reactivities already discussed, the stereochemistry of the amino acid side chains plays an important role in protein folding and intermolecular interactions. The size and shape of the side chain determines the type of packing interactions that can occur with neighboring groups, according to their van der Waals radii. It is the packing of amino acid side chains within the active site of an enzyme molecule that gives overall size and shape to the binding cavity (pocket), which accommodates the substrate molecule; hence these packing interactions help determine the specificity for binding of substrate and inhibitor molecules at these sites. This is a critical aspect of enzyme catalysis; in Chapter 6 we shall discuss further the relationship between the size and shape of the enzyme binding pocket and the structure of ligands.

For the aliphatic amino acids, side chain surface area also influences the overall hydrophobicity of the residue. The hydrophobicity of aliphatic molecules, in general, has been correlated with their exposed surface area. Hansch and Coats (1970) have made the generalization that the $\Delta G_{transfer}$ from a nonpolar solvent, like *n*-octanol, to water increases by about 0.68 kcal/mol for every methylene group added to an aliphatic structure. While this is an oversimplification, it serves as a useful rule of thumb for predicting the relative hydrophobicities of structurally related molecules. This relationship between surface area and hydrophobicity holds not only for the amino acids that line the binding pocket of an enzyme, but also for the substrate and inhibitor molecules that might bind in that pocket.

3.2 The Peptide Bond

The peptide bond is the primary structural unit of the polypeptide chains of proteins. Peptide bonds result from the condensation of two amino acids, as follows:

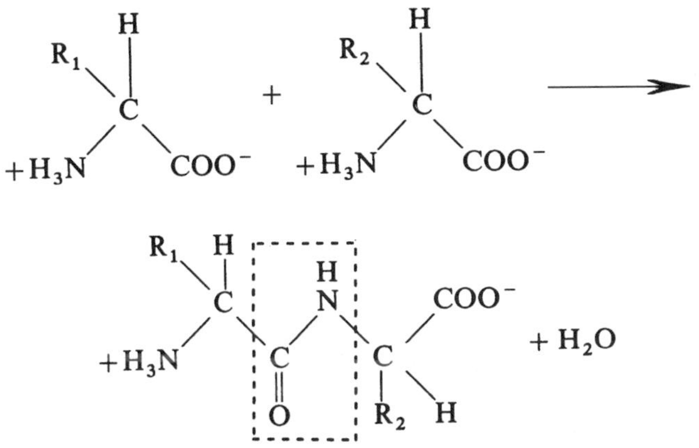

The product of such a condensation is referred to as a dipeptide, because it is composed of two amino acids. A third amino acid could condense with this dipeptide to form a tripeptide, a fourth to form a tetrapeptide, and so on. In this way chains of amino acids can be linked together to form *polypeptides* or *proteins*.

Until now we have drawn the peptide bond as an amide with double-bond character between the peptide carbon and the oxygen, and single-bond character between the peptide carbon and nitrogen atoms. Table 3.2 provides typical bond lengths for carbon–oxygen and carbon–nitrogen double and single bonds. Based on these data, one would expect the peptide carbon–oxygen bond length to be 1.22 Å and the carbon–nitogen bond length to be 1.45 Å. In fact, however, when x-ray crystallography was first applied to small peptides and other amide-containing molecules, it was found that the carbon–oxygen bond length was longer than expected, 1.24 Å, and the carbon–nitrogen bond length was shorter than expected, 1.32 Å (Figure 3.6). These

Table 3.2 Typical bond lengths for carbon–oxygen and carbon–nitrogen bonds

Bond Type	Bond Length (Å)
C—O	1.27
C=O	1.22
C—N	1.45
C=N	1.25

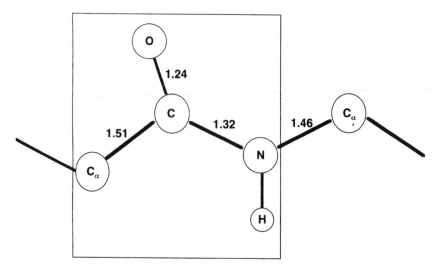

Figure 3.6 Schematic diagram of a typical peptide bond; numbers are typical bond lengths in angstrom units.

values are intermediate between those expected for double and single bonds. These results were rationalized by the chemist Linus Pauling by invoking two resonance structures for the peptide bond; one as we have drawn it at the beginning of this section, with all the double-bond character on the carbon–oxygen bond, and another in which the double bond is between the carbon and the nitrogen, and the oxygen is negatively charged. Thus the π system is actually delocalized over all three atoms, O—C—N. Based on the observed bond lengths, it was concluded that a peptide bond has about 60% C=O character and about 40% C=N character.

The 40% double-bond character along the C–N axis results in about a 20 kcal/mol resonance energy stabilization of the peptide. It also imposes a severe barrier to rotation about this axis. Hence, the six atoms associated with the peptide unit of polypeptides occur in a *planar* arrangement. The planarity of this peptide unit limits the configurations the polypeptide chain can adopt. It also allows for two stereoisomers of the peptide bond to occur: a *trans* configuration, in which the carbonyl oxygen and the nitrogenous proton are on opposite sides of the axis defined by the C—N bond, and a *cis* configuration, in which these groups are on the same side of this axis (Figure 3.7). When proteins are produced on the ribosomes of cells, they are synthesized stereospecifically. They could be synthesized with either all-cis or all-trans peptide bonds. However, because of the steric bulk of the amino acid side chains, polypeptides composed of cis peptide bonds are greatly restricted in terms of the conformational space they can survey. Thus, there is a significant thermodynamic advantage to utilizing trans peptide bonds for proteins and,

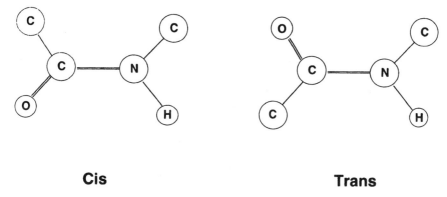

Cis   **Trans**

Figure 3.7 The cis and trans configurations of the peptide bond.

unsurprisingly, almost all the peptide bonds in naturally occurring proteins are present in the trans configuration.

An exception to this general rule is found in prolyl–peptide bonds. Here the energy difference between the cis and trans isomers is much smaller (≈ 2 kcal/ mol), and so the cis isomer can occur without a significant disruption in stability of the protein. Nevertheless, only a very few examples of cis prolyl– peptide bonds have ever been observed in nature. Three examples are known of cis prolyl–peptide bonds within enzymes from x-ray crystallographic studies: in ribonuclease-S (before Pro 93 and before Pro 114), in a subtilisin (before Pro 168), and in staphylococcal nuclease (before Pro 116). Thus, while the cis prolyl–peptide bond is energetically feasible, it is extremely rare. For our purposes, then, we can assume that all the peptide bonds in the proteins we shall be discussing are present in the trans configuration.

3.3 AMINO ACID SEQUENCE OR PRIMARY STRUCTURE

The structure and reactivity of a protein are defined by the identity of the amino acids that make up its polypeptide chain, and by the order in which those amino acids occur in the chain. This information constitutes the *amino acid sequence* or *primary structure* of the protein. Recall that we can link amino acids together through condensation reactions to form poly- peptide chains. We have seen that for most of the amino acids in such a chain, the condensation results in loss of the charged amino and carboxyl- ate moieties. No matter how many times we perform this condensation reaction, however, the final polypeptide will always retain a charged amino group at one end of the chain and a charged carboxylate at the other end. The terminal amino acid that retains the positively charged amino group is referred to as the *N-terminus* or *amino terminus*. The other terminal amino acid, retaining the negatively charged carboxylate group, is referred to as the *C-terminus* or *carboxy terminus*.

The individual amino acids in a protein are identified numerically in sequential order, starting with the N-terminus. The N-terminal amino acid is labeled number 1, and the numbering continues in ascending numerical order, ending with the residue at the carboxy terminus. Thus, when we read in the literature about "active site residue Ser 530," it means that the 530th amino acid from the N-terminus of this enzyme is a serine, and it occurs within the active site of the folded protein.

With the advent of recombinant DNA technologies, it has become commonplace to substitute amino acid residues within proteins (see Davis and Copeland, 1996, for a recent review of these methods). One may read, for example, about a protein in which a His 323—Asn mutation was induced by means of site-directed mutagenesis. This means that in the natural, or *wild-type* protein, a histidine residue occupies position 323, but through mutagenesis this residue has been replaced by an asparagine to create a mutant (or altered) protein. Studies in which mutant proteins have been purposely created through the methods of molecular biology are very common nowadays. It is important to remember, however, that mutations in protein sequences occur naturally as well. For the most part, point mutations of protein sequences occur with little effect on the biological activity of the protein, but in some cases the result is devastating. Consider, for example, the disease sickle cell anemia.

Patients with sickle cell anemia have a point mutation in their hemoglobin molecules. Hemoglobin is a tetrameric protein composed of four polypeptide chains: two identical α chains, and two identical β chains. Together, the four polypeptides of a hemoglobin molecule contain about 600 amino acids. The β chains of normal hemoglobin have a glutamic acid residue at position 6. The crystal structure of hemoglobin reveals that residue 6 of the β chains is at the solvent-exposed surface of the protein molecule, and it is thus not surprising to find a highly polar side chain, like glutamic acid, at this position. In the hemoglobin from sickle cell anemia patients, this glutamic acid is replaced by a valine, a very nonpolar amino acid. When the hemoglobin molecule is devoid of bound oxygen (the deoxy form, which occurs after hemoglobin has released its oxygen supply to the muscles), these valine residues on different molecules of hemoglobin will come together to shield themselves from the polar solvent through protein aggregation. This aggregation leads to long fibers of hemoglobin in the red blood cells, causing the cells to adopt the narrow elongated "sickle" shape that is characteristic of this disease. Thus with only two amino acid changes out of 600 (one residue per β chain), the entire biological activity of the protein is severely altered. (For a very clear and interesting account of the biochemistry of sickle cell anemia, see Stryer, 1989.) Sickle cell anemia was the first human disease that was shown to be caused by mutation of a specific protein (Pauling et al., 1949). We now know that there are many gene-based human diseases. Current efforts in "gene therapy" are aimed at correcting mutation-based diseases of these types.

3.4 SECONDARY STRUCTURE

We mentioned earlier that the delocalization of the peptide π system restricts rotation about the C—N bond axis. While this is true, it should be noted that free rotation is possible about the N—C$_\alpha$ and the C$_\alpha$—C' bond axes (where C' represents the carbonyl carbon of the amino acid residue). The dihedral angles defined by these two rotations are represented by the symbols ϕ and ψ, respectively, and are illustrated for a dipeptide in Figure 3.8.

Upon surveying the observed values for ϕ and ψ for amino acid residues in the crystal structures of proteins, one finds that certain values occur with high frequency. For any amino acid except glycine, a plot of the observed ϕ and ψ angles looks like Figure 3.9. This type of graph is known as a Ramachandran plot, after the scientist who first measured these angles and constructed such plots. The most obvious feature of Figure 3.9 is that the values of ϕ and ψ tend to cluster around two pairs of angles. The two regions of high frequency correspond to the angles associated with two commonly found regular and repeating structural motifs within proteins, the *right-handed α helix* and the *β-pleated sheet*. Both these structural motifs are examples of protein *secondary structure*, an important aspect of the overall conformation of any protein.

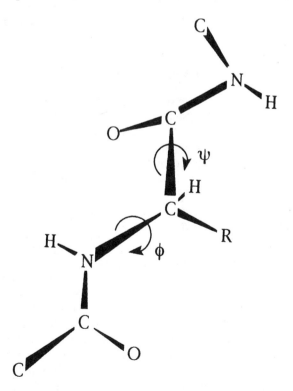

Figure 3.8   The dihedral angles of rotation for an amino acid in a peptide chain.

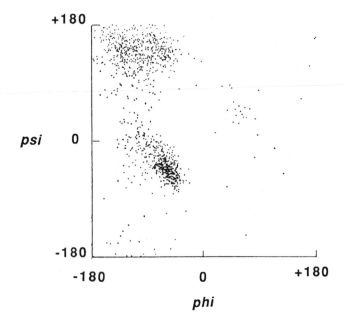

Figure 3.9   A Ramachandran plot for the amino acid alanine.

3.4.1 The Right-Handed α Helix

The right-handed α helix is one of the most commonly found protein secondary structures (Figure 3.10). This structure was first predicted by Linus Pauling, on the basis of the stereochemical properties of polypeptides. The structure is stabilized by a network of hydrogen bonds between the carbonyl oxygen of one residue (i) and the nitrogenous proton of residue i + 4. For most of the residues in the helix, there are thus two hydrogen bonds formed with neighboring peptide bonds, each contributing to the overall stability of the helix. As seen in Figure 3.10, this hydrogen-bonding network is possible because of the arrangement of C=O and N—H groups along the helical axis. The side chains of the amino acid residues all point away from the axis in this structure, thus minimizing steric crowding. The individual peptide bonds are aligned within the α-helical structure, producing, in addition, an overall dipole moment associated with the helix; this too adds some stabilization to the structure. The amino acid residues in an α helix conform to a very precise stereochemical arrangement. Each turn of an α helix requires 3.6 amino acid residues, with a translation along the helical axis of 1.5 Å per residue, or 5.4 Å per turn.

The network of hydrogen bonds formed in an α helix eliminates the possibility of those groups' participating in hydrogen bond formation with solvent–water molecules. For small peptides, the removal of these competing

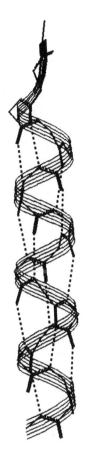

Figure 3.10 The right-handed α helix. (Figure provided by Dr. Steve Betz.)

solvent–hydrogen bonds, by dissolving the peptide in aprotic solvents, tends to promote the formation of α helices. The same tendency is observed when regions of a polypeptide are embedded in the hydrophobic interior of a cell membrane. Within the membrane bilayer, where hydrogen bonding with solvent is not possible, peptides and polypeptides tend to form α-helical structures. The hydrocarbon core cross section of a typical biological membrane is about 30 Å. Since an α helix translates 1.5 Å per residue, one can calculate that it takes, on average, about 20 amino acid residues, arranged in an α-helical structure, to traverse a membrane bilayer. For many proteins that are embedded in cell membranes (known as integral membrane proteins), one or more segments of 20 hydrophobic amino acids are threaded through the membrane bilayer as an α helix. These structures, often referred to as *trans-membrane α helices*, are among the main mechanisms by which proteins associate with cell membranes.

3.4.2 The β-Pleated Sheet

The β-pleated sheet, another very abundant secondary structure found in proteins, is composed of fully extended polypeptide chains linked together through interamide hydrogen bonding between adjacent strands of the sheet (Figure 3.11). Figure 3.11A illustrates a β-pleated sheet composed of two polypeptide chains. Note that the arrangement of peptide C=O and N—H groups permits this structure to be extended in either direction (i.e., to the left or right in the plane of the page as shown in Figure 3.11A) through the same type of interamide hydrogen bonding (Figure 3.11B).

For the moment, let us focus on a two-stranded β sheet, as in Figure 3.11A: the two component polypeptide chains could come from two distinct poly-peptides (an *intermolecular β sheet*) or from two regions of the same contiguous

A

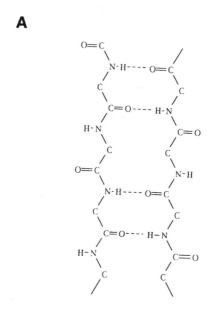

B

Figure 3.11   (A) A β-pleated sheet composed of two segments of polypeptide held together by interchain hydrogen bonding. (B) An extended β-pleated sheet composed of four segments of polypeptide.

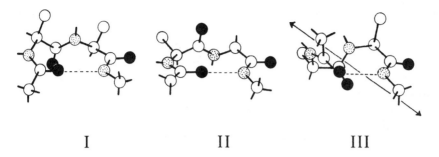

I   II   III

Figure 3.12 Three common forms of β turn.

polypeptide (an *intramolecular* β *sheet*); both types are found in natural proteins. If we imagine a β sheet within the plane of this page, we could have both chains running in the same direction, say from C-terminus at the top of the page to N-terminus at the botton. Alternatively, we could have the two chains running in opposite directions with respect to the placement of their N- and C-termini. These two situations describe structures referred to as *parallel* and *antiparallel* β-pleated sheets, respectively. Again, one finds both types in nature.

3.4.3 β Turns

A third common secondary structure found in natural proteins is the β turn (also known as a reverse turn, hairpin turn, or β bend). The β turns are short segments of the polypeptide chain that allow it to change direction—that is, to *turn* upon itself. Turns are composed of four amino acid residues in a compact configuration in which an interamide hydrogen bond is formed between the first and fourth residue to stabilize the structure. Three types of β turn are commonly found in proteins: types I, II, and III (Figure 3.12). Although turns represent small segments of the polypeptide chain, they occur often in a protein, allowing the molecule to adopt a compact three-dimensional structure. Consider, for example, an intramolecular antiparallel β sheet within a contiguous segment of a protein. To bring the two strands of the sheet into register for the correct hydrogen bonds to form, the polypeptide chain would have to change direction by $180°$. This can be accomplished only by incorporating a type I or type II β turn into the polypeptide chain, between the two segments making up the β sheet. Thus β turns play a very important role in establishing the overall three-dimensional structure of a protein.

3.4.4 Other Secondary Structures

One can imagine other regular repeating structural motifs that are stereochemically possible for polypeptides. In a series of adjacent type III β turns, for example, the polypeptide chain would adopt a helical structure, different from the α helix, that is known as a 3_{10} helix. This structure is indeed found in proteins, but it is rare. Some proteins, composed of high percentages of a single amino acid type, can adopt specialized helical structures, such as the polyproline helices and polyglycine helices. Again, these are special cases, not commonly found in the vast majority of proteins.

Most proteins contain regions of well-defined secondary structures interspersed with segments of nonrepeating, unordered structure in a conformation commonly referred to as *random coil* structure. These regions provide dynamic flexibility to the protein, allowing it to change shape, or conformation. These structural fluctuations can play an important role in facilitating the biological activities of proteins in general. They have particular significance in the cycle of substrate binding, catalytic transformations, and product release that is required for enzymes to function.

3.5 TERTIARY STRUCTURE

The term "*tertiary structure*" refers to the arrangement of secondary structure elements and amino acid side chain interactions that define the three-dimensional structure of the folded protein. Imagine that a newly synthesized protein exists in nature as a fully extended polypeptide chain—it is said then to be *unfolded* (Figure 3.13A) [Actually there is debate over how fully extended the polypeptide chain really is in the unfolded state of a protein; some data suggest that even in the unfolded state, proteins retain a certain amount of structure. However, this is not an important point for our present discussion.] Now suppose that this protein is placed under the set of conditions that will lead to the formation of elements of secondary structure at appropriate locations along the polypeptide chain (Figure 3.13B). Next, the individual elements of second-

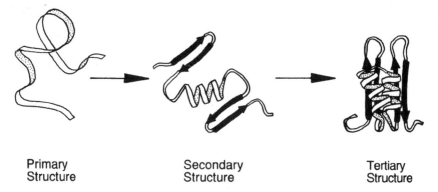

Primary Structure   Secondary Structure   Tertiary Structure

Figure 3.13 The folding of a polypeptide chain illustrating the hierarchy of protein structure from primary structure or amino acid sequence through secondary structure and tertiary structure. [Adapted from Dill et al., *Protein Sci.* **4**, 561 (1995).]

ary structure arrange themselves in three-dimensional space, so that specific contacts are made between amino acid side chains and between backbone groups (Figure 3.13C). The resulting *folded* structure of the protein is referred to as its *tertiary structure*.

What we have just described is the process of protein folding, which occurs naturally in cells as new proteins are synthesized at the ribosomes. The process is remarkable because under the right set of conditions it will also proceed spontaneously outside the cell in a test tube (in vitro). For example, at high concentrations chemicals like urea and guanidine hydrochloride will cause most proteins to adopt an unfolded conformation. In many cases, the subsequent removal of these chemicals (by dialysis, gel filtration chromatography, or dilution) will cause the protein to refold spontaneously into its correct native conformation (i.e., the folded state that occurs naturally and best facilitates the biological activity of the protein). The very ability to perform such experiments in the laboratory indicates that all the information required for the folding of a protein into its proper secondary and tertiary structures is encoded within the amino acid sequence of that protein.

Why is it that proteins fold into these tertiary structures? There are several important advantages to proper folding for a protein. First, folding provides a means of burying hydrophobic residues away from the polar solvent and exposing polar residues to solvent for favorable interactions. In fact, many scientists believe that the shielding of hydrophobic residues from the solvent is one of the strongest thermodynamic forces driving protein folding. Second, through folding the protein can bring together amino acid side chains that are distant from one another along the polypeptide chain. By bringing such groups into close proximity, the protein can form chemically reactive centers, such as the active sites of enzymes. An excellent example is provided by the serine protease chymotrypsin.

Serine proteases are a family of enzymes that cleave peptide bonds in proteins at specific amino acid residues (see Chapter 6 for more details). All these enzymes must have a serine residue within their active sites which functions as the primary nucleophile — that is, to attack the substrate peptide, thereby initiating catalysis. To enhance the nucleophilicity of this residue, the hydroxyl group of the serine side chain participates in hydrogen bonding with an active site histidine residue, which in turn may hydrogen-bond to an active site aspartate as shown in Figure 3.14. This "active site triad" of amino acids is a structural feature common to all serine proteases. In chymotrypsin this triad is composed of Asp 102, His 57, and Ser 195. As the numbering indicates, these three residues would be quite distant from one another along the fully extended polypeptide chain of chymotrypsin. However, the tertiary structure of chymotrypsin is such that when the protein is properly folded, these three residues come together to form the necessary interactions for effective catalysis.

The tertiary structure of a protein will often provide folds or pockets within the protein structure that can accommodate small molecules. We have already used the term "active site" several times, referring, collectively, to the chemically reactive groups of the enzyme that facilitate catalysis. The active site of

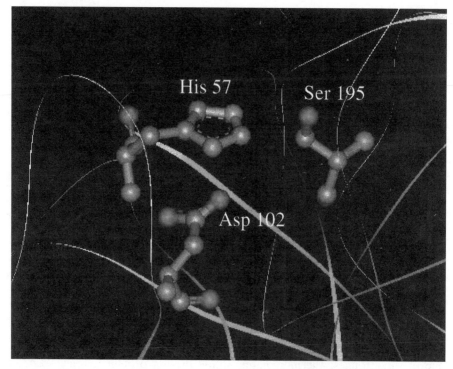

Figure 3.14 The active site triad of the serine protease α-chymotrypsin. [Adapted from the crystal structure reported by Frigerio et al. (1992) *J. Mol. Biol.* **225**, 107.] (See Color Plates.)

an enzyme is also defined by a cavity or pocket into which the substrate molecule binds to initiate the enzymatic reaction; the interior of this binding pocket is lined with the chemically reactive groups from the protein. As we shall see in Chapter 6, there is a precise stereochemical relationship between the structure of the molecules that bind to the enzyme and that of the active site pocket. The same is generally true for the binding of agonists and antagonists to the binding pockets of protein receptors. In all these cases, the structure of the binding pocket is dictated by the tertiary structure of the protein.

While no two proteins have completely identical three-dimensional structures, enzymes that carry out similar functions often adopt similar active site structures, and sometimes similar overall folding patterns. Some arrangements of secondary structure elements, which occur commonly in folded proteins, are referred to by some workers as *supersecondary structure*. Three examples of supersecondary structures are the helical bundle, the β barrel, and the β–α–β loop, illustrated in Figure 3.15.

In some proteins one finds discrete regions of compact tertiary structure that are separated by stretches of the polypeptide chain in a more flexible arrange-

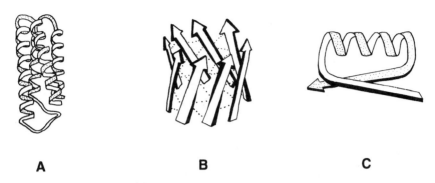

A **B** **C**

Figure 3.15 Examples of supersecondary structures: (A) a helical bundle, (B) a β barrel, and (C) a $\beta–\alpha–\beta$ loop.

ment. These discrete folded units are known as *domains*, and often they define functional units of the protein. For example, many cell membrane receptors play a role in signal transduction by binding extracellular ligands at the cell surface. In response to ligand binding, the receptor undergoes a structural change that results in macromolecular interactions between the receptor and other proteins within the cell cytosol. These interactions in turn set off a cascade of biochemical events that ultimately lead to some form of cellular response to ligand binding. To function in this capacity, such a receptor requires a minimum of three separate domains: an extracellular ligand binding domain, a transmembrane domain that anchors the protein within the cell membrane, and an intracellular domain that forms the locus for protein–protein interactions. These concepts are schematically illustrated in Figure 3.16.

Many enzymes are composed of discrete domains as well. For example, the crystal structure of the integral membrane enzyme prostaglandin synthase was recently solved by Garavito and his coworkers (Picot et al., 1994). The structure reveals three separate domains of the folded enzyme monomer: a β-sheet domain that functions as an interface for dimerization with another molecule of the enzyme, a membrane-incorporated α-helical domain that anchors the enzyme to the biological membrane, and a extramembrane globular (i.e., compact folded region) domain that contains the enzymatic active site and is thus the catalytic unit of the enzyme.

3.6 SUBUNITS AND QUATERNARY STRUCTURE

Not every protein functions as a single folded polypeptide chain. In many cases the biological activity of a protein requires two or more folded polypeptide chains to associate to form a functional molecule. In such cases the individual polypeptides of the active molecule are referred to as *subunits*. The subunits may be multiple copies of the same polypeptide chain (a homomultimer), or

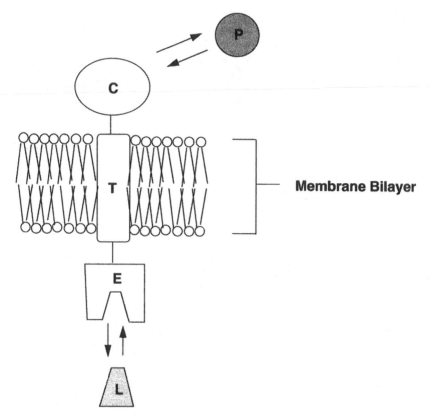

Figure 3.16 Cartoon illustration of the domains of a typical transmembrane receptor. The protein consists of three domains. The extracellular domain (E) forms the center for interaction with the receptor ligand (L). The transmembrane domain (T) anchors the receptor within the phospholipid bilayer of the cellular membrane. The cytosolic domain (C) extends into the intracellular space and forms a locus for interactions with other cytosolic proteins (P), which can then go on to transduce signals within the cell.

they may represent distinct polypeptides (a heteromultimer). In both cases the subunits fold as individual units, acquiring their own secondary and tertiary structures. The association between subunits may be stabilized through non-covalent forces, such as hydrogen bonding, salt bridge formation, and hydrophobic interactions, and may additionally include covalent disulfide bonding between cysteines on the different subunits.

There are numerous examples of multisubunit enzymes in nature, and a few are listed in Table 3.3. In some cases, the subunits act as quasi-independent catalytic units. For example, the enzyme prostaglandin synthase exists as a homodimer, with each subunit containing an independent active site that processes substrate molecules to product. In other cases, the active site of the enzyme is contained within a single subunit, and the other subunits serve to

Table 3.3 Examples of multisubunit enzymes

Enzyme	Number of Subunits
HIV protease	2
Hexokinase	2
Bacterial cytochrome oxidase	3
Lactate dehydrogenase	4
Aspartate carbamoyl transferase	12
Human cytochrome oxidase	13

stabilize the structure, or modify the reactivity of that active subunit. In the cytochrome oxidases, for example, all the active sites are contained in subunit I, and the other 3–12 subunits (depending of species) modify the stability and specific activity of subunit I. In still other cases the active site of the enzyme is formed by the coming together of the individual subunits. A good illustration of this comes from the aspartyl protease of the human immunodeficiency virus, HIV (the causal agent of AIDS). The active sites of all aspartyl proteases require a pair of aspartic acid residues for catalysis. The HIV protease is synthesized as a 99-residue polypeptide chain that dimerizes to form the active enzyme (a homodimer). Residue 25 of each HIV protease monomer is an aspartic acid residue. When the monomers combine to form the active homodimer, the two Asp 25 residues (designated Asp 25 and Asp 25′ to denote their locations on separate polypeptide chains) come together to form the active site structure. Without this subunit association, the enzyme could not perform its catalytic duties.

The arrangement of subunits of a protein relative to one another defines the *quaternary structure* of the protein. Consider a heterotrimeric protein composed of subunits A, B, and C. Each subunit folds into its own discrete tertiary structure. As suggested schematically in Figure 3.17, these three subunits could take up a number of different arrangements with respect to one another in three-dimensional space. This cartoon depicts two particular arrangements, or quaternary structures, that exist in equilibrium with each other. Changes in quaternary structure of this type can occur as part of the activity of many proteins, and these changes can have dramatic consequences.

An example of the importance of protein quaternary structure comes from examination of the biological activity of hemoglobin. Hemoglobin is the protein in blood that is responsible for transporting oxygen from the lungs to the muscles (as well as transporting carbon dioxide in the opposite direction). The active unit of hemoglobin is a heterotetramer, composed of two α subunits and two β subunits. Each of these four subunits contains a heme cofactor (see Section 3.7) that is capable of binding a molecule of oxygen. The affinity of the heme for oxygen depends on the quaternary structure of the protein and on the state of oxygen binding of the heme groups in the other three subunits (a

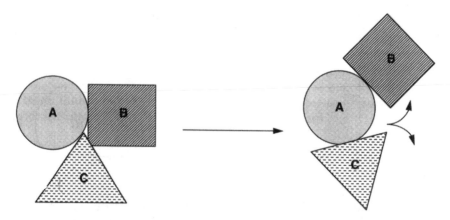

Figure 3.17 Cartoon illustrating the changes in subunit arrangements for a hypothetical heterotrimer that might result from a modification in quaternary structure.

property known as cooperativity). Because of the cooperativity of oxygen binding to the hemes, hemoglobin molecules almost always have all four heme sites bound to oxygen (the oxy form) or all four heme sites free of oxygen (the deoxy form); intermediate forms with one, two, or three oxygen molecules bound are almost never observed.

When the crystal structures of oxy- and deoxyhemoglobin were solved, it was discovered that the two forms differed significantly in quaternary structure. If we label the four subunits of hemoglobin α_1, α_2, β_1, and β_2, we find that at the interface between the α_1 and β_2 subunits, oxygen binding causes changes in hydrogen bonding and salt bridges that lead to a compression of the overall size of the molecule, and a rotation of $15°$ for the $\alpha_1\beta_1$ pair of subunits relative to the $\alpha_2\beta_2$ pair (Figure 3.18). These changes in quaternary structure in part affect the relative affinity of the four heme groups for oxygen, providing a means of reversible oxygen binding by the protein. It is the reversibility of the oxygen binding of hemoglobin that allows it to function as a biological transporter of this important energy source; hemoglobin can bind oxygen tightly in the lungs and then release it in the muscles, thus facilitating cellular respiration in higher organisms. (For a very clear description of all the factors leading to reversible oxygen binding and structural transitions in hemoglobin, see Stryer, 1989.)

3.7   COFACTORS IN ENZYMES

As we have seen, the structures of the 20 amino acid side chains can confer on enzymes a vast array of chemical reactivities. Often, however, the reactions catalyzed by enzymes require the incorporation of additional chemical groups to facilitate rapid reaction. Thus to fulfill reactivity needs

Quaternary Structures of Hemoglobin

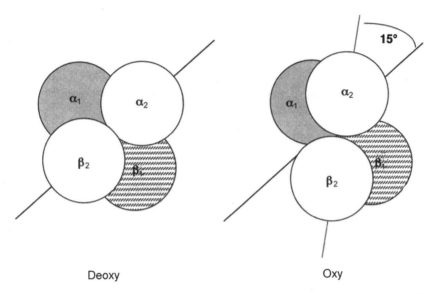

Figure 3.18 Cartoon illustration of the quaternary structure changes that accompany the binding of oxygen to hemoglobin.

that cannot be achieved with the amino acids alone, many enzymes incorporate nonprotein chemical groups into the structures of their active sites. These nonprotein chemical groups are collectively referred to as enzyme *cofactors* or *coenzymes*; Figure 3.19 presents the structures of some common enzyme cofactors.

In most cases, the cofactor and the enzyme associate through noncovalent interactions, such as those described in Chapter 2 (e.g., H-bonding, hydrophobic interactions). In some cases, however, the cofactors are covalently bonded to the polypeptide of the enzyme. For example, the heme group of the electron transfer protein cytochrome *c*, is bound to the protein through thioester bonds with two modified cysteine residues. Another example of covalent cofactor incorporation is the pyridoxal phosphate cofactor of the enzyme aspartate aminotransferase. Here the cofactor is covalently linked to the protein through formation of a Schiff base with a lysine residue in the active site.

In enzymes requiring a cofactor for activity, the protein portion of the active species is referred to as the *apoenzyme*, and the active complex between the protein and cofactor is called the *holoenzyme*. In some cases the cofactors can be removed to form the apoenzyme and be added back later to reconstitute the active holoenzyme. In some of these cases, chemically or isotopically modified versions of the cofactor can be incorporated into the apoenzyme to facilitate structural and mechanistic studies of the enzyme.

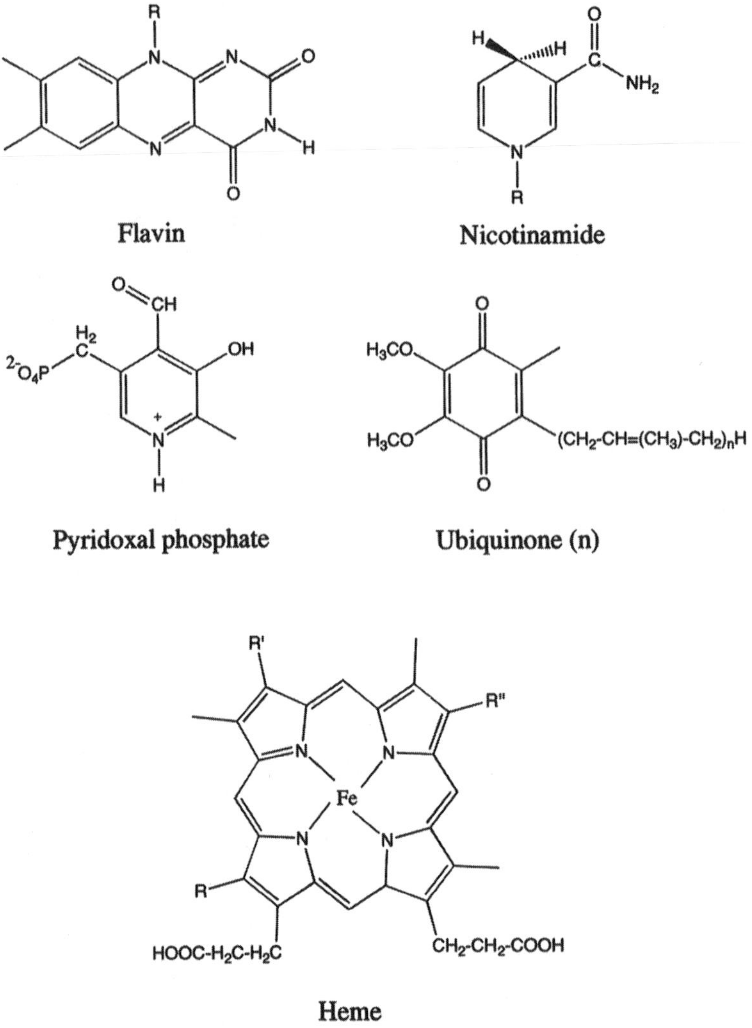

Flavin

Nicotinamide

Pyridoxal phosphate

Ubiquinone (n)

Heme

Figure 3.19 Examples of some common cofactors found in enzymes.

Cofactors fulfill a broad range of reactions in enzymes. One of the more common roles of enzyme cofactors is to provide a locus for oxidation/reduction (redox) chemistry at the active site. An illustrative example of this is the chemistry of flavin cofactors.

Flavins (from the Latin word *flavius*, meaning yellow) are bright yellow (λ_{max} = 450 nm) cofactors common to oxidoreductases, dehydrogenases, and electron transfer proteins. The main structural feature of the flavin cofactor is the highly conjugated isoalloxazine ring system (Figure 3.19). Oxidized flavins readily undergo reversible two-electron reduction to 1,5-dihydroflavin, and thus can act as electron sinks during redox reactions within the enzyme active site. For example, a number of dehydrogenases use flavin cofactors to accept two electrons during catalytic oxidation of NADH (another common enzyme cofactor; see Nicotinamide in Figure 3.19). Alternatively, flavins can undergo discrete one-electron reduction to form a semiquinone radical; this can be further reduced by a second one-electron reduction reaction to yield the fully reduced cofactor. Through this chemistry, flavin cofactors can participate in one-electron oxidations, such as those carried out during respiratory electron transfer in mitochondria. There are actually two stable forms of the flavin semiquinone that interconvert, depending on pH (Figure 3.20) The blue neutral semiquinone occurs at neutral and acidic pH, while the red anionic semiquinone occurs above pH 8.4. Both forms can be stabilized and observed in certain enzymatic reactions.

Additional chemical versatility is demonstrated by flavin cofactors in their ability to form covalent adducts with substrate during redox reactions. The oxidation of dithiols to disulfides by the active site flavin of glutathione reductase is an example of this. Here the thiolate anion adds to the C4a carbon of the isoalloxazine ring system (Figure 3.21A). Likewise, in a number of flavoenzyme oxidases, catalytic reoxidation of reduced flavin by molecular oxygen proceeds with formation of a transient C4a peroxide intermediate (Figure 3.21B).

A variety of other cofactors participate in the catalytic chemistry of the enzyme active site. Some additional examples of these are listed in Table 3.4. This list is, however, far from comprehensive; rather it gives just a hint of the breadth of structures and reactivities provided to enzymes by various cofactors. The texts by Dixon and Webb (1979) Walsh (1979), and Dugas and Penney (1981) give more comprehensive treatments of enzyme cofactors and the chemical reactions they perform.

3.8 SUMMARY

In this chapter we have seen the diversity of chemical reactivities that are imparted to enzymes by the structures of the amino acid side chains. We have described how these amino acids can be linked together to form a polypeptide chain, and how these chains fold into regular patterns of secondary and tertiary structure. The folding of an enzyme into its correct tertiary structure provides a means of establishing the binding pockets for substrate ligands and presents, within these binding pockets, the chemically reactive groups required for catalysis. The active site of the enzyme is defined by these reactive groups, and by the overall topology of the binding pocket. We have seen that the chemically reactive groups used to convert substrate to product molecules are recruited by enzymes, not only from the amino acids that make up the protein, but from cofactor molecules as well; these cofactors are critical components of the biologically active enzyme molecule. In Chapter 6, we shall see how the

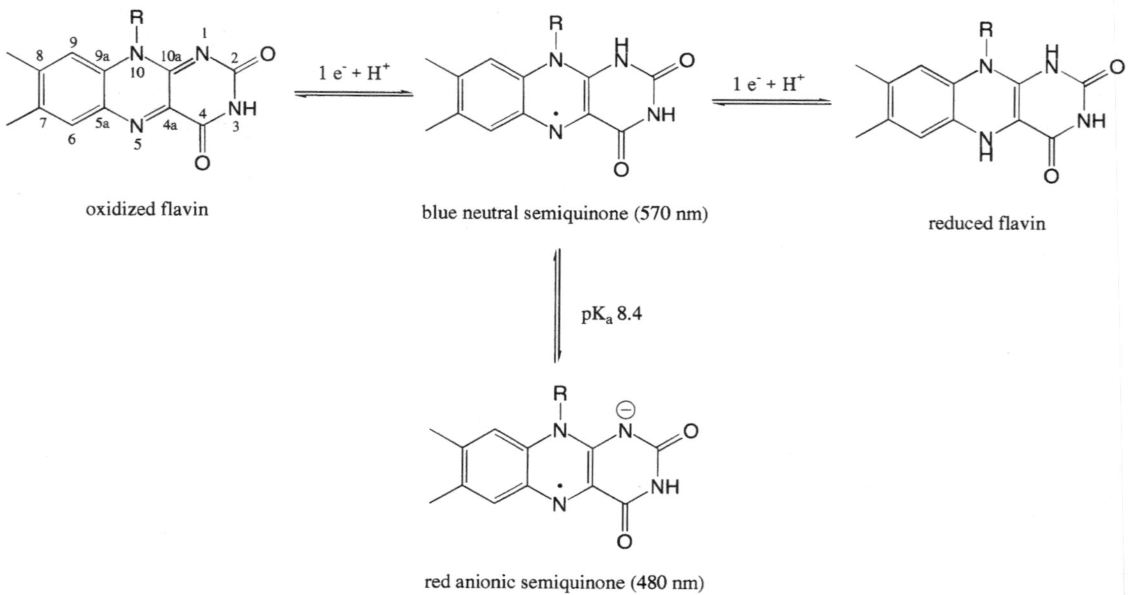

Figure 3.20 Structures of the flavin cofactor in its various oxidation states.

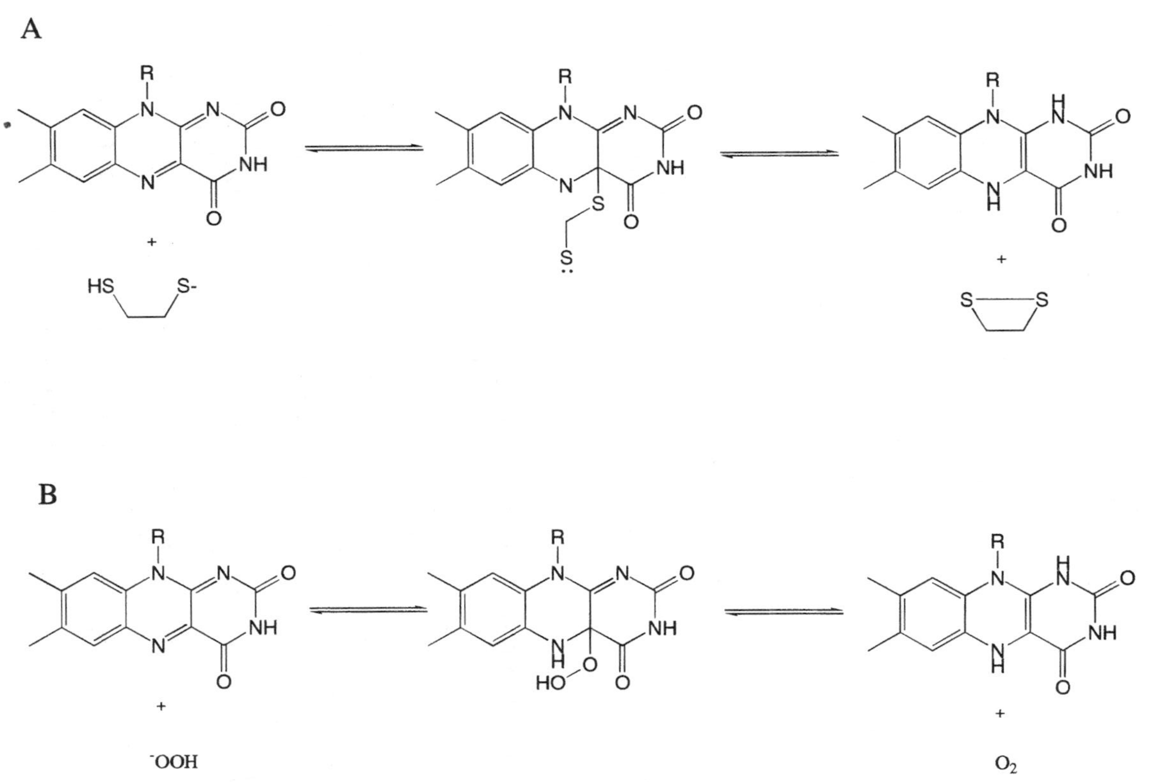

Figure 3.21 Covalent adduct formation by flavin cofactors in enzymes. (A) Oxidation of dithiols to disulfides through formation of a C4a–dithiol complex, as in the reaction of glutathione reductase with dithiols. (B) Reoxidation of reduced flavin by molecular oxygen, with formation of a C4a–peroxide intermediate.

Table 3.4 Some examples of cofactors found in enzymes

Cofactor	Enzymatic Use	Examples of Enzyme
Copper ion	Redox center–ligand binding	Cytochrome oxidase, superoxide
Magnesium ion	Active site electrophile– phosphate binding	dismutase phosphodiesterases, ATP synthases
Zinc ion	Active site electrophile	Matrix metalloproteases, carboxypeptidase A
Flavins	Redox center–proton transfer	Glucose oxidase, succinate dehydrogenase
Hemes	Redox center–ligand binding	Cytochrome oxidase, cytochrome P450s
NAD and NADP	Redox center–proton transfer	Alcohol dehydrogenase, ornithine cyclase
Pyridoxal phosphate	Amino group transfer– stabilizer of intermediate carbanions	Aspartate transaminase, arginine racemase
Quinones	Redox center–hydrogen transfer	Cytochrome b_o, dihydroorotate dehydrogenase
Coenzyme A	Acyl group transfer	Pyruvate dehydrogenase,

structural details of the enzyme active site facilitate substrate binding and the acceleration of reaction rates, which are the hallmarks of enzymatic catalysis.

REFERENCES AND FURTHER READING

Branden, C., and Tooze, J. (1991) *Introduction to Protein Structure*, Garland, New York.

Chotia, C. (1975) *J. Mol. Biol.* **105**, 1.

Copeland, R. A. (1994) *Methods for Protein Analysis: A Practical Guide to Laboratory Protocols*, Chapman & Hall, New York.

Creighton, T. E. (1984) *Proteins, Structure and Molecular Properties*, Freeman, New York.

Davis, J. P., and Copeland, R. A. (1996) Protein engineering, in *Kirk-Othmer Encyclopedia of Chemical Technology*, Vol. 20, 4th ed., Wiley, New York.

Dayoff, M. O. (1978) Atlas of Protein Sequence and Structure, Vol. 5, Suppl. 3, National Biomedical Res. Foundation, Washington, D.C.

Dill, K. A., Bromberg, S., Yue, K., Fiebig, K. M., Yee, D. P., Thomas, P. D., and Chan, H. S. (1995) *Protein Sci.*, **4**, 561.

Dixon, M., and Webb, E. C. (1979) *Enzymes*, 3rd ed., Academic Press, New York.

Dugas, H., and Penny, C. (1981) *Bioorganic Chemistry, A Chemical Approach to Enzyme Action*, Springer-Verlag, New York.

Hansch, C., and Coats, E. (1970) *J. Pharm. Sci.* **59**, 731.

Klapper, M. H. (1977) *Biochem. Biophys. Res. Commun.* **78**, 1018.

Kyte, J., and Doolittle, R. F. (1982) *J. Mol. Biol.* **157**, 105.

Pauling, L., Itano, H. A., Singer, S. J., and Wells, I. C. (1949) *Science*, **110**, 543.

Picot, D., Loll, P., and Garavito, R. M. (1994) *Nature*, **367**, 243.

Stryer, L. (1989) *Molecular Design of Life*, Freeman, New York.

Walsh, C. (1979) *Enzyme Reaction Mechanisms*, Freeman, New York.

4

PROTEIN–LIGAND BINDING EQUILIBRIA

Enzymes catalyze the chemical transformation of one molecule, the substrate, into a different molecular form, the product. For this chemistry to proceed, however, the enzyme and substrate must first encounter one another and form a binary complex through the binding of the substrate to a specific site on the enzyme molecule, the active site. In this chapter we explore the binding interactions that occur between macromolecules, such as proteins (e.g., enzymes) and other molecules, generally of lower molecular weight. These binding events are the initiators of most of the biochemical reactions observed both in vitro and in vivo. Examples of these interactions include agonist and antagonist binding to receptors; protein–protein and protein–nucleic acid complexation; substrate, activator, and inhibitor binding to enzymes; and metal ion and cofactor binding to proteins. We shall broadly define the smaller molecular weight partner in the binding interaction as the *ligand* (L) and the macromolecular binding partner as the *receptor* (R). Mathematical expressions will be derived to describe these interactions quantitatively. Graphical methods for representing experimental data will be presented that allow one to determine the equilibrium constant associated with complex dissociation. The chapter concludes with a brief survey of experimental methods for studying protein–ligand interactions.

4.1 THE EQUILIBRIUM DISSOCIATION CONSTANT, K_d

We begin this chapter by defining the equilibrium between free (i.e., unoccupied) and ligand-bound receptor molecules. We will assume, for now, that the receptor has a single binding site for the ligand, so that any molecule of receptor is either free or ligand bound. Likewise, any ligand molecule must be either free or bound to a receptor molecule. This assumption leads to the following pair of mass conservation equations:

$$[R] = [RL] + [R]_f \tag{4.1}$$

$$[L] = [RL] + [L]_f \tag{4.2}$$

where $[R]$ and $[L]$ are the *total* concentrations of receptor and ligand, respectively, $[R]_f$ and $[L]_f$ are the free concentrations of the two molecules, and $[RL]$ is the concentration of the binary receptor–ligand complex.

Under any specific set of solution conditions, an equilibrium will be established between the free and bound forms of the receptor. The position of this equilibrium is most commonly quantified in terms of the dissociation constant, K_d, for the binary complex at equilibrium:

$$K_d = \frac{[R]_f[L]_f}{[RL]} \tag{4.3}$$

The relative affinities (i.e., strength of binding) of different receptor–ligand complexes are inversely proportional to their K_d values; the tighter the ligand binds, the lower the value of the dissociation constant. Dissociation constants are thus used to compare affinities of different ligands for a particular receptor, and likewise to compare the affinities of different receptors for a common ligand.

The dissociation constant can be related to the Gibbs free energy of binding for the receptor–ligand complex (Table 4.1) as follows:

$$\Delta G_{binding} = RT\ln(K_d) \tag{4.4}$$

The observant reader may have noted the sign change here, relative to the

Table 4.1 Relationship between K_d and $\Delta G_{binding}$ for receptor–ligand complexes at 25°C

K_d (M)	$\Delta G_{binding}$ (kcal/mol)[a]
10^{-3} (mM)	−4.08
10^{-4}	−5.44
10^{-5}	−6.80
10^{-6} (µM)	−8.16
10^{-7}	−9.52
10^{-8}	−10.87
10^{-9} (nM)	−12.23
10^{-10}	−13.59
10^{-11}	−14.95
10^{-12} (pM)	−16.31

[a]Calculated using Equation 4.4.

typical way of expressing the Gibbs free energy. This is because in general chemistry texts, the free energy is more usually presented in terms of the forward reaction, that is, in terms of the *association* constant:

$$\Delta G_{\text{binding}} = - RT \ln \left(\frac{[RL]}{[R]_f [L]_f} \right) \tag{4.5}$$

or

$$\Delta G_{\text{binding}} = - RT \ln(K_a) \tag{4.6}$$

The sign conversion between Equations 4.4 and 4.6 merely reflects the inverse relationship between equilibrium association and dissociation constants:

$$K_d = \frac{1}{K_a} \tag{4.7}$$

At this point the reader may wonder why biochemists choose to express affinity relationships in terms of dissociation constants rather than the more conventional association constants, as found in most chemistry and physics textbooks. The reason is that the dissociation constant has units of molarity and can thus be equated with a specific ligand concentration that leads to half-maximal saturation of the available receptor binding sites; this will become evident in Section 4.3, where we discuss equilibrium binding experiments.

4.2 THE KINETIC APPROACH TO EQUILIBRIUM

Let us begin by considering a simple case of bimolecular binding without any subsequent chemical steps. Suppose that the two molecules, R and L, bind reversibly to each other in solution to form a binary complex, RL. The equilibrium between the free components, R and L, and the binary complex, RL, will be governed by the rate of complex formation (i.e., association of the complex) and by the rate of dissociation of the formed complex. Here we will define the second-order rate constant for complex association as k_{on} and the first-order rate constant for complex dissociation as k_{off}.

$$R + L \underset{k_{off}}{\overset{k_{on}}{\rightleftharpoons}} RL \tag{4.8}$$

The equilibrium dissociation constant for the complex is thus given by the ratio of k_{off} to k_{on}:

$$K_d = \frac{k_{off}}{k_{on}} \tag{4.9}$$

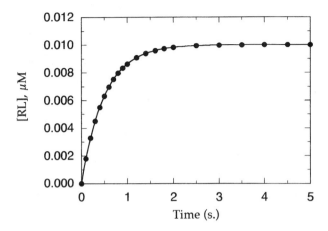

Figure 4.1 Time course for approach to equilibrium after mixing of a receptor and a ligand. The data are fit by nonlinear regression to Equation 4.10, from which an estimate of the observed pseudo-first-order rate constant (k_{obs}) is obtained.

In the vast majority of cases, the strength of interaction (i.e., affinity) between the receptor and ligand is such that a large excess of ligand concentration is required to effect significant binding to the receptor. Hence, under most experimental conditions association to form the binary complex proceeds with little change in the concentration of free (i.e., unbound) ligand; thus the association reaction proceeds with pseudo-first-order kinetics:

$$[RL]_t = [RL]_{eq}[1 - \exp(-k_{obs}t) \tag{4.10}$$

where $[RL]_t$ is the concentration of binary complex at time t, $[RL]_{eq}$ is the concentration of binary complex at equilibrium, and k_{obs} is the experimentally determined value of the pseudo-first-order rate constant for approach to equilibrium. Figure 4.1 illustrates a typical kinetic progress curve for a receptor–ligand pair approaching equilibrium binding. The line drawn through the data in this figure is a nonlinear least-squares best fit to Equation 4.10, from which the researcher can obtain an estimate of k_{obs}.

For reversible binding, it can be shown that the value of k_{obs} is directly proportional to the concentration of ligand present as follows:

$$k_{obs} = k_{off} + k_{on}[L]_f \tag{4.11}$$

Hence, one can determine the value of k_{obs} at a series of ligand concentrations from experiments such as that illustrated in Figure 4.1. A replot of k_{obs} as a function of ligand concentration should then yield a linear fit with slope equal to k_{on} and y intercept equal to k_{off} (Figure 4.2). From these values, the equilibrium dissociation constant can be determined by means of Equation 4.9.

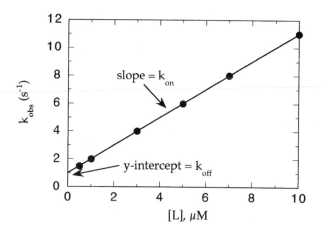

Figure 4.2 Plot of the observed pseudo-first-order rate constant (k_{obs}) for approach to equilibrium as a function of ligand concentration. As illustrated, the data are fit to a linear function from which the values of k_{on} and k_{off} are estimated from the values of the slope and y intercept, respectively.

4.3 BINDING MEASUREMENTS AT EQUILIBRIUM

While in principle the dissociation constant can be determined from kinetic studies, in practice these kinetics usually occur on a short (ca. millisecond) time scale, making them experimentally challenging. Hence, researchers more commonly study receptor–ligand interactions after equilibrium has been established.

4.3.1 Derivation of the Langmuir Isotherm

At equilibrium, the concentration of the RL complex is constant. Hence the rates of complex association and dissociation must be equal. Referring back to the equilibrium scheme shown in Equation 4.3, we see that the rates of association and dissociation are given by Equations 4.12 and 4.13, respectively:

$$\frac{d[\mathrm{RL}]}{dt} = k_{on}[\mathrm{R}]_f[\mathrm{L}]_f \tag{4.12}$$

$$\frac{-d[\mathrm{RL}]}{dt} = k_{off}[\mathrm{RL}] \tag{4.13}$$

Again, at equilibrium, these rates must be equal. Thus:

$$k_{on}[\mathrm{R}]_f[\mathrm{L}]_f = k_{off}[\mathrm{RL}] \tag{4.14}$$

or

$$[RL] = \frac{k_{on}}{k_{off}}[R]_f[L]_f \qquad (4.15)$$

Using the equality that k_{on}/k_{off} is equivalent to the equilibrium association constant K_a, we obtain:

$$[RL] = K_a[R]_f[L]_f \qquad (4.16)$$

Now, in most experimental situations a good measure of the actual concentration of free ligand or protein is lacking. Therefore, we would prefer an equation in terms of the total concentrations of added protein and/or ligand, quantities that are readily controlled by the experimenter. If we take the mass conservation equation, 4.1, divide both sides by $1 + ([RL]/[R]_f)$, and apply a little algebra, we obtain:

$$[R]_f = \frac{[R]}{1 + \frac{[RL]}{[R]_f}} \qquad (4.17)$$

Using the equality that $K_a = [RL]/[R]_f[L]_f$, and a little more algebra, we obtain:

$$[RL] = K_a[L]_f \frac{[R]}{1 + K_a[L]_f} \qquad (4.18)$$

Rearranging and using the equality $K_d = 1/K_a$, leads to:

$$[RL] = \frac{[R][L]_f}{K_d + [L]_f} \qquad (4.19)$$

Again we consider that under most conditions the concentration of receptor is far less than that of the ligand. Hence, formation of the binary complex does not significantly diminish the concentration of free ligand. We can thus make the approximation that the free ligand concentration is about the same as the total ligand concentration added:

$$[L]_f \approx [L] \qquad (4.20)$$

With this approximation, Equation 4.19 can be rewritten as follows:

$$[RL] = \frac{[R][L]}{K_d + [L]} = \frac{[R]}{1 + \frac{K_d}{[L]}} \qquad (4.21)$$

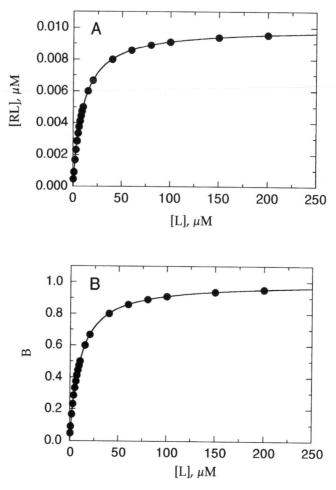

Figure 4.3 (A) Langmuir isotherm for formation of the binary RL complex as a function of ligand concentration. (B) The data from (A) expressed in terms of fractional receptor occupancy, B. The data in plots (A) and (B) are fit to Equations 4.21 and 4.23, respectively.

This equation describes a square hyperbola that is typical of saturable binding in a variety of chemical, physical, and biochemical situations. It was first derived by Langmuir to describe adsorption of gas molecules on a solid surface as a function of pressure at constant temperature (i.e., under isothermal conditions). Hence, the equation is known as the *Langmuir isotherm equation*. Because the data are well described by this equation, plots of [RL] as a function of total ligand concentration are sometimes referred to as *binding isotherms*. Figure 4.3 illustrates two formulations of a typical binding isotherm for a receptor–ligand binding equilibrium; the curve drawn through the data in each case represents the nonlinear least-squares best fit of the data to Equation 4.21, from which the researcher can obtain estimates of both the K_d and the total receptor concentration. Such binding isotherms are commonly used in a variety of biochemical studies; we shall see examples of the use of such plots in subsequent chapters when we discuss enzyme–substrate and enzyme–inhibitor interactions.

It is not necessary, and often not possible, to know the concentration of total receptor in an experiment calling for the use of the Langmuir isotherm equation. As long as one has some experimentally measurable signal that is unique to the receptor–ligand complex (e.g., radioactivity associated with the macromolecule after separation from free radiolabeled ligand), one can construct a binding isotherm. Let us say that we have some signal, Y, that we can follow as a measure of [RL]. In terms of signal, the Langmuir isotherm equation can be recast as follows:

$$Y = \frac{Y_{max}}{1 + \dfrac{K_d}{[L]}} \qquad (4.22)$$

The values of Y_{max} and K_d are then determined from fitting a plot of Y as a function of ligand concentration to Equation 4.22.

Note that the ratio Y/Y_{max} is equal to [RL]/[R] for any concentration of ligand. The ratio [RL]/[R] is referred to as the *fractional occupancy* of the receptor, and is often represented by the symbol B (for *bound* receptor). Equations 4.21 and 4.22 can both be cast in terms of fractional occupancy by dividing both sides of the equations by [R] and Y_{max}, respectively:

$$\frac{[RL]}{[R]} = \frac{Y}{Y_{max}} = B = \frac{1}{1 + \dfrac{K_d}{[L]}} \qquad (4.23)$$

This form of the Langmuir isotherm equation is useful in normalizing data from different receptor samples, which may differ slightly in their total receptor concentrations, for the purposes of comparing K_d values and overall binding isotherms.

4.3.2 Multiple Binding Sites

4.3.2.1 Multiple Equivalent Binding Sites   In the discussion above we assumed the simplest model in which each receptor molecule had a single specific binding site for the ligand. Hence the molarity of specific ligand binding sites was identical to the molarity of receptor molecules. There are examples, however, of receptors that are multivalent for a ligand: that is, each receptor

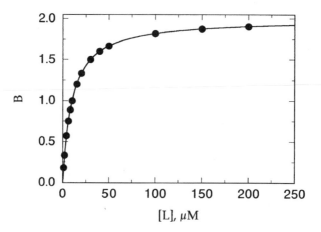

Figure 4.4 Langmuir isotherm for a receptor with two equivalent ligand binding sites (n). The data in this plot are fit to Equation 4.24.

molecule contains more than one specific binding pocket for the ligand. In the simplest case of multivalency, each ligand binding site behaves independently and identically with respect to ligand binding. That is, the affinity of each site for ligand is the same as all other specific binding sites (identical K_d values), and the binding of ligand at one site does not influence the affinity of other binding sites on the same receptor molecule (independent binding). In this case a binding isotherm constructed by plotting the concentration of bound ligand as a function of total ligand added will be indistinguishable from that in Figure 4.3B. The Langmuir isotherm equation still describes this situation, but now the molarity of specific binding sites available for the ligand will be $n[R]$, where n is the number of specific binding sites per molecule of receptor:

$$[RL] = \frac{n[R]}{1 + \dfrac{K_d}{[L]}} \tag{4.24}$$

Again, a plot of fractional occupancy as a function of total ligand concentration will be qualitatively indistinguishable from Figure 4.3B, except that the maximum value of B is now n, rather than 1.0 (Figure 4.4):

$$B = \frac{n}{1 + \dfrac{K_d}{[L]}} \tag{4.25}$$

4.3.2.2 Multiple Nonequivalent Binding Sites It is also possible for a single receptor molecule to have more than one type of independent specific

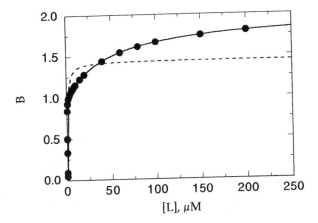

Figure 4.5 Langmuir isotherm for a receptor with multiple, nonequivalent ligand binding sites. In this simulation the receptor has two ligand binding sites, one with a K_d of 0.1 μM and the second with a K_d of 50 μM. The solid line through the data is the best fit to Equation 4.27, while the dashed line is the best fit to the standard Langmuir isotherm equation, Equation 4.23.

binding site for a particular ligand, with each binding site type having a different dissociation constant. In this case the binding isotherm for the receptor will be a composite of the individual binding isotherms for each type of binding site population (Feldman, 1972; Halfman and Nishida, 1972):

$$B = \sum_{i=1}^{i=j} \frac{n_i}{1 + \frac{K_d^i}{[L]}} \qquad (4.26)$$

In the simplest of these cases, there would be binding sites of two distinct types on the receptor molecule, with n_1 and n_2 copies of each type per receptor molecule, respectively. The individual dissociation constants for each type of binding site are defined as K_d^1 and K_d^2. The overall binding isotherm is thus given by:

$$B = \frac{n_1}{1 + \frac{K_d^1}{[L]}} + \frac{n_2}{1 + \frac{K_d^2}{[L]}} \qquad (4.27)$$

If K_d^1 and K_d^2 are very different (by about a factor of 100), it is possible to observe the presence of multiple, nonequivalent binding sites in the shape of the binding isotherm. This is illustrated in Figure 4.5 for a receptor with two types of binding site with dissociation constants of 0.1 and 50 μM, respectively.

Note that the presence of multiple, nonequivalent binding sites is much more apparent in other graphical representations of the binding isotherm, as will be discussed below.

4.3.2.3 Cooperative Interactions Among Multiple Binding Sites   Thus far we have assumed that the multiple binding sites on a receptor molecule (whether equivalent or nonequivalent) behave independently of one another with respect to ligand binding. This, however, is not always the case. Sometimes the binding of a ligand at one site on a receptor will influence the affinity of other sites on the same molecule; when the affinity of one site on a receptor is modulated by ligand occupancy at another site, the two sites are said to display *cooperative binding*. If binding of ligand at one site increases the affinity of another site, the sites are said to display *positive cooperativity*. On the other hand, if the binding of ligand at one site decreases the affinity of another site, the sites are said to display *negative cooperativity*. Both positive and negative forms of cooperativity are observed in nature, and this is a common mechanism for controlling biochemical processes in, for example, metabolic pathways. We shall discuss cooperativity as it applies to enzyme catalysis in Chapter 12. For a more in-depth discussion of receptor cooperativity in general, the reader is referred to the excellent text on this subject by Perutz (1990).

4.3.3 Correction for Nonspecific Binding

In the preceding discussion the binary complex [RL] referred to the molecular situation of a ligand that interacts at a specific binding pocket (or binding site) within the receptor molecule. This *specific binding* is what leads to biological activity and is thus the focus of research on receptor–ligand interactions. Experimentally, however, one often observes additional *nonspecific binding* of ligands to the receptor molecule and to other components of the reaction system (e.g., binding to container surfaces, filter membranes, detergent micelles). Nonspecific binding of ligands to the receptor itself occur adventitiously to sites that do not confer any biological activity. Such interactions are artifactual and must be corrected for in the data analysis if one is to obtain a true measure of the specific binding interactions.

In the vast majority of ligand binding experiments, one is measuring some unique signal associated with the ligand. Typically this is radioisotope incorporation, fluorescence, or some unique optical absorbance. For the purposes of this discussion, let us say that the primary ligand of interest has been synthesized with a radiolabel (e.g., ^{3}H, ^{14}C, ^{35}S, ^{33}P, ^{125}I; the ligand containing such a label is often referred to as a "hot ligand"). We can then measure ligand binding by somehow separating the receptor–ligand complex from the free ligand and ascertaining the amount of radioactivity associated with the bound and free fractions. In such an experiment the total radioactivity associated with the receptor will be the sum of specific and nonspecific binding. Being associated with a specific binding site on the receptor, the specific radioligand binding could be displaced by the addition of a large molar excess of a non-

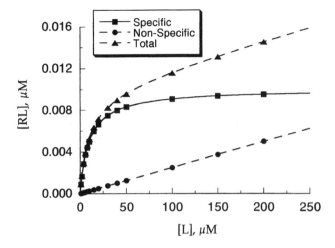

Figure 4.6 Typical results from a ligand binding experiment; the data show the effects of nonspecific ligand binding on the titration data. Triangles illustrate the typical experimental data for total ligand binding, reflecting the sum of both specific and nonspecific binding. Circles represent the nonspecific binding component, determined by measurements in the presence of excess unlabeled (cold) ligand. Squares represent the specific ligand binding, which is determined by subtracting the value of the nonspecific binding component from the total binding value at each concentration of labeled (hot) ligand.

labeled version of the same ligand (i.e., the same ligand lacking the radioactivity, often referred to as "cold ligand"). Nonspecific binding sites, however, are not affected by the addition of cold ligand, and the radioactivity associated with this nonspecific binding will thus be unaffected by the presence of excess cold ligand. This then, is how one experimentally determines the extent of nonspecific binding of ligand to receptor molecules. A good rule of thumb is that to displace all the specific binding, the unlabeled ligand concentration should be about 100- to 1000-fold above the K_d for the receptor–ligand complex (Hulme, 1992; Klotz, 1997). The remaining radioactivity measured under these conditions is then defined as the nonspecific binding.

Unlike specific binding, nonspecific binding is typically nonsaturable, increasing linearly with the total concentration of radioligand. If, in a binding experiment, one varied the total radioligand concentration over a range that bracketed the receptor K_d, one would typically obtain data similar to the curve indicated by triangles in Figure 4.6. If this experiment were repeated with a constant excess of cold ligand present at every point, the nonspecific binding could be measured (circles in Figure 4.6). From these data, the specific binding at each ligand concentration could be determined by subtracting the nonspecific binding from the total binding measured:

$$[RL]_{specific} = [RL]_{total} - [RL]_{nonspecific} \qquad (4.28)$$

The specific binding from such an experiment (squares in Figure 4.6) displays the saturable nature expected from the Langmuir isotherm equation.

4.4 GRAPHIC ANALYSIS OF EQUILIBRIUM LIGAND BINDING DATA

Direct plots of receptor–ligand binding data provide the best measure of the binding parameters K_d and n (the total number of binding sites per receptor molecule). Over the years, however, a number of other graphical methods have been employed in data analysis for receptor–ligand binding experiments. Here we briefly review some of these methods.

4.4.1 Direct Plots on Semilog Scale

In the semilog method one plots the direct data from a ligand binding experiment, as in Figure 4.3, but with the x axis (i.e., [L]) on a $\log_{10}$ rather than linear scale. The results of such plotting are illustrated in Figure 4.7, where the data are again fit to the Langmuir isotherm equation. The advantages of this type of plot, reviewed recently by Klotz (1997), are immediately obvious from inspection of Figure 4.7. The isotherm appears as an "S-shaped" curve, with the ligand concentration corresponding to the K_d at the midpoint of the quasi-linear portion of the curve.

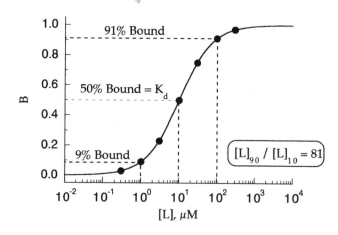

Figure 4.7 Langmuir isotherm plot on a semilog scale, illustrating the relationship between ligand concentration and percent of occupied receptor molecules. At [L] = K_d, 50% of the ligand binding sites on the receptor are occupied by ligand. At a ligand concentration 10-fold below the K_d, only 9% of the binding sites are occupied, while at a ligand concentration 10-fold above the K_d, 91% of the binding sites are occupied. It is clear that several decades of ligand concentration, bracketing the K_d, must be used to characterize the isotherm fully. Plots of this type make visual inspection of the binding data much easier and are highly recommended for presentation of such data.

Semilog plots allow for rapid visual estimation of K_d from the midpoint of the plot; as illustrated in Figure 4.7, the K_d is easily determined by dropping a perpendicular line from the curve to the x value corresponding to $B = n/2$. These plots also give a clearer view of the extent of binding site saturation achieved in a ligand titration experiment. In Figure 4.3, our highest ligand concentration results in about 95% saturation of the available binding sites on the receptor. Yet, the direct plot already demonstrates some leveling off, which could be easily mistaken for complete saturation. In the semilog plot, the limited level of saturation is much more clearly illustrated. Today, of course, the values of both K_d and n are determined by nonlinear least-squares curve fitting of either form of direct plot (linear or semilog), not by visual inspection. Nevertheless, the visual clarity afforded by the semilog plots is still of value. For these reasons Klotz (1997) strongly recommends the use of these semilog plots as the single most preferred method of plotting experimental ligand binding data.

Figure 4.7 illustrates more clearly another aspect of idealized ligand binding isotherms: the dependence of binding site saturation on ligand concentration. Note from Figure 4.7 that when the ligand concentration is one decade below the K_d we achieve about 9% saturation of binding sites, and when the ligand concentration is one decade above the K_d we achieve about 91% saturation. Put another way, one finds (in the absence of cooperativity: see Chapter 12) that the ratio of ligand concentrations needed to achieve 90 and 10% saturation of binding sites is an 81-fold change in [L]:

$$\frac{[L]_{90}}{[L]_{10}} = 81 \tag{4.29}$$

For example, to cover just the range of 20–80% receptor occupancy, one needs to vary the ligand concentration from $0.25K_d$ to $5.0K_d$. Hence, to cover a reasonable portion of the binding isotherm, experiments should be designed to cover *at least* a 100-fold ligand concentration range bracketing the K_d value. Since the K_d is usually unknown prior to the experiment (in fact, the whole point of the experiment is typically to determine the value of K_d), it is best to design experiments to cover as broad a range of ligand concentrations as is feasible.

Another advantage of the semilog plot is the ease with which nonequivalent multivalency of receptor binding sites can be discerned. Compare, for example, the data plotted in Figures 4.5 and 4.8. While careful inspection of Figure 4.5 reveals a deviation from a simple binding isotherm, the biphasic nature of the binding events is much more clearly evident in the semilog plot.

4.4.2 Linear Transformations of Binding Data: The Wolff Plots

Before personal computers were so widespread, nonlinear curve-fitting methods could not be routinely applied to experimental data. Hence scientists

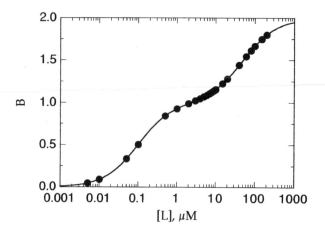

Figure 4.8 Semilog plot of a Langmuir isotherm for a receptor with multiple, nonequivalent ligand binding sites. The data and fit to Equation 4.27 are the same as for Figure 4.5.

in all fields put a great deal of emphasis on finding mathematical transformations of data that would lead to linearized plots, so that the data could be graphed and analyzed easily. During the 1920s B. Wolff developed three linear transformation methods for ligand binding data and graphic forms for their representation (Wolff, 1930, 1932). We shall encounter all three of these transformation and plotting methods again in Chapter 5, where they will be applied to the analysis of steady state enzyme kinetic data. As discussed by Klotz (1997), despite Wolff's initial introduction of these graphs, each of the three types is referred to by the name of another scientist, who reintroduced the individual method in a more widely read publication. The use of these linear transformation methods is no longer recommended, since today nonlinear fitting of direct data plots is so easily accomplished digitally. These methods are presented here mainly for historic perspective.

One method for linearizing ligand binding data is to present a version of Equation 4.25 rearranged in reciprocal form. Taking the reciprocal of both sides of Equation 4.25 leads to the following expression:

$$\frac{1}{B} = \left(\frac{K_d}{n}\right)\frac{1}{[L]} + \frac{1}{n} \tag{4.30}$$

Comparing Equation 4.30 with the standard equation for a straight line, we have:

$$y = (mx) + b \tag{4.31}$$

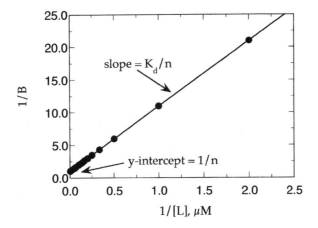

Figure 4.9 Double-reciprocal plot for the data in Figure 4.3. The data in this type of plot are fit to a linear function from which estimates of $1/n$ and K_d/n can be obtained from the values of the y intercept and slope, respectively.

where m is the slope and b is the y intercept. Hence we see that Equation 4.30 predicts that a plot of $1/B$ as a function of $1/[L]$ will yield a linear plot with slope equal to K_d/n and y intercept equal to $1/n$ (Figure 4.9). Plots of this type are referred to as double-reciprocal or Langmuir plots. We shall see in Chapter 5 that similar plots are used for representing steady state enzyme kinetic data, and in this context these plots are often referred to as Lineweaver–Burk plots.

A second linearization method is to multiply both sides of Equation 4.23 by $(1 + K_d/[L])$ to obtain:

$$B\left(1 + \frac{K_d}{[L]}\right) = n \tag{4.32}$$

This can be rearranged to yield:

$$B = -K_d\left(\frac{B}{[L]}\right) + n \tag{4.33}$$

From Equation 4.33 we see that a plot of B as a function of $B/[L]$ will yield a straight line with slope equal to $-K_d$ and intercept equal to n (Figure 4.10). Plots of the type illustrated in Figure 4.10, known as Scatchard plots, are the most popular linear transformation used for ligand binding data. We shall encounter this same plot in Chapter 5 for use in analyzing steady state enzyme kinetic data; in this context these plots are often referred to as Eadie–Hofstee plots.

Note that when the receptor contains multiple, equivalent binding sites, the Scatchard plot remains linear, but with intercept $n > 1$. If, however, the receptor

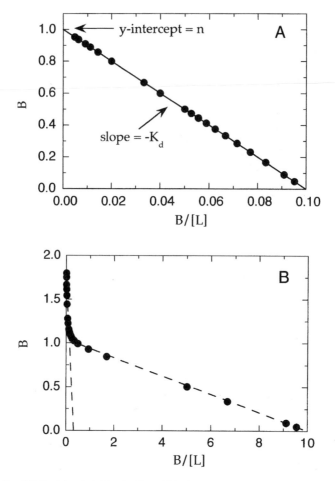

Figure 4.10 (A) Scatchard plot for the ligand binding data presented in Figure 4.3. In this type of plot the data are fit to a linear function from which estimates of n and $-K_d$ can be obtained from the values of the y intercept and the slope, respectively. (B) Scatchard plot for a receptor with multiple, nonequivalent ligand binding sites; the data are from Figure 4.5. Dashed lines illustrate an attempt to fit the data to two independent linear functions. This type of analysis of nonlinear Scatchard plots is inappropriate and can lead to significant errors in determinations of the individual K_d values (see text for further details).

contains multiple, *nonequivalent* binding sites, the Scatchard plot will no longer be linear. Figure 4.10B illustrates this for the case of two nonequivalent binding sites on a receptor: note that the curvilinear plot appears to be the superposition of two straight lines. Many researchers have attempted to fit data like these to two independent straight lines to determine K_d and n values for each binding site type. This, however, can be an oversimplified means of data analysis that does not account adequately for the curvature in these plots. Additionally, nonlinear Scatchard plots can arise for reasons other than nonequivalent binding sites (Klotz, 1997). The reader who encounters nonlinear data like that in Figure 4.10B is referred to the more general discussion of such nonlinear effects in the texts by Klotz (1997) and Hulme (1992).

The third transformation that is commonly used is obtained by multiplying both sides of Equation 4.30 by [L] to obtain:

$$\frac{[L]}{B} = \frac{1}{n}[L] + \frac{K_d}{n} \tag{4.34}$$

From this we see that a plot of [L]/B as a function of [L] also yields a straight line with slope equal to $1/n$ and intercept equal to K_d/n (Figure 4.11). Again, this plot is used for analysis of both ligand binding data and steady state enzyme kinetic data. In both contexts these plots are referred to as Hanes–Wolff plots.

A word of caution is in order with regard to these linear transformations. Any mathematical transformation of data can introduce errors due to the transformation itself (e.g., rounding errors during calculations). Also, some of the transformation methods give uneven weighting to certain experimental data points; this artifact is discussed in greater detail in Chapter 5 with respect to Lineweaver–Burk plots. Today almost all researchers have at their disposal computer graphics software capable of performing nonlinear curve fitting. Hence, the use of these linear transformation methods is today antiquated, and the errors in data analysis introduced by their use are no longer counterbal-

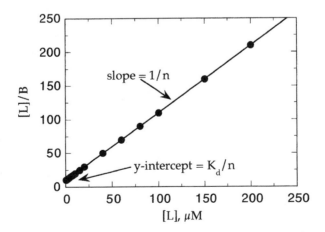

Figure 4.11 Hanes—Wolff plot for the ligand binding data from Figure 4.3. In this type of plot the data are fit to a linear function from which estimates of $1/n$ and K_d/n can be obtained from the values of the slope and the y intercept, respectively.

anced by ease of analysis. Therefore, it is strongly recommended that the reader refrain from the use of these plots, relying instead on the direct plotting methods described above.

4.5 EQUILIBRIUM BINDING WITH LIGAND DEPLETION (TIGHT BINDING INTERACTIONS)

In all the equations derived until now, we have assumed that the receptor concentration used in the experimental determination of K_d was much less than K_d. This allowed us to use the simplifying approximation that the free ligand concentration $[L]_{free}$ was equal to the total ligand concentration added to the binding mixture, $[L]$. If, however, the affinity of the ligand for the receptor is very strong, so that the K_d is similar in magnitude to the concentration of receptor, this assumption is no longer valid. In this case binding of the ligand to form the RL complex significantly depletes the free ligand concentration. Use of the simple Langmuir isotherm equation to fit the experimental data would lead to errors in the determination of K_d. Instead, we must develop an equation that explicitly accounts for the depletion of free ligand and receptor concentrations due to formation of the binary RL complex. We again begin with the mass conservation Equations 4.1 and 4.2. Using these we can recast Equation 4.3 in terms of the concentrations of free receptor and ligand as follows:

$$K_d = \frac{([R] - [RL])([L] - [RL])}{[RL]} \qquad (4.35)$$

If we multiply both side of Equation 4.35 by $[RL]$ and then subtract the term $K_d[RL]$ from both sides, we obtain:

$$0 = ([R] - [RL])([L] - [RL]) - K_d[RL] \qquad (4.36)$$

which can be rearranged to:

$$0 = [RL]^2 - ([R] + [L] + K_d)[RL] + [R][L] \qquad (4.37)$$

Equation 4.37 is a quadratic equation for $[RL]$. From elementary algebra we know that such an equation has two potential solutions; however, in our case only one of these would be physically meaningful:

$$[RL] = \frac{([R] + [L] + K_d) - \sqrt{([R] + [L] + K_d)^2 - 4[R][L]}}{2} \qquad (4.38)$$

The quadratic equation above is always rigorously correct for any receptor–ligand binding interaction. When ligand or receptor depletion is not significant, the much simpler Langmuir isotherm equation, 4.21, is used for convenience, but the reader should consider carefully the assumptions, outlined above, that go into use to this equation in any particular experimental design. As we shall see in Chapter 9, inhibitors that mimic the transition state of the enzymatic reaction can bind very tightly to enzyme molecules. In such cases, the dissociation constant for the enzyme–inhibitor complex can be accurately determined only by use of a quadratic equation similar to Equation 4.38.

4.6 COMPETITION AMONG LIGANDS FOR A COMMON BINDING SITE

In many experimental situations, one wishes to compare the affinities of a number of ligands for a particular receptor. For example, given a natural ligand for a receptor (e.g., a substrate for an enzyme), one may wish to design nonnatural ligands (e.g., enzyme inhibitors) to compete with the natural ligand for the receptor binding site. As discussed in Chapter 1, this is a common strategy for the design of pharmacological agents that function by blocking the activity of key enzymes that are critical for a particular disease process.

If one of the ligands (e.g., the natural ligand) has some special physicochemical property that allows for its detection and for the detection of the RL complex, then competition experiments can be used to determine the affinities for a range of other ligands. This first ligand (L) might be radiolabeled by synthetic incorporation of a radioisotope into its structure. Alternatively, it may have some unique optical or fluorescent signal, or it may stimulate the biological activity of the receptor in a way that is easily followed (e.g., enzymatic activity induced by substrate). For the sake of illustration, let us say that this ligand is radiolabeled and that we can detect the formation of the RL complex by measuring the radioactivity associated with the protein after the protein has been separated from free ligand by some method (see Section 4.7). In a situation like this, we can readily determine the K_d for the receptor by titration with the radioligand, as described above. Having determined the value of K_d^L we can fix the concentrations of receptor and of radioligand to result in a certain concentration of receptor–ligand complex [RL]. Addition of a second, nonlabeled ligand (A) that binds to the same site as the radioligand will cause a competition between the two ligands for the binding site on the receptor. Hence, as the concentration of ligand A is increased, less of the radioligand effectively competes for the binding sites, and the concentration of the [RL] that is formed will be decreased. A plot of [RL] or B as a function of [A] will appear as an inverted Langmuir isotherm (Figure 4.12) from which the dissociation constant for the nonlabeled ligand (K_d^A) can be determined from fitting of the experimental data to the following equation:

$$\frac{[RL]}{[R]} = \frac{n}{1 + \frac{K_d^L}{[L]}\left(1 + \frac{[A]}{K_d^A}\right)} \tag{4.39}$$

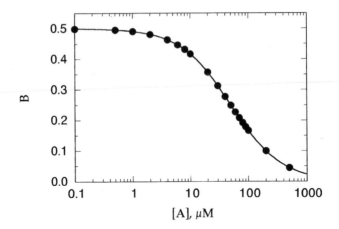

Figure 4.12   Binding isotherm for competitive ligand binding. B, the fractional occupancy of receptor with a labeled ligand (L) is measured as a function of the concentration of a second, unlabeled ligand (A). In this simulation $[L] = K_d^L = 10\,\mu M$, and $K_d^A = 25\ \mu M$. The line through the data represents the best fit to Equation 4.36.

We shall see a form of this equation in Chapter 8 when we discuss competitive inhibition of enzymes. Fitting of the data in Figure 4.12 to Equation 4.39 provides a good estimate of K_d^A assuming that the researcher has previously determined the values of $[L]$ and K_d^L. Note, however, that the midpoint value of $[A]$ in Figure 4.12 does not correspond to the dissociation constant K_d^A. The presence of the competing ligand results in the displacement of the midpoint in this plot from the true dissociation constant value. In situations like this, where the plot midpoint is not a direct measure of affinity, the midpoint value is often referred to as the EC_{50} or IC_{50}. This terminology, and its use in data analysis, will be discussed more fully in Chapter 8.

4.7   EXPERIMENTAL METHODS FOR MEASURING LIGAND BINDING

The focus of this text is on enzymes and their interactions with ligands, such as substrates and inhibitors. The most common means of studying these interactions is through the use of steady state kinetic measurements, and these methods are discussed in detail in Chapter 7. A comprehensive discussion of general experimental methods for measuring ligand binding to proteins is outside the scope of this text; these more general methods are covered in great detail elsewhere (Hulme, 1992).

Here we shall briefly describe some of the more common methods for direct measurement of protein–ligand complexation. These methods fall into two general categories. In the first, the ligand has some unique physicochemical property, such as radioactivity, fluorescence, or optical signal. The concentration of receptor–ligand complex is then quantified by measuring this unique signal after the free and protein-bound fractions of the ligand have been physically separated. In the second category of techniques, it is the receptor–ligand complex itself that has some unique spectroscopic signal associated with it. Methods within this second category require no physical separation of the free and bound ligand, since only the bound ligand leads to the unique signal.

4.7.1 Equilibrium Dialysis

Equilibrium dialysis is one of the most well-established methods for separating the free and protein-bound ligand populations from each other. Use of this method relies on some unique physical signal associated with the ligand; this is usually radioactivity or fluorescence signal, although the recent application of sensitive mass spectral methods has eliminated the need for such unique signals.

A typical apparatus for performing equilibrium dialysis measurements is illustrated in Figure 4.13. The apparatus consists of two chambers of equal

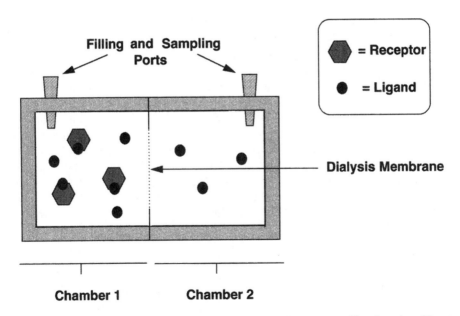

Figure 4.13 Schematic illustration of an equilibrium dialysis apparatus. Chambers 1 and 2 are separated by a semipermeable dialysis membrane. At the beginning of the experiment, chamber 1 is filled with a solution containing receptor molecules and chamber 2 is filled with an equal volume of the same solution, with ligand molecules instead of receptor molecules. Over time, the ligand equilibrates between the two chambers. After equilibrium has been reached, the ligand concentration in chamber 1 will represent the sum of free and receptor-bound ligands, while the ligand concentration in chamber 2 will represent the free ligand concentration.

volume that are separated by a semipermeable membrane (commercially available dialysis tubing). The membrane is selected so that small molecular weight ligands will readily diffuse through the membrane under osmotic pressure, while the larger protein molecules will be unable to pass through. At the beginning of an experiment one chamber is filled with a known volume of solution containing a known concentration of protein. The other chamber is filled with a known volume of the same solution containing the ligand but no protein (actually, in many cases both protein-containing and protein-free solutions have equal concentrations of ligand at the initiation of the experiment). The chambers are then sealed, and the apparatus is place on an orbital rocker, or a mechanical rotator, in a thermostated container (e.g., an incubator or water bath) to ensure good mixing of the solutions in each individual chamber and good thermal equilibration. The solutions are left under these conditions until equilibrium is established, as described later. At that time a sample of known volume is removed from each chamber, and the concentration of ligand in each sample is quantified by means of the unique signal associated with the ligand.

The concentration of ligand at equilibrium in the protein-free chamber represents the free ligand concentration. The concentration of ligand at equilibrium in the protein-containing chamber represents the sum of the free and bound ligand concentrations. Thus, from these two measurements both the free and bound concentrations can be simultaneously determined, and the dissociation constant can be calculated by use of Equation 4.3.

Accurate determination of the dissociation constant from equilibrium dialysis measurements depends critically on attainment of equilibrium by the two solutions in the chambers of the apparatus. The time required to reach equilibrium must thus be well established for each experimental situation. A common way of establishing the equilibration time is to perform an equilibrium dialysis measurement in the absence of protein. Ligand is added to one of two identical solutions that are placed in the two chambers of the dialysis apparatus, set up as described above. Small samples are removed from each chamber at various times, and the ligand concentration of each solution is determined. These data are plotted as illustrated in Figure 4.14 to determine the time required for full equilibration. It should be noted that the equilbration time can vary dramatically among different ligands, even for two ligands of approximately equal molecular weight. It is well known, for example, that relative to other molecules of similar molecular weight, adenosine triphosphate (ATP) takes an unusually long time to equilibrate across dialysis tubing. This is because in solution ATP is highly hydrated, and it is the ATP–water complex, not the free ATP molecule, that must diffuse through the membrane. Hence, equilibration time must be established empirically for each particular protein–ligand combination.

A number of other factors must be taken into account to properly analyze data from equilibrium dialysis experiments. Protein and ligand binding to the membrane and chamber surfaces, for example, must be corrected for. Likewise,

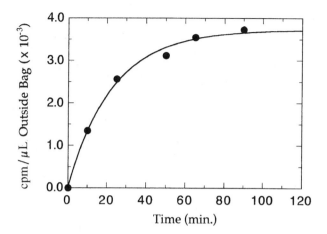

Figure 4.14 Time course for approach to equilibrium between two chambers of an equilibrium dialysis apparatus. The data are for a radiolabeled deoxyribonucleoside 5′-triphosphate. At the beginning of the experiment all of the radioligand was inside the dialysis bag. At various time points the researcher sampled the radioactivity of the solution outside the dialysis bag. The data are fit to a first-order approach to equilibrium. This plot was constructed using data reported by Englund et al. (1969).

any net charge on the protein and/or ligand molecules themselves can affect the osmotic equilibrium between the chambers (referred to as the Donnan effect) and must therefore be accounted for. These and other corrections are described in detail in a number of texts (Segel, 1976; Bell and Bell, 1988; Klotz, 1997), and the interested reader is referred to these more comprehensive treatments. An interesting Internet site provides a detailed discussion of experimental methods for equilibrium dialysis.* Visiting this Web site would be a good starting point for designing such experiments.

4.7.2 Membrane Filtration Methods

Membrane filtration methods generally make use of one of two types of membrane to separate protein-bound ligand from free ligand. One membrane type, which binds proteins through various hydrophobic and/or electrostatic forces, allows one to wash away the free ligand from the adhered protein molecules. Filter materials in this class include nitrocellulose and Immobilon P. The second membrane type consists of porous semipermeable barriers, with nominal molecular weight cutoffs (Paulus, 1969), that allow the passage of small molecular weight species (i.e., free ligand) but retain macromolecules (i.e., receptor and receptor-bound ligands). Applications of both these membrane types for enzyme activity measurements are discussed in Chapter 7.

* (*http://biowww.chem.utoledo.edu/eqDial/intro*)

Experiments utilizing the first membrane type are performed as follows. The receptor and ligand are mixed together at known concentrations in a known volume of an appropriate buffer solution and allowed to come to equilibrium. The sample, or a measured portion thereof, is applied to the surface of the membrane, and binding to the membrane is allowed to proceed for some fixed period of time. When membrane binding is complete, the membrane is washed several times with an appropriate buffer (typically containing a low concentration of detergent to remove adventitiously bound materials). These binding and washing steps are greatly facilitated by use of vacuum manifolds of various types that are now commercially available (Figure 4.15A). With this apparatus, the membrane sits above a vacuum source and a liquid reservoir. Binding and wash buffers are efficiently removed from the membranes by application of a vacuum from below. As illustrated in Figure 4.15A, protein binding membranes and accompanying vacuum manifolds are now available in 96-well plate formats, making these experiments very simple and efficient. Upon completion of the binding and washing steps, signal from the bound ligand is measured on the membrane surface. In the most typical application, radioligands are used, and the membrane is immersed in scintillation fluid and quantified by means of a scintillation counter. From such measurements the concentration of bound ligand can be determined.

Molecular weight cutoff filters are today available in centrifugation tube units, as illustrated in Figure 4.15B, and these are the most common means of utilizing such filters for ligand binding experiments (Freundlich and Taylor, 1981; see also product literature from vendors such as Amicon and Millipore). Here the receptor and ligand are again allowed to reach equilibrium in an appropriate buffer solution, and the sample is loaded into the top of the centrifugation filtration unit (i.e., above the membrane). A small volume aliquot is removed from the sample, and the total concentration of ligand is quantified from this aliquot by whatever signal the researcher can measure (ligand radioactivity, ligand fluorescence, etc.). The remainder of the sample is then centrifuged for a brief period (according to the manufacturer's instructions), so that a small volume of the sample is filtered through the membrane and collected in the filtrate reservoir. By keeping the centrifugation time brief (on the order of ca. 30 seconds), one avoids large displacements from equilibrium during the centrifugation period. A known volume of the filtrate is then taken and quantified. This provides a measure of the free ligand concentration at equilibrium. From such an experiment both the total and free ligand concentrations are experimentally determined, and thus the bound ligand concentration can be determined by subtraction (i.e., using a rearranged form of Equation 4.2).

Note that the reliability of these measurements depends on the off rate of the receptor–ligand complex. If the off rate is long compared to the time required for membrane washing or centrifugation, the measurements of bound and free ligand will reflect accurately the equilibrium concentrations. If, however, the complex has a short half-life, the measured concentrations may deviate significantly from the true equilibrium levels.

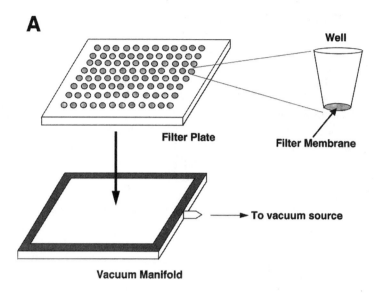

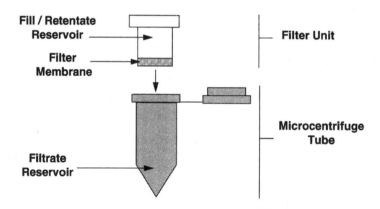

Figure 4.15 (A) Schematic illustration of a 96-well filter binding apparatus. The individual wells on the plate have a filter membrane at the bottom. After binding has reached equilibrium, the reaction mixture is place in the wells of this plate, and the solution is drawn through the filter by placing the plate into the vacuum manifold and applying a vacuum. The manifold has space for a second 96-well plate below the filter plate to capture the filtrate (not shown). Such 96-well filter binding plates are available commercially with a number of membrane compositions. (B) Schematic illustration of a centrifugal filtration device. A molecular weight cutoff filter at the bottom of the fill/retentate reservoir is filled with reaction mixture, and the reservoir is placed into a microcentrifuge tube. The apparatus is centrifuged and the filtrate is collected into the bottom of the microcentrifuge tube. See the text for further information on the use of these devices.

4.7.3 Size Exclusion Chromatography

Size exclusion chromatography is commonly used to separate proteins from small molecular weight species in what are referred to as protein desalting methods (see Copeland, 1994, and Chapter 7). Because of the nature of the stationary phase in these columns, macromolecules are excluded and pass through the columns in the void volume. Small molecular weight species, such as salts or free ligand molecules, are retained longer within the stationary phase. Traditional size exclusion chromatography requires tens of minutes to hours to perform, and is thus usually inappropriate for ligand binding measurements. Two variations of size exclusion chromatography are, however, quite useful for this purpose.

In the first variation that is useful for ligand binding measurements, spin columns are employed for size exclusion chromatography (Penefsky, 1977; Zeeberg and Caplow, 1979; Anderson and Vaughan, 1982; Copeland, 1994). Here a small bed volume size exclusion column is constructed within a column tube that fits conveniently into a microcentrifuge tube. Separation of excluded and retained materials is accomplished by centrifugal force, rather than by gravity or peristaltic pressure, as in conventional chromatography. After the column has been equilibrated with buffer, a sample of the equilibrated receptor–ligand mixture is applied to the column. A separate sample of the mixture is retained for measurement of total ligand concentration. The column is then centrifuged according to the manufacturer's instructions, and the excluded material is collected at the bottom of the microcentrifuge tube. This excluded material contains the protein-bound ligand population. By quantifying the ligand concentration in the sample before centrifugation and in the excluded material, one can determine the total and bound ligand concentrations, respectively. Again, by subtraction, one can also calculate the free ligand concentration and thus determine the dissociation constant. Prepacked spin columns, suitable for these studies are now commercially available from a number of manufacturers (e.g., BioRad, AmiKa Corporation).

The second variation of size exclusion chromatography that is applicable to ligand binding measurements is known as Hummel–Dreyer chromatography (HDC: Hummel and Dreyer, 1962; Ackers, 1973; Cann and Hinman, 1976). In HDC the size exclusion column is first equilibrated with ligand at a known concentration. A receptor solution is equilibrated with ligand at the same concentration as the column, and this solution is applied to the column. The column is then run with isocratic elution using buffer containing the same concentration of ligand. Elution is typically followed by measuring some unique signal from the ligand (e.g., radioactivity, fluorescence, a unique absorption signal). If there is no binding of ligand to the protein, the signal measured during elution should be constant and related to the concentration of ligand with which the column was equilibrated. If, however, binding occurs, the total concentration of ligand that elutes with the protein will be the sum of the

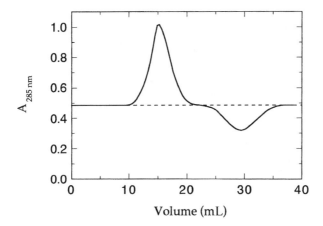

Figure 4.16 Binding of 2′-cytidylic acid to the enzyme ribonuclease as measured by Hummel—Dreyer chromatography. The positive peak of ligand absorbance is coincident with the elution of the enzyme. The trough at latter time results from free ligand depletion from the column due to the binding events. [Data redrawn from Hummel and Dreyer (1962).]

bound and free ligand concentrations. Hence, during protein elution the net signal from ligand elution will increase by an amount proportional to the bound ligand concentration. The ligand that is bound to the protein is recruited from the general pool of free ligand within the column stationary and mobile phases. Hence, some ligand depletion will occur subsequent to protein elution. This results in a period of diminished ligand concentration during the chromatographic run. The degree of ligand diminution in this phase of the chromatograph is also proportional to the concentration of bound ligand.

Figure 4.16 illustrates the results of a typical chromatographic run for an HDC experiment. From generation of a standard curve (i.e., signal as a function of known concentration of ligand), the signal units can be converted into molar concentrations of ligand. From the baseline measurement, one determines the free ligand concentration (which also corresponds to the concentration of ligand used to equilibrate the column), while the bound ligand concentration is determined from the signal displacements that are observed during and after protein elution (Figure 4.16). Because the column is equilibrated with ligand throughout the chromatographic run, displacement from equilibrium is not a significant concern in HDC. This method is considered by many to be one of the most accurate measures of protein–ligand equilibria. Oravcova et al. (1996) have recently reviewed HDC and other methods applicable to protein–ligand binding measurements; their paper provides a good starting point for acquiring a more in-depth understanding of many of these methods.

4.7.4 Spectroscopic Methods

The receptor–ligand complex often exhibits a spectroscopic signal that is distinct from the free receptor or ligand. When this is the case, the spectroscopic signal can be utilized to follow the formation of the receptor–ligand complex, and thus determine the dissociation constant for the complex. Examples exist in the literature of distinct changes in absorbance, fluorescence, circular dichroism, and vibrational spectra (i.e., Raman and infrared spectra) that result from receptor–ligand complex formation. The bases for these spectroscopic methods are not detailed here because they have been presented numerous times (see Chapter 7 of this text; Campbell and Dwek, 1984; Copeland, 1994). Instead we shall present an overview of the use of such methods for following receptor–ligand complex formation.

Because of its sensitivity, fluorescence spectroscopy is often used to follow receptor–ligand interactions, and we shall use this method as an example. Often a ligand will have a fluorescence signal that is significantly enhanced or quenched (i.e., diminished) upon interaction with the receptor. For example, warfarin and dansylsulfonamide are two fluorescent molecules that are known to bind to serum albumin. In both cases the fluorescence signal of the ligand is significantly increased upon complex formation, and knowledge of this behavior has been used to measure the interactions of these ligands with albumin (Epps et al., 1995). In contrast, ligand fluorescence can also often be quenched by interaction with the receptor. For example, my group synthesized a tripeptide, Lys-Cys-Lys, which we expected to bind to the kringle domains of plasminogen (Balciunas et al., 1993). We then chemically modified the peptide with a stilbene–maleimide derivative to impart a fluorescence signal (via covalent modification of the cysteine thiol). The stilbene-labeled peptide was highly fluorescent in solution, but it displayed significant fluorescence quenching upon complex formation with plasminogen and other kringle-containing proteins (Figure 4.17)

Even when the fluorescence intensity of the ligand is not significantly perturbed by binding to the receptor, it is often possible to follow receptor–ligand interaction by a technique known as fluorescence polarization. Fluorescence occurs when light of an appropriate wavelength excites a molecule from its ground electronic state to an excited electronic state (Copeland, 1994). One means of relaxation back to the ground state is by emission of light energy (fluorescence). The transitions between the ground and excited states are accompanied by a redistribution of electron density within the molecule, and this usually occurs mainly along one axis of the molecule (Figure 4.18). The axis along which electron density is perturbed between the ground and excited state is referred to as the transition dipole moment of the molecule.

If the excitation light beam is plane-polarized (by passage through a polarizing filter), the efficiency of fluorescence will depend on the alignment of the plane of light polarization with the transition dipole moment. Suppose that for a particular molecule the transition dipole moment is aligned with the plane

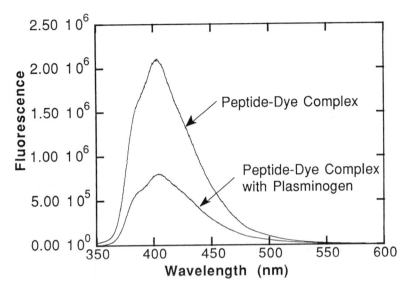

Figure 4.17 Fluorescence spectra of a fluorescently labeled peptide (Lys-Cys-Lys) free in solution (peptide—dye complex) and bound to the protein plasminogen. Note the significant quenching of the probe fluorescence upon peptide—plasminogen binding. [Data from Balciunas et al. (1993).]

of light polarization at the moment of excitation (i.e., light absorption by the molecule). In this case the light emitted from the molecule will also be plane-polarized and will thus pass efficiently through a properly oriented polarization filter placed between the sample and the detector. In this sequence (Figure 4.18A), the molecule has not rotated in space during its excited state lifetime, and so the plane of polarization remains the same. This is not always the case, however. If the molecule rotates during the excited state, less fluorescent light will pass through the oriented polarization filter between the sample and the detector: the faster the rotation, the less light passes (Figure 4.18B). Hence, as the rotational rate of the molecule is slowed down, the efficiency of fluorescence polarization increases. Small molecular weight ligands rotate in solution much faster than macromolecules, such as proteins. Hence, when a fluorescent ligand binds to a much larger protein, its rate of rotation in solution is greatly diminished, and a corresponding increase in fluorescence polarization is observed. This is the basis for measuring protein–ligand interactions by fluorescence polarization. A more detailed description of this method can be found in the texts by Campbell and Dwek (1984) and Lackowicz (1983). The PanVera Corporation (Madison, WI) also distributes an excellent primer and applications guide on the use of fluorescence polarization measurements for studying protein–ligand interactions.

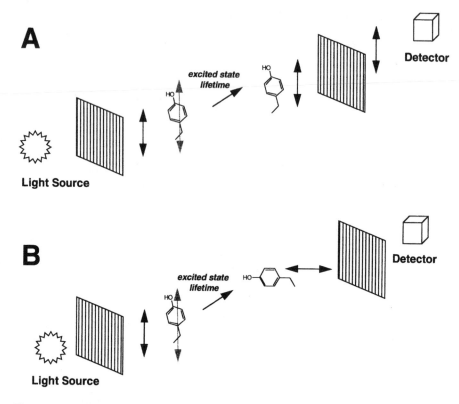

Figure 4.18 Schematic illustration of fluorescence polarization, in which a plane polarizing filter between the light source and the sample selects for a single plane of light polarization. The plane of excitation light polarization is aligned with the transition dipole moment (illustrated by the gray double-headed arrow) of the fluorophore there, the amino acid tyrosine. The emitted light is also plane-polarized and can thus pass through a polarizing filter, between the sample and detector, only if the plane of the emitted light polarization is aligned with the filter. (A) The molecule does not rotate during the excited state lifetime. Hence, the plane of polarization of the emitted light remains aligned with that of the excitation beam. (B) The molecule has rotated during the excited state lifetime so that the polarization planes of the excitation light and the emitted light are no longer aligned. In this latter case, the emitted light is said to have undergone depolarization.

Proteins often contain the fluorescent amino acids tryptophan and tyrosine (Campbell and Dwek, 1984; Copeland, 1994), and in some cases the intrinsic fluoresence of these groups is perturbed by ligand binding to the protein. There are a number of examples in the literature of proteins containing a tryptophan residue at or near the binding site for some ligand. Binding of the ligand in these cases often results in a change in fluorescence intensity and/or wavelength maximum for the affected tryptophan. Likewise, tyrosine-containing proteins often display changes in tyrosine fluorescence intensity upon complex formation with ligand. A number of DNA binding proteins, for example, display dramatic quenching of tyrosine fluorescence when DNA is bound to them.

Any spectroscopic signal that displays distinct values for the bound and free versions of the spectroscopically active component (either ligand or receptor), can be used as a measure of protein–ligand complex formation. Suppose that some signal δ has one distinct value for the free species δ_{free} and another value for the bound species δ_{bound}. If the spectroscopically active species is the receptor, then the concentration of receptor can be fixed, and the signal at any point within a ligand titration will be given by:

$$\delta = [RL]\,\delta_{bound} + [R]_{free}(\delta_{free}) \tag{4.40}$$

Since $[R]_{free}$ is equivalent to $[R] - [RL]$, we can rearrange this equation to:

$$\delta = [RL](\delta_{bound} - \delta_{free}) + [R](\delta_{free}) \tag{4.41}$$

Equation 4.41 can be rearranged further to give the fraction of bound receptor at any point in the ligand titration as follows:

$$\frac{[RL]}{[R]} = \frac{\delta - \delta_{free}}{\delta_{bound} - \delta_{free}} \tag{4.42}$$

Similarly, if the spectroscopically active species is the ligand, a fixed concentration of ligand can be titrated with receptor, and the fraction of bound ligand can be determined as follows:

$$\frac{[RL]}{[L]} = \frac{\delta - \delta_{free}}{\delta_{bound} - \delta_{free}} \tag{4.43}$$

The dissociation constant for the receptor–ligand complex can then be determined from isothermal analysis of the spectroscopic titration data as described above.

4.8 SUMMARY

In this chapter we have described methods for the quantitative evaluation of protein–ligand binding interactions at equilibrium. The Langmuir binding isotherm equation was introduced as a general description of protein–ligand equilibria. From fitting of experimental data to this equation, estimates of the equilibrium dissociation constant K_d and the concentration of ligand binding sites n, can be obtained. We shall encounter the Langmuir isotherm equation in different forms throughout the remainder of this text in our discussions of enzyme interactions with ligands such as substrates inhibitors and activators.

The basic concepts described here provide a framework for understanding the kinetic evaluation of enzyme activity and inhibition, as discussed in these subsequent chapters.

REFERENCES AND FURTHER READING

Ackers, G. K. (1973) *Methods Enzymol.* **27**, 441.

Anderson, K. B., and Vaughan, M. H. (1982) *J. Chromatogr.* **240**, 1.

Balciunas, A., Fless, G., Scanu, A., and Copeland, R. A. (1993) *J. Protein Chem.* **12**, 39.

Bell, J. E., and Bell, E. T. (1988) *Proteins and Enzymes*, Prentice-Hall, Englewood Cliffs, NJ.

Campbell, I. D., and Dwek, R. A. (1984) *Biological Spectroscopy*, Benjamin/Cummings, Menlo Park, CA.

Cann, J. R., and Hinman, N. D. (1976) *Biochemistry*, **15**, 4614.

Copeland, R. A. (1994) *Methods for Protein Analysis: A Practical Guide to Laboratory Protocols*, Chapman & Hall, New York.

Englund, P. T., Huberman, J. A., Jovin, T. M., and Kornberg, A. (1969) *J. Biol. Chem.* **244**, 3038.

Epps, D. E., Raub, T. J., and Kezdy, F. J. (1995) *Anal. Biochem.* **227**, 342.

Feldman, H. A. (1972) *Anal. Biochem.* **48**, 317.

Freundlich, R. and Taylor, D. B. (1981) *Anal. Biochem.* **114**, 103.

Halfman, C. J., and Nishida, T. (1972) *Biochemistry*, **18**, 3493.

Hulme, E. C. (1992) *Receptor–Ligand Interactions: A Practical Approach*, Oxford University Press, New York.

Hummel, J. R., and Dreyer, W. J. (1962) *Biochim. Biophys. Acta*, **63**, 530.

Klotz, I. M. (1997) *Ligand–Receptor Energetics: A Guide for the Perplexed*, Wiley, New York.

Lackowicz, J. R. (1983) *Principle of Fluorescence Spectroscopy*, Plenum Press, New York.

Oravcova, J., Böhs, B., and Lindner, W. (1996) *J. Chromatogr. B* **677**, 1.

Paulus, H. (1969) *Anal. Biochem.* **32**, 91.

Penefsky, H. S. (1977) *J. Biol. Chem.* **252**, 2891.

Perutz, M. (1990) *Mechanisms of Cooperativity and Allosteric Regulation in Proteins*, Cambridge University Press, New York.

Segel, I. H. (1976) *Biochemical Calculations*, 2nd ed., Wiley, New York.

Wolff, B. (1930) In *Enzymes*, J. B. S. Haldane, Ed., Longmans, Green & Co., London.

Wolff, B. (1932) In *Allgemeine Chemie der Enzyme*, J. B. S. Haldane and K. G. Stern, Eds., Steinkopf, Dresden, pp. 119ff.

Zeeberg, B., and Caplow, M. (1979) *Biochemistry*, **18**, 3880.

5

KINETICS OF SINGLE-SUBSTRATE ENZYME REACTIONS

Enzyme-catalyzed reactions can be studied in a variety of ways to explore different aspects of catalysis. Enzyme–substrate and enzyme–inhibitor complexes can be rapidly frozen and studied by spectroscopic means. Many enzymes have been crystallized and their structures determined by x-ray diffraction methods. More recently, enzyme structures have been determined by multidimensional NMR methods. Kinetic analysis of enzyme-catalyzed reactions, however, is the most commonly used means of elucidating enzyme mechanism and, especially when coupled with protein engineering, identifying catalytically relevant structural components. In this chapter we shall explore the use of steady state and transient enzyme kinetics as a means of defining the catalytic efficiency and substrate affinity of simple enzymes. As we shall see, the term *steady state* refers to experimental conditions in which the enzyme–substrate complex can build up to an appreciable "steady state" level. These conditions are easily obtained in the laboratory, and they allow for convenient interpretation of the time courses of enzyme reactions. All the data analysis described in this chapter rests on the ability of the scientist to conveniently measure the initial velocity of the enzyme-catalyzed reaction under a variety of conditions. For our discussion, we shall assume that some convenient method for determining the initial velocity of the reaction exists. In Chapter 7 we shall address specifically how initial velocities are measured and describe a variety of experimental methods for performing such measurements.

5.1 THE TIME COURSE OF ENZYMATIC REACTIONS

Upon mixing an enzyme with its substrate in solution and then (by some convenient means) measuring the amount of substrate remaining and/or the

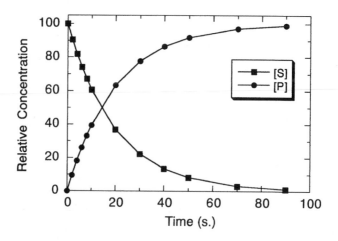

Figure 5.1 Reaction progress curves for the loss of substrate [S] and production of product [P] during an enzyme-catalyzed reaction.

amount of product produced over time, one will observe *progress curves* similar to those shown in Figure 5.1. Note that the substrate depletion curve is the mirror image of the product appearance curve. At early times substrate loss and product appearance change rapidly with time but as time increases these rates diminish, reaching zero when all the substrate has been converted to product by the enzyme. Such time courses are well modeled by first-order kinetics, as discussed in Chapter 2:

$$[S] = [S_0]e^{-kt} \qquad (5.1)$$

where [S] is the substrate concentration remaining at time t, $[S_0]$ is the starting substrate concentration, and k is the pseudo-first-order rate constant for the reaction. The velocity v of such a reaction is thus given by:

$$v = -\frac{d[S]}{dt} = \frac{d[P]}{dt} = k[S_0]e^{-kt} \qquad (5.2)$$

Let us look more carefully at the product appearance profile for an enzyme-catalyzed reaction (Figure 5.2). If we restrict our attention to the very early portion of this plot (shaded area), we see that the increase in product formation (and substrate depletion as well) tracks approximately linear with time. For this limited time period, the *initial velocity* v_0 can be approximated as the slope (change in y over change in x) of the linear plot of [S] or [P] as a function of time:

$$v_0 = -\frac{\Delta[S]}{\Delta t} = \frac{\Delta[P]}{\Delta t} \qquad (5.3)$$

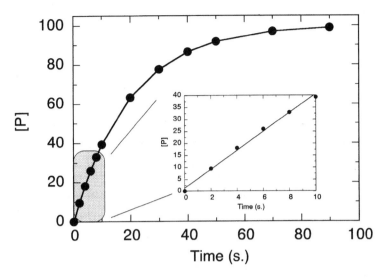

Figure 5.2   Reaction progress curve for the production of product during an enzyme-catalyzed reaction. Inset highlights the early time points at which the initial velocity can be determined from the slope of the linear plot of [P] versus time.

Experimentally one finds that the time course of product appearance and substrate depletion is well modeled by a linear function up to the time when about 10% of the initial substrate concentration has been converted to product (Chapter 2). We shall see in Chapter 7 that by varying solution conditions, we can alter the length of time over which an enzyme-catalyzed reaction will display linear kinetics. For the rest of this chapter we shall assume that the reaction velocity is measured during this early phase of the reaction, which means that from here $v = v_0$, the initial velocity.

5.2   EFFECTS OF SUBSTRATE CONCENTRATION ON VELOCITY

From Equation 5.2, one would expect the velocity of a pseudo-first-order reaction to depend linearly on the initial substrate concentration. When early studies were performed on enzyme-catalyzed reactions, however, scientists found instead that the reactions followed the substrate dependence illustrated in Figure 5.3. Figure 5.3A illustrates the time course of the enzyme-catalyzed reaction observed at different starting concentrations of substrate; the velocities for each experiment are measured as the slopes of the plots of [P] versus time. Figure 5.3B replots these data as the initial velocity v as a function of [S], the starting concentration of substrate. Rather than observing the linear relationship expected for first-order kinetics, we find the velocity apparently saturable at high substrate concentrations. This behavior puzzled early enzymologists.

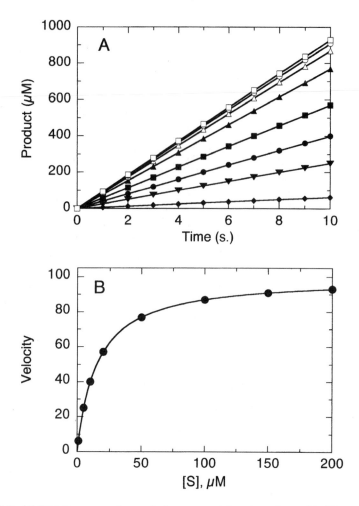

Figure 5.3 (A) Progress curves for a set of enzyme-catalyzed reactions with different starting concentrations of substrate [S]. (B) Plot of the reaction velocities, measured as the slopes of the lines from (A), as a function of [S].

Three distinct regions of this curve can be identified: at low substrate concentrations the velocity appears to display first-order behavior, tracking linearly with substrate concentration; at very high concentrations of substrate, the velocity switches to zero-order behavior, displaying no dependence on substrate concentration; and in the intermediate region, the velocity displays a curvilinear dependence on substrate concentration. How can one rationalize these experimental observations?

A qualitative explanation for the substrate dependence of enzyme-catalyzed reaction velocities was provided by Brown (1902). At the same time that the kinetic characteristics of enzyme reactions were being explored, evidence for complex formation between enzymes and their substrates was also accumulating. Brown thus argued that enzyme-catalyzed reactions could best be described by the following reaction scheme:

$$E + S \underset{k_{-1}}{\overset{k_1}{\rightleftharpoons}} ES \overset{k_2}{\longrightarrow} E + P$$

This scheme predicts that the reaction velocity will be proportional to the concentration of the ES complex as: $v = k_2[ES]$. Suppose that we held the total enzyme concentration constant at some low level and varied the concentration of S. At low concentrations of S the concentration of ES would be directly proportional to [S]; hence the velocity would depend on [S] in an apparent first-order fashion. At very high concentrations of S, however, practically all the enzyme would be present in the form of the ES complex. Under such conditions the velocity depends of the rate of the chemical transformations that convert ES to EP and the subsequent release of product to re-form free enzyme. Adding more substrate under these conditions would not effect a change in reaction velocity; hence the slope of the plot of velocity versus [S] would approach zero (as seen in Figure 5.3B). The complete [S] dependence of the reaction velocity (Figure 5.3B) predicted by the model of Brown resembles the results seen from the Langmuir isotherm Equation (Chapter 4) for equilibrium binding of ligands to receptors. This is not surprising, since in the model of Brown, catalysis is critically dependent on initial formation of a binary ES complex through equilibrium binding.

5.3 THE RAPID EQUILIBRIUM MODEL OF ENZYME KINETICS

Although the model of Brown provided a useful qualitative picture of enzyme reactions, to be fully utilized by experimental scientists, it needed to be put into a rigorous mathematical framework. This was accomplished first by Henri (1903) and subsequently by Michaelis and Menten (1913). Ironically, Michaelis and Menten are more widely recognized for this contribution, although they themselves acknowledged the prior work of Henri. The basic rate equation derived in this section is commonly referred to as the Michaelis–Menten equation. Several writers have recently taken to referring to the equation as the Henri–Michaelis–Menten equation, in an attempt to correct this neglect of Henri's contributions. The reader should be aware, however, that the majority of the scientific literature continues to use the traditional terminology.

The Henri–Michaelis–Menten approach assumes that a rapid equilibrium is established between the reactants (E + S) and the ES complex, followed by slower conversion of the ES complex back to free enzyme and product(s); that is, this model assumes that $k_2 \ll k_{-1}$ in the scheme presented in Section 5.2. In this model, the free enzyme E_f first combines with the substrate S to form the binary ES complex. Since substrate is present in large excess over enzyme, we can use the assumption that the free substrate concentration $[S]_f$ is well approximated by the total substrate concentration added to the reaction $[S]$. Hence, the equilibrium dissociation constant for this complex is given by:

$$K_S = \frac{[E]_f[S]}{[ES]} \tag{5.4}$$

Similar to the treatment of receptor–ligand binding in Chapter 4, here the free enzyme concentration is given by the difference between the total enzyme concentration $[E]$ and the concentration of the binary complex $[ES]$:

$$[E]_f = [E] - [ES] \tag{5.5}$$

and therefore,

$$K_S = \frac{([E] - [ES])[S]}{[ES]} \tag{5.6}$$

This can be rearranged to give an expression for $[ES]$:

$$[ES] = \frac{[E][S]}{K_S + [S]} \tag{5.7}$$

Next, the ES complex is transformed by various chemical steps to yield the product of the reaction and to recover the free enzyme. In the simplest case, a single chemical step, defined by the first-order rate constant k_2, results in product formation. More likely, however, there will be a series of rapid chemical events following ES complex formation. For simplicity, the overall rate for these collective chemical steps can be described by a single first-order rate constant k_{cat}. Hence:

$$E + S \underset{K_S}{\rightleftharpoons} ES \xrightarrow{k_{cat}} E + P$$

and the rate of product formation is thus given by the first-order equation:

$$v = k_{cat}[ES] \tag{5.8}$$

Combining Equations 5.7 and 5.8, we obtain:

$$v = \frac{k_{cat}[E][S]}{K_S + [S]} \tag{5.9}$$

Equation 5.9 is similar to the equation for a Langmuir isotherm, as derived in Chapter 4 (Equation 4.21). This, then, describes the reaction velocity as a hyperbolic function of [S], with a maximum value of $k_{cat}[E]$ at infinite [S]. We refer to this value as the maximum reaction velocity, or V_{max}.

$$V_{max} = k_{cat}[E] \tag{5.10}$$

Combining this definition with Equation 5.9, we obtain:

$$v = \frac{V_{max}[S]}{K_S + [S]} = \frac{V_{max}}{1 + \dfrac{K_S}{[S]}} \tag{5.11}$$

Equation 5.11 is the final equation derived independently by Henri and Michaelis and Menten to describe enzyme kinetic data. Note the striking similarity between this equation and the forms of the Langmuir isotherm equation presented in Chapter 4 (Equations 4.21 and 4.22). Thus, much of enzyme kinetics can be explained in terms of a simple equilibrium model involving rapid equilibrium between free enzyme and substrate to form the binary ES complex, followed by chemical transformation steps to produce and release product.

5.4 THE STEADY STATE MODEL OF ENZYME KINETICS

The original derivations by Henri and by Michaelis and Menten depended on a rapid equilibrium approach to enzyme reactions. This approach is quite useful in rapid kinetic measurements, such as single-turnover reactions, as described later in this chapter. The majority of experimental measurements of enzyme reactions, however, occur when the ES complex is present at a constant, steady state concentration (as defined below). Briggs and Haldane (1925) recognized that the equilibrium-binding approach of Henri and Michaelis and Menten could be described more generally by a steady state approach that did not require $k_2 \ll k_{-1}$. The following discussion is based on this description by Briggs and Haldane. As we shall see, the final equation that results from this treatment is very similar to Equation 5.11, and despite the differences between the rapid equilibrium and steady state approaches, the final steady state equation is commonly referred to as the Henri–Michaelis–Menten equation.

Steady state refers to a time period of the enzymatic reaction during which the rate of formation of the ES complex is exactly matched by its rate of decay to free enzyme and products. This kinetic phase can be attained when the concentration of substrate molecules is in great excess of the free enzyme concentration. To achieve a steady state, certain condition must be met, and these conditions allow us to make some reasonable assumption, which greatly simplify the mathematical treatment of the kinetics. These assumptions are as follows:

1. During the initial phase of the reaction progress curve (i.e., conditions under which we are measuring the linear initial velocity), there is no appreciable buildup of any intermediates other than the ES complex. Hence, all the enzyme molecules can be accounted for by either the free enzyme or by the enzyme–substrate complex. The total enzyme concentration [E] is therefore given by:

$$[E] = [E]_f + [ES] \qquad (5.12)$$

2. As in the rapid equilibrium treatment, we assume that the enzyme is acting catalytically, so that it is present in very low concentration relative to substrate, that is, $[S] \gg [E]$. Hence, formation of the ES complex does not significantly diminish the concentration of free substrate. We can therefore make the approximation: $[S]_f \sim [S]$, where $[S]_f$ is the free substrate concentration and $[S]$ is the total substrate concentration).

3. During the initial phase of the progress curve, very little product is formed relative to the total concentration of substrate. Hence, during this early phase $[P] \sim 0$ and therefore depletion of $[S]$ is minimal. At the initiation of the reaction there will be a rapid burst of formation of the ES complex followed by a kinetic phase in which the rate of formation of new ES complex is balanced by the rate of its decomposition back to free enzyme and product. In other words, during this phase the concentration of ES is constant. We refer to this kinetic phase as the *steady state*, which is defined by:

$$\frac{d[ES]}{dt} = 0 \qquad (5.13)$$

Figure 5.4 illustrates the development and duration of the steady state for the enzyme cytochrome c oxidase interacting with its substrates cytochrome c and molecular oxygen. As soon as the substrates and enzyme are mixed, we see a rapid pre–steady state buildup of ES complex, followed by a long time window in which the concentration of ES does not change (the steady state phase), and finally a post–steady state phase characterized by significant depletion of the starting substrate concentration.

With these assumptions made, we can now work out an expression for the enzyme velocity under steady state conditions. As stated previously, for the simplest of reaction schemes, the pseudo-first-order progress curve for an enzymatic reaction can be described by:

$$v = k_2[ES] \qquad (5.14)$$

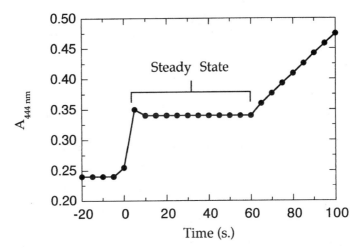

Figure 5.4 Development of the steady state for the reaction of cytochrome c oxidase with its substrates, cytochrome c and molecular oxygen. The absorbance at 444 nm reflects the ligation state of the active site heme cofactor of the enzyme. Prior to substrate addition (time < 0) the heme group is in the Fe^{3+} oxidation state and is ligated by a histidine group from the enzyme. Upon substrate addition, the active site heme iron is reduced to the Fe^{2+} state and rapidly reaches a steady state phase of substrate utilization in which the iron is ligated by some oxygen species. The steady state phase ends when a significant portion of the molecular oxygen in solution has been used up. At this point the heme iron remains reduced (Fe^{2+}) but is no longer bound to a ligand at its sixth coordination site; this heme species has a much larger extinction coefficient at 444 nm; hence the rapid increase in absorbance at this wavelength following the steady state phase. [Data adapted and redrawn from Copeland (1991).]

Now, [ES] is dependent on the rate of formation of the complex (governed by k_1) and the rate of loss of the complex (governed by k_{-1} and k_2). The rate equations for these two processes are thus given by:

$$\frac{d[ES]}{dt} = k_1[E]_f[S]_f \quad \text{and} \quad \frac{-d[ES]}{dt} = (k_{-1} + k_2)[ES] \quad (5.15)$$

Under steady state conditions these two rates must be equal, hence:

$$k_1[E]_f[S]_f = (k_{-1} + k_2)[ES] \quad (5.16)$$

This can be rearranged to obtain an expression for [ES]:

$$[ES] = \frac{[E]_f[S]_f}{\dfrac{k_{-1} + k_2}{k_1}} \quad (5.17)$$

At this point let us define the term K_m as an abbreviation for the kinetic constants in the denominatior of the right-hand side of Equation 5.17:

$$K_m = \frac{k_{-1} + k_2}{k_1} \tag{5.18}$$

For now we will consider K_m to be merely an abbreviation to make our subsequent mathematical expressions less cumbersome. Later, however, we shall see that K_m has a more significant meaning. Substituting Equation 5.18 into Equation 5.17 we obtain:

$$[ES] = \frac{[E]_f [S]_f}{K_m} \tag{5.19}$$

Now, since substrate depletion is insignificant during the steady state phase, we can replace the term $[S]_f$ by the total substrate concentration $[S]$ (which is much more easily measured in real experimental situations). We can also use the equality of Equation 5.12 to replace $[E]_f$ by $([E] - [ES])$. With these substitutions, Equation 5.19 can be recast as follows:

$$[ES] = [E] \frac{[S]}{[S] + K_m} \tag{5.20}$$

If we now combine this expression for $[ES]$ with the velocity expression of Equation 5.14, we obtain:

$$v = k_2 [E] \frac{[S]}{[S] + K_m} \tag{5.21}$$

Or, we can generalize Equation 5.21 for more complex reaction schemes by substituting k_{cat} for k_2:

$$v = k_{cat} [E] \frac{[S]}{[S] + K_m} \tag{5.22}$$

As described earlier, as the concentration of substrate goes towards infinity, the velocity reaches a maximum value that we have defined as V_{max}. Under these conditions, the K_m term is a very small contribution to Equation 5.22. Therefore:

$$\lim_{[S] \to \infty} \frac{[S]}{[S] + K_m} \cong \frac{[S]}{[S]} = 1 \tag{5.23}$$

and thus we again arrive at Equation 5.10: $V_{max} = k_{cat}[E]$. Combining this with

Equation 5.23 we finally arrive at an expression very similar to that first described by Henri and Michaelis and Menten (i.e., similar to Equation 5.11):

$$v = \frac{V_{max}[S]}{K_m + [S]} = \frac{V_{max}}{1 + \dfrac{K_m}{[S]}} \tag{5.24}$$

This is the central expression for steady state enzyme kinetics. While it differs from the equilibrium expression derived by Henri and by Michaelis and Menten, it is nevertheless universally referred to as the Michaelis–Menten or Henri–Michaelis–Menten equation.

In our definition of K_m (Equation 5.18), we combined first-order rate constants (k_{-1} and k_2, which have units of reciprocal time) with a second-order rate constant (k_1, which has units of reciprocal molarity, reciprocal time) in such a way that the resulting K_m has units of molarity, as does [S]. If we set up our experimental system so that the concentration of substrate exactly matches K_m, Equation 5.24 will reduce to:

$$v = \frac{V_{max}[S]}{[S] + [S]} = \frac{V_{max}}{2} \tag{5.25}$$

This provides us with a working definition of K_m: *The K_m is the substrate concentration that provides a reaction velocity that is half of the maximal velocity obtained under saturating substrate conditions.* The K_m value is often referred to in the literature as the *Michaelis constant*. In comparing Equation 5.24 for steady state kinetics with Equation 5.11 for the rapid equilibrium treatment, we see that the equations are identical except for the substitution of K_m for K_S in the steady state treatment. It is therefore easy to confuse these terms and to treat K_m as if it were the thermodynamic dissociation constant for the ES complex. However, the two constants are not always equal, even in considerations of the simplest of reactions schemes, as here. Recall that K_S can be defined by the rato of the reverse and forward reaction rate constants:

$$K_S = \frac{k_{-1}}{k_1} \tag{5.26}$$

This value is not identical to the expression for K_m given in Equation 5.18. Only under the specific conditions that $k_2 \ll k_{-1}$ are K_m and K_S equivalent. For more complex reaction schemes one would replace the k_2 term in Equation 5.18 by k_{cat}. Recall that k_{cat} reflects a summation of multiple chemical steps in catalysis. Hence, depending on the details of the reaction mechanism, and the values of the individual rate constants, situations can arise in which the value of K_m is less than, greater than, or equal to K_S. Therefore, K_m should generally be considered as a kinetic, not thermodynamic, constant.

5.5 THE SIGNIFICANCE OF k_{cat} AND K_m

We have gone to great lengths in this chapter to define and derive expressions for the kinetic constants k_{cat} and K_m. What value do these constants add to our understanding of the enzyme under study?

5.5.1 K_m

The value of K_m varies considerably from one enzyme to another, and for a particular enzyme with different substrates. We have already defined K_m as the substrate concentration that results in half-maximal velocity for the enzymatic reaction. An equivalent way of stating this is that the K_m represents the substrate concentration at which half of the enzyme active sites in the sample are filled (i.e., saturated) by substrate molecules in the steady state. Hence, while K_m is not equivalent to K_S under most conditions, it can nevertheless be used as a relative measure of substrate binding affinity. In some instances, changes in solution conditions (pH, temperature, etc.) can have selective effects on the value of K_m. Also, one sometimes observes effects on the value of K_m in the course of comparing different mutants or isoforms of an enzyme, or different substrates with a common enzyme. In these cases one can reasonably relate the changes to effects on the stability (i.e., affinity) of the ES complex. As we shall see below, however, the ratio k_{cat}/K_m is generally a better measure of effects on substrate binding.

5.5.2 k_{cat}

Considering Equations 5.22–5.24, we see that if one knows the concentration of enzyme used experimentally, the value of k_{cat} can be directly calculated by dividing the experimentally determined value of V_{max} by [E]. The value of k_{cat} is sometimes referred to as the *turnover number* for the enzyme, since it defines the number of catalytic turnover events that occur per unit time. The units of k_{cat} are reciprocal time (e.g., min^{-1}, s^{-1}). Turnover numbers, however, are typically reported in units of molecules of product produced per unit time per molecules of enzyme present. As long as the same units are used to express the amount of product produced and the amount of enzyme present, these units will cancel and, as expected, the final units will be reciprocal time. It is important, however, that the units of product and enzyme concentration be expressed in molar or molarity units. In crude enzyme samples, such as cell lysates and other nonhomogeneous protein samples, it is often impossible to know the concentration of enzyme in anything other than units of total protein mass. The literature is thus filled with enzyme activity values expressed as number of micrograms of product produced per minute per microgram of protein in the enzyme sample. While such units may be useful in comparing one batch of crude enzyme to another (see the discussion of specific activity measurements in Chapter 7), it is difficult to relate these values to kinetic constants, such as k_{cat}.

In the laboratory we can easily determine the turnover number as k_{cat}, by measuring the reaction velocity under conditions of $[S] \gg K_m$ so that v approaches V_{max}. The rate of enzyme turnover under most physiological conditions in vivo, however, is very different from our laboratory situation. In vivo, the concentration of substrate is more typically 0.1–$1.0 K_m$. When $[S] \ll K_m$ we must change our expression for velocity to:

$$v = \frac{k_{cat}}{K_m} [E][S]_f \qquad (5.27)$$

(Since $[S] \ll K_m$ here, the free enzyme concentration is well approximated by the total enzyme concentration $[E]$; thus we used this term in Equation 5.27 in place of $[E]_f$.) Recalling our definition of K_m, we note that:

$$\frac{k_{cat}}{K_m} = \frac{k_{cat} k_1}{k_{-1} + k_{cat}} \qquad (5.28)$$

Thus, under our laboratory conditions, where $[S] \gg K_m$, formation of the ES complex is rapid and often is not the rate-limiting step. In vivo, however, where $[S] \ll K_m$, the overall reaction may be limited by the diffusional rate of encounter of the free enzyme with substrate, which is defined by k_1. The rate constant for diffusional encounters between molecules like enzymes and substrates is typically in the range of 10^8–$10^9 \ M^{-1} s^{-1}$. Thus we must keep in mind that the rate-limiting step in catalysis is not always the same in vivo as in vitro. Nevertheless, measurement of k_{cat} (i.e., velocity under saturating substrate concentration) gives us the most consistent means of comparing rates for different enzymatic reactions.

The significance of k_{cat} is that it defines for us the maximal velocity at which an enzymatic reaction can proceed at a fixed concentration of enzyme and infinite availability of substrate. Because k_{cat} relates to the chemical steps subsequent to formation of the ES complex, changes in k_{cat}, brought about by changes in the enzyme (e.g., mutagenesis of specific amino acid residues, or comparison of different enzymes), in solution conditions (e.g., pH, ionic strength, temperature, etc.), or in substrate identity (e.g., structural analogues or isotopically labeled substrates), define perturbations that affect the chemical steps in enzymatic catalysis. In other words, changes in k_{cat} reflect perturbations of the chemical steps subsequent to initial substrate binding. Since k_{cat} reflects multiple chemical steps, it does not provide detailed information on the rates of any of the individual steps subsequent to substrate binding. Instead k_{cat} provides a lower limit on the first-order rate constant of the slowest (i.e., rate-determining) step following substrate binding that leads eventually to product release.

5.5.3 k_{cat}/K_m

The catalytic efficiency of an enzyme is best defined by the ratio of the kinetic constants, k_{cat}/K_m. This ratio has units of a second-order rate constant and is generally used to compare the efficiencies of different enzymes to one another.

The values of k_{cat}/K_m are also used to compare the utilization of different substrates for a particular enzyme. As we shall see in Chapter 6, in comparisons of different substrates for an enzyme, the largest differences often are seen in the values of k_{cat}, rather than in K_m. This is because substrate specificity often results from differences in transition state, rather than ground state binding interactions (see Chapter 6 for more details). Hence, the ratio k_{cat}/K_m captures the effects of differing substrate on either kinetic constant and provides a lower limit for the second-order rate constant of productive substrate binding (i.e., substrate binding leading to $ES^{\ddagger}$ complex formation and eventual product formation); this ratio is therefore considered to be the best measure of substrate specificity.

The ratio k_{cat}/K_m is also used to compare the efficiency with which an enzyme catalyzes a particular reaction in the forward and reverse directions. Enzymatic reactions are in principle reversible, although for many enzymes the reverse reaction is thermodynamically unfavorable. The presence of an enzyme in solution does not alter the equilibrium constant K_{eq} between the free substrate and free product concentrations. Hence, the value of K_{eq} is fixed for specific solution conditions, and this constrains the values of k_{cat}/K_m that can be achieved in the forward (f) and reverse (r) directions. At equilibrium the forward and reverse reactions occur with equal frequency so that:

$$\left(\frac{k_{cat}}{K_m}\right)_f [E][S] = \left(\frac{k_{cat}}{K_m}\right)_r [E][P] \tag{5.29}$$

hence,

$$K_{eq} = \frac{(k_{cat}/K_m)_f}{(k_{cat}/K_m)_r} \tag{5.30}$$

Equation 5.30, known as the Haldane relationship, provides a useful measure of the directionality of an enzymatic reaction under a specific set of solution conditions.

In either direction, the ratio k_{cat}/K_m can be related to the free energy difference between the free reactants (E and S, in the forward direction) and the transition state complex ($ES^{\ddagger}$). If we normalize the free energy of the reactant state to zero, the free energy difference is defined by:

$$\Delta G_{ES^{\ddagger}} = -RT \ln\left(\frac{k_{cat}}{K_m}\right) + RT \ln\left(\frac{k_B T}{h}\right) \tag{5.31}$$

where k_B is the Boltzmann constant, T is temperature in degrees Kelvin, and

h is Planck's constant, so that at a fixed temperature the term $RT \ln(k_B T/h)$ is a constant. This relationship holds because generally attainment of the transition state is the most energetically costly component of the multiple steps contributing to k_{cat}. If we compare different substrates for a single enzyme, or different enzymes or mutants versus a common substrate, we can calculate the difference in transition state energies ($\Delta\Delta G_{ES^\ddagger}$) from experimentally measured values of k_{cat}/K_m at constant temperature:

$$\Delta\Delta G_{ES^\ddagger} = -RT \ln \left[\frac{(k_{cat}/K_m)^1}{(k_{cat}/K_m)^2} \right] \tag{5.32}$$

where the superscripts 1 and 2 refer to the different substrates or enzymes being compared. By this type of analysis, one can quantitate the thermodynamic contributions of particular structural components to catalysis. For example, suppose one suspected that an active site tyrosine residue was forming a critical hydrogen bond with substrate in the transition state of the enzymatic reaction. Through the tools of molecular biology, one could replace this tyrosine with a phenylalanine (which would be incapable of forming an H bond) by site-directed mutagenesis and measure the value of k_{cat}/K_m for both the wild type and the Tyr→Phe mutant. Suppose these values turned out to be $88\ \text{M}^{-1}\,\text{s}^{-1}$ for the wild-type enzyme and $0.1\ \text{M}^{-1}\,\text{s}^{-1}$ for the mutant. The ratio of these k_{cat}/K_m would be 880 and, from Equation 5.30, this would correspond to a difference in transition state free energy of 4 kcal/mol, consistent with a strong H-bonding interaction of the tyrosine (of course this does not prove that the exact role of the tyrosine OH group is H-bonding, but the data do prove that this OH group plays a critical role in catalysis). A good example of the use of this approach can be found in the paper by Wilkinson et al. (1983).

5.5.4 Diffusion-Controlled Reactions and Kinetic Perfection

For an enzyme in solution, the rate-determining step in catalysis will be either k_1, the rate of ES formation, or one of the multiple steps contributing to k_{cat}. If k_{cat} is rate limiting, the catalytic events that occur after substrate binding are slower that the rate of formation of the ES complex. If, however, k_1 is rate limiting, the enzyme turns over essentially instantaneously once the ES complex has formed. In either case we see that the fastest rate of catalysis for an enzyme in solution is limited by the rate of diffusion of molecules in the solution. Some enzymes, such as carbonic anhydrase, display k_{cat}/K_m values of 10^8–$10^9\ \text{M}^{-1}\,\text{s}^{-1}$, which is at the diffusion limit. Such enzymes are said to have achieved *kinetic perfection*, because they convert substrate to product as fast as the substrate is delivered to the active site of the enzyme!

The diffusion limit would seem to set an upper limit on the value of k_{cat}/K_m that an enzyme can achieve. This is true for most enzymes in solution. However, some enzyme systems have overcome this limit by compartmentalizing themselves and their substrates within close proximity in subcellular locals where three-dimensional diffusion no longer comes into play. This can be accomplished by assembling enzymes and substrates into organized systems such as multienzyme complexes or cellular membranes. Two examples are presented.

We first consider the respiratory electron transfer system of the inner mitochondrial membrane. Here enzymes in a cascade are localized in close proximity to one another within the membrane bilayer. The product of one enzyme is the substrate for the next in the cascade. Because of the proximity of the enzymes in the membrane, the product leaves the active site of one enzyme and is presented to the active site of the next enzyme without the need for diffusion through solution.

The second example comes from the de novo synthetic pathway for pyrimidines. The first three steps in the synthesis of uridine monophosphate are performed by a supercomplex of three enzymes that are noncovalently associated as a multiprotein complex. This supercomplex, referred to as CAD, comprises the enzymes carbamyl phosphate synthase, aspartate transcarbamylase, and dihydroorotase. Because the active sites of the three enzymes are compartmentalized inside the supercomplex, the product of the first enzyme is immediately in proximity to the active site of the second enzyme, and so on. In this way, the supercomplex can overcome the diffusion barrier to rapid catalysis.

5.6 EXPERIMENTAL MEASUREMENT OF k_{cat} AND K_m

5.6.1 Graphical Determinations from Untransformed Data

The kinetic constants V_{max} and K_m are determined graphically with initial velocity measurements obtained at varying substrate concentrations. The graphical methods are best illustrated by working through examples with some numerical data. The quality of the estimates of V_{max} and K_m depend on covering a substrate concentration range that spans a significant portion of the binding isotherm. Experimentally, a convenient method for choosing substrate concentrations is to first make a stock solution of substrate at the highest concentration that is experimentally reasonable. Then, twofold serial dilutions can be made from this stock to produce a range of lower substrate concentrations. For example, let us say that the highest concentration of substrate to be used in an enzymatic reaction is 250 μM. We could make a 2.5 mM stock solution of the substrate that would be diluted 10-fold into the final assay reaction mixture (i.e., to give a final concentration of 250 μM). We could then take a portion of this stock solution and dilute it with an equal volume of buffer to yield a 1.25 mM solution, which upon dilution into the assay reaction mixture would give a final substrate concentration of 125 μM. A portion of this solution could also be diluted in half with buffer, and so on, to yield a series

Table 5.1   Initial velocity (with random error added) as a function of substrate concentration for a model enzymatic reaction

[S](μM)[a]	v(μM product formed s^{-1})	$1/v$(μM$^{-1}\cdot$s)	$1/$[S](μM^{-1})
0.98	10	0.100	1.024
1.95	12	0.083	0.512
3.91	28	0.036	0.256
7.81	40	0.025	0.128
15.63	55	0.018	0.064
31.25	75	0.013	0.032
62.50	85	0.012	0.016
125.00	90	0.011	0.008
250.00	97	0.010	0.004

[a]Substrate concentrations reflect a twofold serial dilution starting with an initial solution that provides 250 μM substrate to the final assay reaction mixture.

of solutions of diminishing substrate concentrations. The final substrate concentrations presented in Table 5.1 illustrate the use of such a twofold *serial dilution* scheme. Let us suppose that we are studying a simple enzymatic reaction for which the true values of V_{max} and K_m are 100 μM/s and 12 μM, respectively. In Table 5.1 we have listed experimental values for the initial velocity v at each of the substrate concentrations used. In generating this table, some random error has been added to each of the velocity values to better simulate real experimental conditions. The largest percent errors in this table occur at the lowest substrate concentrations, where in real experiments one encounters the greatest difficulty in obtaining accurate velocity measurements.

The first and most straightforward way of graphing the data is as a direct plot of velocity as a function of [S]; we shall refer to such a plot as a Michaelis–Menten plot. Figure 5.5 is a Michaelis–Menten plot for the data in Table 5.1; the line drawn through the data was generated by a nonlinear least-squares fit of the data to Equation 5.24. With modern computer graphics programs, the reader has a wide choice of options for performing such curve fitting; some programs that are particularly well suited for enzyme studies are listed in Appendix II. The plots in this book, for example, were generated with the program Kaleidagraph (from Abelbeck Software), which contains a built-in iterative method for performing nonlinear curve fitting to user-generated equations. For the data in Figure 5.5 both V_{max} and K_m were set as unknowns that were simultaneously solved for by the curve-fitting routine. The estimates of V_{max} and K_m determined in this way were 100.36 μM/s and 11.63 μM, respectively, in excellent agreement with the true values of these constants. Such direct fits of the untransformed data provide the most reliable estimates of both kinetic constants.

With the widespread availability of computer curve-fitting programs, what limitations are there on our ability to estimate V_{max} and K_m from experimental

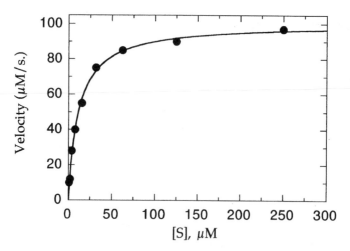

Figure 5.5 Michaelis–Menten plot for the velocity data in Table 5.1. The solid line through the data points represents the nonlinear least-squares best fit to Equation 5.24.

data? As mentioned above, the accuracy of such estimates will depend on the range of substrate concentrations over which the initial velocity has been determined. If measurements are made only at low substrate concentrations, the data will appear to be first-ordered (i.e., v will appear to be a linear function of [S]). This is illustrated in Figure 5.6A for the data in Table 5.1 between substrate concentrations of 0.98 and 3.91 μM (i.e., $\leq 0.33 K_m$). In this concentration range, the enzyme active sites never reach saturation, and graphically, both V_{max} and K_m appear to be infinite (but see Section 5.8). On the other hand, Figure 5.6B illustrates what happens when measurements are made at very high substrate concentrations only; here the data for substrate concentrations above 60 μM are considered (i.e., [S] $\geq 5K_m$). In this saturating substrate concentration range, the velocity appears to be almost independent of substrate concentration. While a rough estimate of V_{max} might be obtained from these data (although the reader should note that the true V_{max} is only reached at infinite substrate concentration; hence any experimentally measured velocity at high [S] may approach, but never fully reach V_{max}), there is no way to determine the K_m value here.

The plots in Figure 5.6 emphasize the need for exploring a broad range of substrate concentrations to accurately determine the kinetic constants for the enzyme of interest. Again, there may be practical limits on the range of substrate concentrations over which such measurements can be performed. In Chapter 4 we suggested that to best characterize a ligand binding isotherm, it is necessary to cover a ligand concentration range that resulted in 20–80% receptor saturation. Likewise, in determining the steady state kinetic constants for an enzymatic system, it is best to at least cover substrate concentrations that yield velocities of 20–80% of V_{max}; this corresponds to [S] of 0.25–5.0K_m.

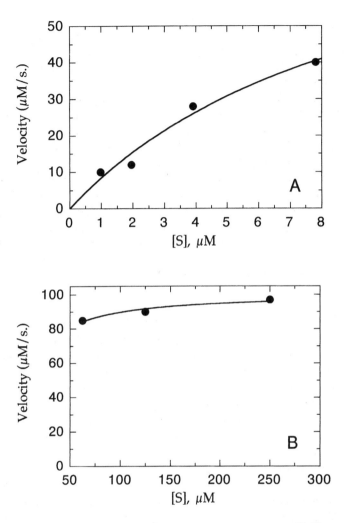

Figure 5.6 Michaelis–Menten plots for restricted data from Table 5.1. (A) The range of [S] values is inappropriately low ($\leqslant 0.33K_m$), hence K_m and V_{max} appear to be infinite. (B) The range of [S] values is inappropriately high, with the result that every data point represents near-saturating conditions; one may be able to approximate V_{max}, but K_m cannot be determined.

Since the kinetic constants are unknowns prior to these experiments, it is common to perform initial experiments with a limited number of data points that span as broad a range of substrate concentrations as possible (at least a 100-fold substrate concentration range) to obtain a rough estimate of K_m and V_{max}. Improved estimates can then be obtained by narrowing the substrate concentration range between 0.25 and 5.0K_m and obtaining a larger number of data points within this range. Table 5.1 illustrates an ideal situation for determining V_{max} and K_m. The data span a 250-fold range of substrate concentrations that cover the range from 0.08 to $20.8K_m$. Hence, use of such a twofold serial dilution setup, starting with the highest substrate concentration that is feasible, is highly recommended.

Alternatively, one could perform a limited number of experiments by using a fivefold serial dilution setup starting at our maximum substrate concentration of $250\,\mu M$. With only five experiments we would then cover the substrate concentrations 250, 50, 10, 2, and $0.4\,\mu M$. Data from such a hypothetical experiment are shown in Figure 5.7A and would yield an estimate of K_m of about $10\,\mu M$. With this initial estimate in hand, one might then chose to expand the number of data points within the narrower range of $0.25–5.0K_m$ to obtain better estimates of the kinetic constants. Figure 5.7B, for example, illustrates the type of data one might obtain from velocity measurements at substrate concentrations of 2.5, 5, 10, 15, 20, 25, 30, 35, 40, 45, and $50\,\mu M$. From this second set of measurements, values of 102 M/s and $12\,\mu M$ for V_{max} and K_m, respectively, would be obtained.

5.6.2 Lineweaver–Burk Plots of Enzyme Kinetics

The widespread availability of user-friendly nonlinear curve-fitting programs is a relatively recent development. In the past, determination of the kinetic constants for an enzyme from the untransformed data was not so routine. To facilitate work in this area, scientists searched for means of transforming the data to produce linear plots from which the kinetic constants could be determined simply with graph paper and a straightedge. While today many of us have nonlinear curve-fitting programs at our disposal (and this is the preferred means of determining the values of V_{max} and K_m), there is still considerable value in linearized plots of enzyme kinetic data. As we shall see in subsequent chapters, these plots are extremely useful in diagnosing the mechanistic details of multisubstrate enzymes and for determining the mode of interaction between an enzyme and an inhibitor.

The most commonly used method for linearizing enzyme kinetic data is that of Lineweaver and Burk (1934). We start with the same steady state assumption described earlier. Applying some simple algebra, we can rewrite Equation 5.24 in the following form:

$$v = V_{max}\left(\cfrac{1}{1 + \cfrac{K_m}{[S]}}\right) \qquad (5.33)$$

Now we simply take the reciprocal of this equation and rearrange to obtain:

$$\frac{1}{v} = \left(\frac{K_m}{V_{max}}\frac{1}{[S]}\right) + \frac{1}{V_{max}} \qquad (5.34)$$

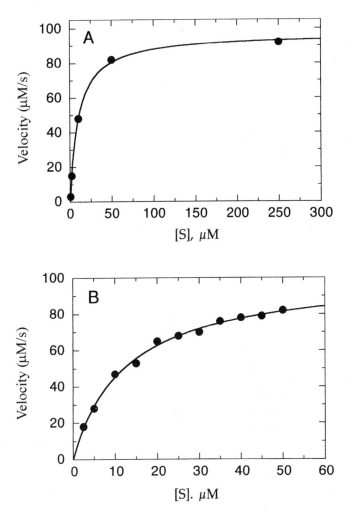

Figure 5.7 Experimental strategy for estimating K_m and V_{max}. (A) A limited data set is collected over a broad range of [S] to get a rough estimate of the kinetic constants. (B) Once a rough estimate of K_m has been determined, a second set of experiments is performed with more data within the range of 0.25–$5.0K_m$ to obtain more precise estimates of the kinetic constants.

Comparing Equation 5.34 with the standard equation for a straight line, we have

$$y = mx + b \tag{5.35}$$

where m is the slope and b is the y intercept. We see that Equation 5.34 is an equation for a straight line with slope of K_m/V_{max} and y intercept of $1/V_{max}$.

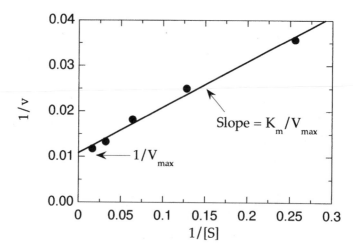

Figure 5.8 Lineweaver–Burk double-reciprocal plot for selected data from Table 5.1 within the range of [S] = 0.25–5.0K_m.

Thus if the reciprocal of initial velocity is plotted as a function of the reciprocal of [S], we would expect from Equation 5.34 to obtain a linear plot. For the same reasons described earlier for untransformed data, these plots work best when the substrate concentration covers the range of 0.25–5.0K_m. Within this range, good linearity is observed, as illustrated in Figure 5.8 for the data between [S] = 3.91 and [S] = 62.50 μM in Table 5.1. A plot like that in Figure 5.8 is known as a Lineweaver–Burk plot.

The kinetic constants K_m and V_{max} can be determined from the slope and intercept values of the linear fit of the data in a Lineweaver–Burk plot. Since the x axis is reciprocal substrate concentration, the value of $x = 0$ (i.e., 1/[S] = 0) corresponds to [S] = infinity. Hence, the extrapolated value of the y intercept corresponds to the reciprocal of V_{max}. The value of K_m can be determined from a Lineweaver–Burk plot in two ways. First we note from Equation 5.34 that the slope is equal to K_m divided by V_{max}. If we therefore divide the slope of our best fit line by the y-intercept value (i.e., by 1/V_{max}), the product will be equal to K_m. Alternatively, we could extrapolate our linear fit to the point of intersecting the x axis. This x intercept is equal to $-1/K_m$; thus we could determine K_m from the absolute value of the reciprocal of the x intercept of our plot.

We have noted several times that the preferred way to determine K_m and V_{max} values is from nonlinear fitting of untransformed data to the Michaelis–Menten equation. Figure 5.8 illustrates why we have stressed this point. In real experimental data, small errors in the measured values of v are amplified by the mathematical transformation of taking the reciprocal. The greatest percent error is likely to be associated with velocity values at low substrate concentration. Unfortunately, in the reciprocal plot, the lowest values of [S] correspond

Table 5.2   Estimates of the kinetic constants V_{max} and K_m from various graphical treatments of the data from Table 5.1

Graphical Method	$K_m(\mu M)$	Deviation from True $K_m(\%)$	$V_{max}(\mu M/s)$	Deviation from True $V_{max}(\%)$
True values	12.00		100.00	
Michaelis–Menten	11.63	3.08	100.36	0.36
Lineweaver–Burk (full data set)	7.57	36.92	79.28	20.72
Lineweaver–Burk ([S] = 0.25–5.0K_m only)	9.17	23.58	91.84	8.16
Eadie–Hofstee	9.66	19.50	94.45	5.55
Hanes–Wolff	11.84	1.33	100.97	0.97
Eisenthal–Cornish-Bowden	11.53	3.92	100.64	0.64

to the highest values of 1/[S], and because of the details of linear regression, these data points are weighted more heavily in the analysis. Hence the experimental error is amplified and unevenly weighted in this analysis, resulting in poor estimates of the kinetic constants even when the experimental error is relatively small. To illustrate this, let us compare the estimates of V_{max} and K_m obtained for the data in Table 5.1 by various graphical methods; this is summarized in Table 5.2. The true values of V_{max} and K_m for the hypothetical data in Table 5.1 were $100 \, \mu M/s$ and $12 \, \mu M$, respectively. The fitting of the untransformed data to the Michaelis–Menten equation provided estimates of 100.36 and 11.63 for the two kinetic constants, with deviations from the true values of only 0.36 and 3.08%, respectively. The linear fitting of the data in Figure 5.8, on the other hand, yields estimates of V_{max} and K_m of 91.84 and 9.17, with deviations from the true values of 8.16 and 23.58%, respectively. The errors are even greater when the double-reciprocal plots are used for the full data set in Table 5.1, as illustrated in Figure 5.9 and Table 5.2. Here the inclusion of the low substrate data values ($<[S] = 3.91 \, \mu M$) are very heavily weighted in the linear regression and further limit the precision of the kinetic constant estimates. The values of V_{max} and K_m derived from this fitting are 79.28 and 7.57, representing deviations from the true values of 20.72 and 36.92%, respectively.

The foregoing example, should convince the reader of the limitations of using linear transformations of the primary data for determining the values of the kinetic constants. Nevertheless, the Lineweaver–Burk plots are still commonly used by many researchers and, as we shall see in later chapters, are valuable tools for certain purposes. In these situations (described in detail in Chapters 8 and 11), we make the following recommendation. Rather than using linear regression to fit the reciprocal data in Lineweaver–Burk plots, one should determine the values of V_{max} and K_m by nonlinear regression analysis of the untransformed data fit to the Michaelis–Menten equation. These values are then inserted as constants into Equation 5.34 to create a line through the

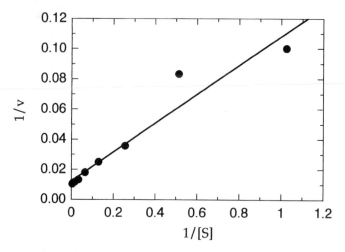

Figure 5.9 Lineweaver–Burk double-reciprocal plot for the full data set from Table 5.1. Note the strong influence of the data points at low [S] (high 1/[S] values) on the best fit line from linear regression.

reciprocal data on the Lineweaver–Burk plot. The line drawn by this method may not appear to fit the reciprocal data as well as a linear regression fit, but it will be a much more accurate reflection of the kinetic behavior of the enzyme. The use of this method will be more clear when it is applied in Chapters 8 and 11 to studies of enzyme inhibition and multisubstrate enzyme mechanisms, respectively.

If one is to ultimately present experimental data in the form of a double-reciprocal plot, it is desirable to chose substrate concentrations that will be evenly spaced along a reciprocal x axis (i.e., 1/[S]). This is easily accomplished experimentally as follows. One picks a maximum value of [S] ($[S_{max}]$) to work with and makes a stock solution of substrate that will give this final concentration after dilution into the assay reaction mixture. Additional initial velocity measurements are then made by adding the same final volume to the enzyme reaction mixture from stock substrate solutions made by diluting the original stock solution by 1:2, 1:3, 1:4, 1:5, and so on. In this way, the data points will fall along the 1/[S] axis at intervals of 1, 2, 3, 4, 5, ..., units.

For example, let us say that we have decided to work with a maximum substrate concentration of 60 μM in our enzymatic reaction. If we prepare a 600 μM stock solution of substrate for this data point, we might dilute it 1:10 into our assay reaction mixture to obtain the desired final substrate concentration. If, for example, our total reaction volume were 1.0 mL, we could start our reaction by mixing 100 μL of substrate stock, with 900 μL of the other components of our reaction system (enzyme, buffer, cofactors, etc.). Table 5.3 summarizes the additional stock solutions that would be needed to prepare final substrate concentrations evenly spaced along a 1/[S] axis.

Table 5.3 Setup for an experimental determination of enzyme kinetics using a Lineweaver–Burk plot

Stock [S] (μM)	Final [S] in Reaction Mixture (μM)	1/[S] (μM^{-1})
600	60.0	0.017
300	30.0	0.033
200	20.0	0.050
150	15.0	0.067
120	12.0	0.083
100	10.0	0.100
86	8.6	0.116
75	7.5	0.133
67	6.7	0.149
60	6.0	0.167
55	5.5	0.182
50	5.0	0.200

5.7 OTHER LINEAR TRANSFORMATIONS OF ENZYME KINETIC DATA

Despite the errors associated with this method, the Lineweaver–Burk double-reciprocal plot has become the most popular means of graphically representing enzyme kinetic data. There are, however, a variety of other linearizing transformations. Again, the use of these transformation methods is no longer necessary because most researchers have access to computer-based nonlinear curve-fitting methods, and the direct fitting of untransformed data by these methods is highly recommended. For the sake of historic persepective, however, we shall describe three other popular graphical methods for presenting enzyme kinetic data: Eadie–Hofstee, Hanes–Wolff, and Eisenthal–Cornish-Bowden direct plots. These linear transformation methods, which are here applied to enzyme kinetic data, are identical to the Wolff transformations described in Chapter 4 for receptor–ligand binding data.

5.7.1 Eadie–Hofstee Plots

If we multiply both sides of Equation 5.24 by $K_m + [S]$, we obtain:

$$v(K_m + [S]) = V_{max}[S] \tag{5.36}$$

If we now divide both sides by [S] and rearrange, we obtain:

$$v = V_{max} - K_m \left(\frac{v}{[S]} \right) \tag{5.37}$$

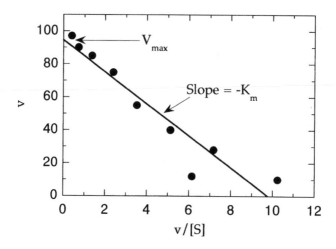

Figure 5.10 Eadie–Hofstee plot of enzyme kinetic data. Data taken from Table 5.1.

Hence, if we plot v as a function of $v/[S]$, Equation 5.37 would predict a straight-line relationship with slope of $-K_m$ and y intercept of V_{max}. Such a plot, referred to as an Eadie–Hofstee plot, is illustrated in Figure 5.10.

5.7.2 Hanes–Wolff Plots

If one multiplies both sides of the Lineweaver–Burk Equation (Equation 5.34) by $[S]$, one obtains:

$$\frac{[S]}{v} = [S]\left(\frac{1}{V_{max}}\right) + \frac{K_m}{V_{max}} \tag{5.38}$$

This treatment also leads to linear plots when $[S]/v$ is plotted as a function of $[S]$. Figure 5.11 illustrates such a plot, which is known as a Hanes–Wolff plot. In this plot the slope is $1/V_{max}$, the y intercept is K_m/V_{max}, and the x intercept is $-K_m$.

5.7.3 Eisenthal–Cornish-Bowden Direct Plots

In our final method, pairs of v, $[S]$ data (as in Table 5.1) are plotted as follows: values of v along the y axis and the negative values of $[S]$ along the x axis (Eisenthal and Cornish-Bowden, 1974). For each pair, one then draws a straight line connecting the points on the two axes and extrapolates these lines past their point of intersection (Figure 5.12). When a horizontal line is drawn from the point of intersection of these line to the y axis, the value at which this horizontal line crosses the y axis is equal to V_{max}. Similarly, when a vertical line is dropped from the point of intersection to the x axis, the value at which this

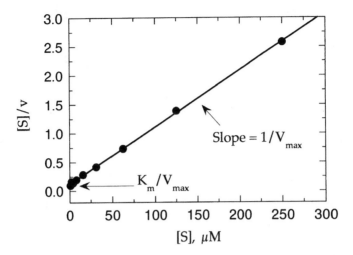

Figure 5.11 Hanes–Wolff plot of enzyme kinetic data. Data taken from Table 5.1.

vertical line crosses the x axis defines K_m. Plots like Figure 5.12, are referred to as Eisenthal–Cornish-Bowden direct plots and are considered to give the best estimates of K_m and V_{max} of any of the linear transformation methods. Hence they are highly recommended when it is desired to determine these kinetic parameters but nonlinear curve fitting to Equation 5.24 is not feasible.

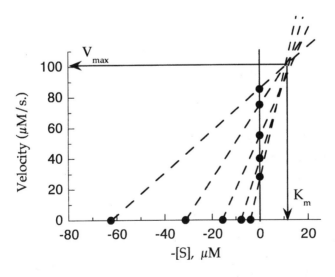

Figure 5.12 Eisenthal–Cornish-Bowden direct plot of enzyme kinetic data. Selected data taken from Table 5.1.

5.8   MEASUREMENTS AT LOW SUBSTRATE CONCENTRATIONS

In some instances the concentration range of substrates suitable for experimental measurements is severely limited because of poor solubility or some physicochemical property of the substrate that interferes with the measurements above a critical concentration. If one is limited to measurements in which the substrate concentration is much less than the K_m, the reaction will follow pseudo-first-order kinetics, and it may be difficult to find a time window over which the reaction velocity can be approximated by a linear function. Even if quasi-linear progress curves can be obtained, a plot of initial velocity as a function of [S] cannot be used to determine the individual kinetic constants k_{cat} and K_m, since the substrate concentration range that is experimentally attainable is far below saturation (as in Figure 5.6A). In such situations one can still derive an estimate of k_{cat}/K_m by fitting the reaction progress curve to a first-order equation at some fixed substrate concentration.

Suppose that we were to follow the loss of substrate as a function of time under first-order conditions (i.e., where $[S] \ll K_m$). The progress curve could be fit by the following equation:

$$[S] = [S_0]e^{-kt} \tag{5.39}$$

where [S] is the substrate concentration remaining after time t, $[S_0]$ is the starting concentration of substrate, and k is the observed first-order rate constant. When $[S] \ll K_m$, the [S] term can be ignored in the denominator of Equation 5.24. Combining this with our definition of V_{max} from Equation 5.10 we obtain:

$$-\frac{d[S]}{dt} = \frac{k_{cat}}{K_m}[E][S] \tag{5.40}$$

Rearranging Equation 5.40 and integrating, we obtain:

$$[S] = [S_0]\exp\left(\frac{k_{cat}}{K_m}[E]t\right) \tag{5.41}$$

Comparing Equation 5.41 with Equation 5.39, we see that:

$$k = \frac{k_{cat}}{K_m}[E] \tag{5.42}$$

Thus if the concentration of enzyme used in the reaction is known, an estimate of k_{cat}/K_m can be obtained from the measured first-order rate constant of the reaction progress curve when $[S] \ll K_m$ (Chapman et al., 1993; Wahl, 1994).

5.9 DEVIATIONS FROM HYPERBOLIC KINETICS

In most cases enzyme kinetic measurements fit remarkably well to the Henri–Michaelis–Menten behavior discussed in this chapter. However, occasional deviations from the hyperbolic dependence of velocity on substrate concentration are seen. Such anomalies occur for several reasons. Some physical methods of measuring velocity, such as optical spectroscopies, can lead to experimental artifacts that have the appearance of deviations from the expected behavior, and we shall discuss these in detail in Chapter 7.

Nonhyperbolic behavior can also be caused by the presence of certain types of inhibitor as well. In the most often encountered case, substrate inhibition, a second molecule of substrate can bind to the ES complex to form an inactive ternary complex, SES. Because formation of the ES complex must precede formation of the inhibitory ternary complex, substrate inhibition is usually realized only at high substrate concentrations, and it is detected as a lower than expected value for the measured velocity at these high substrate concentrations. Figure 5.13 illustrates the type of behavior one might see for an enzyme that exhibits substrate inhibition. At low substrate concentrations, the kinetics follow simple Michaelis–Menten behavior. Above a critical substrate concentration, however, the data deviate significantly from the expected behavior. The binding of the second, inhibitory, molecule of substrate can be accounted for by the following equation:

$$ v = \frac{V_{max}[S]}{K_m + [S]\left(1 + \dfrac{[S]}{K_i}\right)} \tag{5.43}$$

or dividing the top and bottom of the right-hand side of Equation 5.43 by [S], we obtain:

$$ v = \frac{V_{max}}{1 + \dfrac{K_m}{[S]} + \dfrac{[S]}{K_i}} \tag{5.44}$$

where the term K_i in Equations 5.43 and 5.44 represents the dissociation constant for the inhibitory SES ternary complex. Inhibition effects at very high substrate concentrations also can be readily detected as nonlinearity in the Lineweaver–Burk plots of the data. Here one observes a sudden and dramatic curving up of the data near the y-axis intercept.

Another cause of nonhyperbolic kinetics is the presence of more than one enzyme acting on the same substrate (see also Chapter 4, Section 4.3.2.2). Many enzyme studies are performed with only partially purified enzymes, and many clinical diagnostic tests that rely on measuring enzyme activities are performed on crude samples (of blood, tissue homogenates, etc.). When the

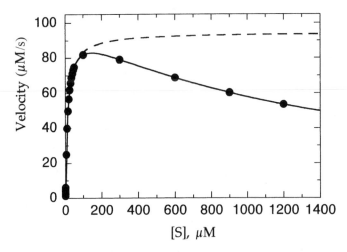

Figure 5.13 Michaelis–Menten plot for an enzyme reaction displaying substrate inhibition at high substrate concentrations: dashed line, best fit of the data at low substrate concentrations to Equation 5.24; solid line, fit of all the data to Equation 5.44. The constant K_i (Equation 5.44) will be described further in subsequent chapters.

substrate for the reaction is unique to the enzyme of interest, these crude samples can be used with good results. If, however, the sample contains more than one enzyme that can act on the substrate, deviations from the expected kinetic results occur. Suppose that our sample contains two enzymes; both can convert the substrate to product, but they display different kinetic constants. Suppose further that for one of the enzymes $V_{max} = V_1$ and $K_m = K_1$, and for the second enzyme $V_{max} = V_2$ and $K_m = K_2$. The velocity of the overall mixture then is given by:

$$v = \frac{V_1[S]}{K_1 + [S]} + \frac{V_2[S]}{K_2 + [S]} \tag{5.45}$$

This can be rearranged to give the following expression (Schulz, 1994):

$$v = \frac{(V_1K_2 + V_2K_1)[S] + (V_1 + V_2)[S]^2}{K_1K_2 + (K_1 + K_2)[S] + [S]^2} \tag{5.46}$$

Equation 5.46 is a polynomial expression, which yields behavior very different from the rectangular hyperbolic behavior we expect; this is illustrated in Figure 5.14.

One last example of deviation from hyperbolic kinetics is that of enzymes displaying cooperativity of substrate binding. In the derivation of Equation 5.24 we assumed that the active sites of the enzyme molecules behave independently of one another. As we saw in Chapter 3 and 4, sometimes proteins occur as multimeric assemblies of subunits. Some enzymes occur as

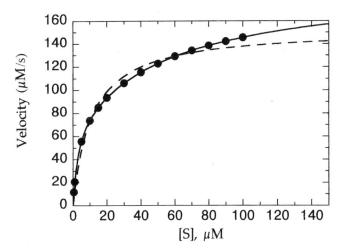

Figure 5.14 Effects of multiple enzymes acting on the same substrate. The dashed line represents the fit of the data to Equation 5.24 for a single enzyme, while the solid line represents the fit to Equation 5.46 for two enzymes acting on the same substrate with $V_1 = 120 \, \mu M/s$, $V_2 = 75 \, \mu M/s$, $K_1 = 65 \, \mu M$, and $K_2 = 3 \, \mu M$.

homomultimers, each subunit containing a separate active site. It is possible that the binding of a substrate molecule at one of these active sites could influence the affinity of the other active sites in the multisubunit assembly (see Chapter 12 for more details). This effect is known as *cooperativity*. It is said to be *positive* when the binding of a substrate molecule to one active site increases the affinity for substrate of the other active sites. On the other hand, when the binding of substrate to one active site lowers the affinity of the other active sites for the substrate, the effect is called *negative cooperativity*. The number of potential substrate binding sites on the enzyme and the degree of cooperativity among them can be quantified by the Hill coefficient, h. The influence of cooperativity on the measured values of velocity can be easily taken into account by modifying Equation 5.24 as follows:

$$v = \frac{V_{max}[S]^h}{K' + [S]^h} \qquad (5.47)$$

where K' is related to K_m but also contains terms related to the effect of substrate occupancy at one site on the substrate affinity of other sites (see Chapter 12). Figure 5.15 illustrates how positive cooperativity can affect the Michaelis–Menten and Lineweaver–Burk plots of an enzyme reaction.

The velocity data for cooperative enzymes can be presented in a linear form by use of Equation 5.48:

$$\log\left(\frac{v}{V_{max} - v}\right) = h \log[S] - \log(K') \qquad (5.48)$$

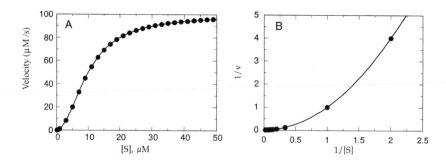

Figure 5.15 Effects of positive cooperativity on the kinetics of an enzyme-catalyzed reaction: (A) data graphed as a Michaelis–Menten (i.e., direct) plot and (B) data from (A) replotted as a Lineweaver–Burk double-reciprocal plot.

Thus, a plot of $\log(v/(V_{max} - v))$ as a function of $\log[S]$ should yield a straight line with slope of h and a y intercept of $-\log(K')$, as illustrated in Figure 5.16. The utility of such plots is limited, however, by the need to know V_{max} a priori and because the linear relationship described by Equation 5.48 holds over only a limited range of substrate concentrations (in the region of $[S] = K'$). Hence, whenever possible, it is best to determine V_{max}, h, and K_m for cooperative enzymes from direct nonlinear curve fits to Equation 5.47.

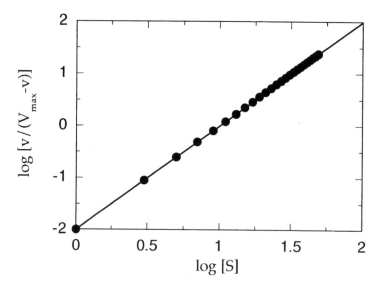

Figure 5.16 Hill plots for the data from Figure 5.15: $\log[v/(V_{max} - v)]$ is plotted as a function of $\log[S]$. The slope of the best fit line provides an estimated of the Hill coefficient h, and the y intercept provides an estimate of $-\log(K')$.

These examples illustrate the more commonly encountered deviations from hyperbolic kinetics. A number of other causes of deviations are known, but they are less common. A more comprehensive discussion of such deviations can be found in the texts by Segel (1975) and Bell and Bell (1988).

5.10 TRANSIENT STATE KINETIC MEASUREMENTS

Much of the enzymology literature describes studies of the steady state kinetics of enzymatic reactions, and a great deal of biochemical insight can be derived from such studies. Steady state kinetics, however, does have some limitations. The steady state kinetic constants k_{cat} and K_m are complex functions that combined rate constants from multiple steps in the overall enzymatic reaction. Hence, these constants do not provide rate information on any individual steps in the reaction pathway. For example, let us again consider a simple-single substrate enzymatic reaction (as discussed above), but this time let us define the individual rate constants for each step:

$$E + S \underset{k_{-1}}{\overset{k_1}{\rightleftharpoons}} ES \underset{k_{-2}}{\overset{k_2}{\rightleftharpoons}} EX \underset{k_{-3}}{\overset{k_3}{\rightleftharpoons}} EP \underset{k_{-4}}{\overset{k_4}{\rightleftharpoons}} E + P$$

In this scheme, EX represents some transient intermediate in the enzymatic reaction. This could be a distinct chemical species (e.g., an acyl–enzyme intermediate as in peptide hydrolysis by serine proteases; see Chapter 6) or a kinetically significant conformational state that is formed by a change in structure of the ES complex prior to the chemical steps of catalysis. A steady state kinetic study would define the overall reaction in terms of k_{cat} and K_m but would not provide much information on the rates and nature of the individual steps in the reaction. Steady state kinetics also limits the mechanistic detail that one can derive. For example, in Chapter 11 we shall see how steady state kinetic measurements can define the order of substrate binding and product release for multisubstrate enzymes. These studies do not, however, give information on the specific reactions of the various enzyme species involved in catalysis, nor can they identify the rate-limiting step. The steady state kinetic constant K_m is often confused with the dissociation constant for the ES complex (K_S), and k_{cat} is often mistakenly thought of as a specific rate constant for a rate-limiting step in catalysis (i.e., k_2); we have already discussed the correct interpretation of K_m and k_{cat}.

To overcome some of these limitations, researchers turn to rapid kinetic methods that allow them to make measurements on a time scale (i.e., milliseconds) consistent with the approach to steady state (pre–steady state kinetics) and to measure the kinetics of transient species that occur after initial substrate binding. These methods are collectively referred to as transient state kinetics, and their application to enzymatic systems provides much richer kinetic detail than simple steady state measurements (Johnson, 1992).

Specialized apparatus must be used to measure kinetic events on a millisecond time scale. A variety of instruments have been designed for this purpose (Fersht, 1985), but the two most commonly used methods for measuring transient kinetics are stopped-flow and rapid reaction quenching.

In a stopped-flow experiment, the researcher is measuring the formation of a transient species by detecting a unique optical signal (typically absorbance or fluorescence). Figure 5.17 illustrates a typical design for a stopped-flow apparatus. The instrument consists of two syringes, one holding a enzyme solution and the other holding a substrate solution. Both syringes are attached to a common drive bar that compresses the plungers of both syringes at a steady and common rate, forcing the solutions from each syringe to mix in the mixing chamber and flow through the detection tube. A third syringe is located at the end of the detection tube. As liquid is forced into this third syringe, its plunger is pushed back until it is stopped by contact with a stopping bar. The stopping bar has attached to it a microswitch, which triggers the controlling computer to initiate observation of the optical signal from the solution trapped in the detection tube. Measurements of the optical signal are then made over time as the solution ages in the detection tube.

In a rapid quench apparatus (Figure 5.18), three syringes are compressed by a common driving bar. The first two syringes contain enzyme and substrate solutions, respectively. These solutions flow into the first mixing chamber and then through a reaction aging tube to the second mixing chamber. The length of reaction time is controlled by the length of the reaction aging tube, or by the rate of flow through this tube. In the second mixing chamber, the reaction mixture is combined with a third solution containing the quenching material, which rapidly stops (or quenches) the reaction. The quenching solution must be able to instantaneously stop the reaction by denaturing the enzyme or sequestering a critical cofactor or other component of the reaction mixture. For example, strong acids are commonly used to quench enzyme reactions in this way. Also, enzymes that rely on divalent metal ions as necessary cofactors can be effectively quenched by mixing with EDTA (ethylenediaminetetraacetic acid) or other chelators. Once the reaction has been quenched, the mixed solution flows into a collection container, from which it can be retrieved by the scientist. Detection is performed off-line by any convenient method, including spectroscopy, radiometric chromatography (including thin-layer chromatography), or electrophoretic separation of substrates and products (see Chapter 7 for details).

With instruments like stopped-flow and rapid quench apparatus, one can measure the formation of products or intermediates on a millisecond time scale. A number of kinetic schemes can be studied by these rapid kinetic techniques. We describe two common situations. The first is simple, reversible association of the subtrate and enzyme to form the ES complex:

$$E + S \underset{k_{-1}}{\overset{k_1}{\rightleftharpoons}} ES$$

If the experiment is performed under conditions where [S] $\gg$ [E], formation of

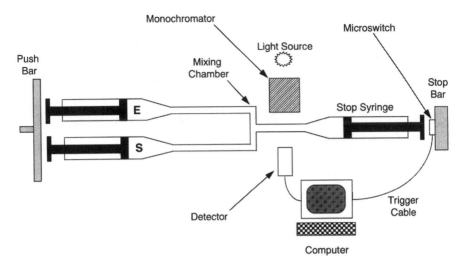

Figure 5.17 Schematic diagram of a typical stopped-flow instrument for rapid kinetic measurements.

the ES complex will follow pseudo-first-order kinetics and will be equivalent to the approach to equilibrium for receptor–ligand complexes, as discussed in Chapter 4. Hence, the observed rate of formation k_{obs} will depend on substrate concentration as follows:

$$k_{obs} = k_1[S] + k_{-1} \tag{5.49}$$

A plot of k_{obs} as a function of [S] will be linear with y intercept of k_{-1} and slope of k_1, as illustrated earlier (Figure 4.2).

In the second situation, formation of an intermediate species EX is rate limiting. Here initial substrate binding comes to equilibrium on a time scale

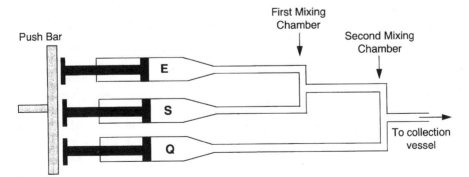

Figure 5.18 Schematic diagram of a typical rapid quench instrument for rapid kinetic measurements.

much faster than the subsequent first-order "isomerization" step:

$$E + S \underset{k_{-1}}{\overset{k_1}{\rightleftharpoons}} ES \underset{k_{-2}}{\overset{k_2}{\rightleftharpoons}} EX$$

Three forms of the enzyme appear in this reaction scheme, the free enzyme E_f, the ES complex, and the intermediate EX. The time dependence for each of these can be derived, yielding the following equations for the mole fraction of each species (Johnson, 1992):

$$\frac{[E]_f}{[E]} = \frac{1 - (K_1[S] + K_1 K_2[S])}{\alpha}(1 - e^{\lambda t}) \tag{5.50}$$

$$\frac{[ES]}{[E]} = \frac{K_1[S]}{\alpha}(1 - e^{\lambda t}) \tag{(5.51)}$$

$$\frac{[EX]}{[E]} = \frac{K_1 K_2[S]}{\alpha}(1 - e^{\lambda t}) \tag{5.52}$$

where

$$\alpha = 1 + K_1[S] + K_1 K_2[S] \tag{5.53}$$

The preexponential term in Equations 5.50–5.52 represents in each case an amplitude term that corresponds to the concentration of that enzyme species at equilibrium. The observed rate constant for formation of the EX complex, λ, follows a hyperbolic dependence on substrate concentration, similar to velocity in the Michaelis–Menten equation:

$$\lambda = \frac{K_1 k_2[S]}{K_1[S] + 1} + k_{-2} \tag{5.54}$$

Three distinctions between Equations 5.54 and 5.24 can be made. First, λ is a hyperbolic function of the true dissociation constant for the ES complex (i.e., $K_S = 1/K_1$), not of the kinetic constant K_m. Second, the maximal rate observed is equal to the sum of $k_2 + k_{-2}$. Third, the y intercept is nonzero in this case, and equal the rate constant k_{-2}. Thus, from fitting of the rapid kinetic data to Equation 5.54, one can simultaneously determine the values of K_S, k_2, and k_{-2}.

The application of transient kinetics to the study of enzymatic reactions, and more generally to protein–ligand binding events, is widespread throughout the biochemical literature. The reader should be aware of the power of these methods for determining individual rate constants and of the value of such information for the development of detailed mechanistic models of catalytic turnover. Because of space limitations, and because these methods require specialized equipment that beginners may not have at their disposal, we shall suspend further discussion of these methods. Several noteworthy reviews on the methods of transient kinetics (Gibson, 1969; Johnson, 1992; Fierke and Hammes, 1995) are highly recommended to the reader who is interested in learning more about these techniques.

5.11 SUMMARY

This chapter focused on steady state kinetic measurements, since these are easiest to perform in a standard laboratory. These methods provide important kinetic and mechanistic information, mainly in the form of two kinetic constants, k_{cat} and K_m. Graphical methods for determining the values for these kinetic constants were presented. We also briefly discussed the application of rapid kinetic techniques to the study of enzymatic reactions. These methods provide even more detailed information on the individual rate constants for different steps in the reaction sequence, but they require more specialized instrumentation and analysis methods. The chapter provided references to more advanced treatments of rapid kinetic methods to aid the interested reader in learning more about these powerful techniques.

REFERENCES AND FURTHER READING

Bell, J. E., and Bell, E. T. (1988) *Proteins and Enzymes*, Prentice-Hall, Englewood Cliffs, NJ.

Briggs, G. E., and Haldane, J. B. S. (1925) *Biochem. J.* **19**, 383.

Brown, A. J. (1902) *J. Chem. Soc.* **81**, 373.

Chapman, K. T., Kopka, I. E., Durette, P. I., Esser, C. K., Lanza, T. J., Izquierdo-Martin, M., Niedzwiecki, L., Chang, B., Harrison, R. K., Kuo, D. W., Lin, T.-Y., Stein, R. L., and Hagmann, W. K. (1993) *J. Med. Chem.* **36**, 4293.

Cleland, W. W. (1967) *Adv. Enzymol.* **29** 1–65.

Copeland, R. A. (1991) *Proc. Natl. Acad. Sci. USA* **88**, 7281.

Cornish-Bowden, A., and Wharton, C. W. (1988) *Enzyme Kinetics*, IRL Press, Oxford.

Eisenthal, R., and Cornish-Bowden, A. (1974) *Biochem. J.* **139**, 715.

Fersht, A. (1985) *Enzyme Structure and Mechanism*, Freeman, New York.

Fierke, C. A., and Hammes, G. G. (1995) *Methods Enzymol.* **249**, 3–37.

Gibson, Q. H. (1969) *Methods Enzymol.* **16**, 187.

Henri, V. (1903) *Lois Générales de l'action des diastases*, Hermann, Paris.

Johnson, K. A. (1992) *Enzymes*, **XX**, 1–61.

Lineweaver, H., and Burk, J. (1934) *J. Am. Chem. Soc.* **56**, 658.

Michaelis, L., and Menten, M. L. (1913) *Biochem. Z.* **49** 333.

Schulz, A. R. (1994) *Enzyme Kinetics from Diastase to Multi-enzyme Systems*, Cambridge University Press, New York.

Segel, I. H. (1975) *Enzyme Kinetics*, Wiley, New York.

Wahl, R. C. (1994) *Anal. Biochem.* **219**, 383.

Wilkinson, A. J., Fersht, A. R., Blow, D. M., and Winter, G. (1983) *Biochemistry*, **22**, 3581.

6

CHEMICAL MECHANISMS IN ENZYME CATALYSIS

The essential role of enzymes in almost all physiological processes stems from two key features of enzymatic catalysis: (1) enzymes greatly *accelerate* the rates of chemical reactions; and (2) enzymes act on *specific* molecules, referred to as substrates, to produce *specific* reaction products. Together these properties of rate acceleration and substrate specificity afford enzymes the ability to perform the chemical conversions of metabolism with the efficiency and fidelity required for life. In this chapter we shall see that both substrate specificity and rate acceleration result from the precise three-dimensional structure of the substrate binding pocket within the enzyme molecule, known as the *active site*. Enzymes are (almost always) proteins, hence the chemically reactive groups that act upon the substrate are derived mainly from the natural amino acids. The identity and arrangement of these amino acids within the enzyme active site define the active site topology with respect to stereochemistry, hydrophobicity, and electrostatic character. Together these properties define what molecules may bind in the active site and undergo catalysis. The active site structure has evolved to bind the substrate molecule in such a way as to induce strains and perturbations that convert the substrate to its transition state structure. This transition state is greatly stabilized when bound to the enzyme; its stability under normal solution conditions is much less. Since attainment of the transition state structure is the main energetic barrier to the progress of any chemical reaction, we shall see that the stabilization of the transition state by enzymes results in significant acceleration of the reaction rate.

6.1 SUBSTRATE–ACTIVE SITE COMPLEMENTARITY

When a protein and a ligand combine to form a binary complex, the complex must result in a net stabilization of the system relative to the free protein and ligand; otherwise binding would not be thermodynamically favorable. We discussed in Chapter 4 the main forces involved in stabilizing protein–ligand interactions: hydrogen bonding, hydrophobic forces, van der Waals interactions, electrostatic interactions, and so on. All these contribute to the overall binding energy of the complex and must more than compensate for the lose of rotational and translational entropy that accompanies binary complex formation.

These same forces are utilized by enzymes in binding their substrate molecules. It is clear today that formation of an enzyme–substrate binary complex is but the first step in the catalytic process used in enzymatic catalysis. Formation of the initial encounter complex (also referred to as the enzyme–substrate, ES, or Michaelis complex; see Chapter 5) is followed by steps leading sequentially to a stabilized enzyme–transition state complex ($ES^{\ddagger}$), an enzyme–product complex (EP), and finally dissociation to reform the free enzyme with liberation of product molecules. Initial ES complex formation is defined by the dissociation constant K_s, which is the quotient of the rate constants k_{off} and k_{on} (see Chapters 4 and 5). As discussed in Chapter 5, the rates of the chemical steps following ES complex formation are, for simplicity, often collectively described by a single kinetic constant, k_{cat}. As we shall see, k_{cat} is most often limited by the rate of attainment of the transition state species $ES^{\ddagger}$. Hence, a minimalist view of enzyme catalysis is captured in the scheme illustrated in Figure 6.1.

To understand the rate enhancement and specificity of enzymatic reactions, we must consider the structure of the reactive center of these molecules, the active site, and its relationship to the structures of the substrate molecule in its ground and transition states in forming the ES and the $ES^{\ddagger}$ binary complexes. While the active site of every enzyme is unique, some generalizations can be made:

1. The active site of an enzyme is small relative to the total volume of the enzyme.

2. The active site is three-dimensional — that is, amino acids and cofactors in the active site are held in a precise arrangement with respect to one another and with respect to the structure of the substrate molecule. This active site three-dimensional structure is formed as a result of the overall tertiary structure of the protein.

3. In most cases, the initial interactions between the enzyme and the substrate molecule (i.e., the binding events) are noncovalent, making use of hydrogen bonding, electrostatic, hydrophobic interactions, and van der Waals forces to effect binding.

$$E + S \underset{K_S}{\xrightarrow{\Delta G_{ES}}} ES \xrightarrow[\Delta G_{kcat}]{k_{cat}} E + P$$

$$E + S \underset{E_a = \Delta G_{ES\ddagger}}{\xrightarrow{k_{cat}/K_S}} ES^{\ddagger}$$

Figure 6.1 Generic scheme for an enzyme-catalyzed reaction showing the component free energy terms that contribute to the overall activation energy of reaction.

4. The active sites of enzymes usually occur in clefts and crevices in the protein. This design has the effect of excluding bulk solvent (water), which would otherwise reduce the catalytic activity of the enzyme. In other words, the substrate molecule is desolvated upon binding, and shielded from bulk solvent in the enzyme active site. Solvation by water is replaced by the protein.

5. The specificity of substrate utilization depends on the well-defined arrangement of atoms in the enzyme active site that in some way complements the structure of the substrate molecule.

Experimental evidence for the existence of a binary ES complex rapidly accumulated during the late nineteenth and early twentieth centuries. This evidence, some of which was discussed in Chapter 5, was based generally on studies of enzyme stability, enzyme inhibition, and steady state kinetics. During this same time period, scientists began to appreciate the selective utilization of specific substrates that is characteristic of enzyme-catalyzed reactions. This cumulative information led to the general view that substrate specificity was a result of selective binding of substrate molecules by the enzyme at its active site. The selection of particular substrates reflected a structural complementarity between the substrate molecule and the enzyme active site. In the late nineteenth century Emil Fisher formulated these concepts into the *lock and key* model, as illustrated in Figure 6.2. In this model the enzyme active site and the substrate molecule are viewed as static structures that are stereochemically complementary. The insertion of the substrate into the static enzyme active site is analogous to a key fitting into a lock, or a jigsaw piece fitting into the rest of the puzzle: the best fits occur with the substrates that best complement the structure of the active site; hence these molecules bind most tightly.

Active site–substrate complementarity results from more than just stereochemical fitting of the substrate into the active site. The two structures must also be electrostatically complementary, ensuring that charges are

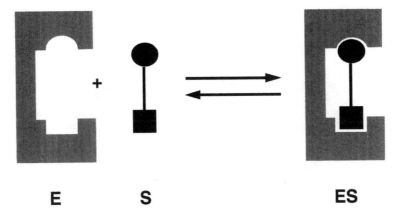

E   **S**   **ES**

Figure 6.2   Schematic illustration of the lock and key model of enzyme–substrate interactions.

counterbalanced to avoid repulsive effects. Likewise, the two structures must complement each other in the arrangement of hydrophobic and hydrogen-bonding interactions to best enhance binding interactions.

Enzyme catalysis is usually stereo-, regio-, and enantiomerically selective. Hence substrate recognition must result from a minimum of three contact points of attachment between the enzyme and the substrate molecule. Consider the example of the alcohol dehydrogenases (Walsh, 1979) that catalyze the transfer of a methylene hydrogen of ethyl alcohol to the carbon at the 4-position of the NAD^+ cofactor, forming NADH and acetaldehyde. Studies in which the methylene hydrogens of ethanol were replaced by deuterium demonstrated that alcohol dehydrogenases exclusively transferred the pro-*R* hydrogen to NAD^+ (Loewus et al., 1953; Fersht, 1985). This stereospecificity implies that the alcohol bind to the enzyme active site through specific interactions of its methyl, hydroxyl, and pro-*R* hydrogen groups to form a *three-point attachment* with the reactive groups within the active site; this concept is illustrated in Figure 6.3. Having anchored down the methyl and hydroxyl groups as depicted in Figure 6.3, the enzyme is committed to the transfer of the specific pro-*R* hydrogen atom because of its relative proximity to the NAD^+ cofactor. The three-point attachment hypothesis is often invoked to explain the stereospecificity commonly displayed by enzymatic reactions.

The concepts of the lock and key and three-point attachment models help to explain substrate selectivity in enzyme catalysis by invoking a structural complementarity between the enzyme active site and substrate molecule. We have not, however, indicated the form of the substrate molecule to which the enzyme active site shows structural complementarity. Early formulations of these hypotheses occurred before the development of transition state theory (Pauling, 1948), hence viewed the substrate ground state as the relevant configuration. Today, however, there is clear evidence that enzyme active sites

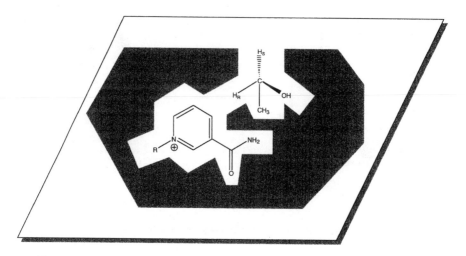

Figure 6.3 Illustration of three-point attachment in enzyme–substrate interactions.

have in fact evolved to best complement the substrate transition state structure, rather than the ground state. For example, it is well known that inhibitor molecules that are designed to mimic the structure of the reaction transition state bind much more tightly to the target enzyme than do the substrate or product molecules. Some scientists have, in fact, argued that "the sole source of catalytic power is the stabilization of the transition state; reactant state interactions are by nature inhibitory and only waste catalytic power" (Schowen, 1978). Others argue that some substrate ground state affinity is required for initial complex formation and to utilize the accompanying binding energy to drive transition state formation (see, e.g., Menger, 1992). Indeed some evidence from site-directed mutagenesis studies suggests that the structural determinants of substrate specificity can at least in part be distinguished from the mechanism of transition state stabilization (Murphy and Benkovic, 1989; Wilson and Agard, 1991). Nevertheless, the bulk of the experimental evidence strongly favors active site–transition state complementarity as the primary basis for substrate specificity and catalytic power in most enzyme systems. There are, for example, numerous studies of specificity in enzyme systems measured through steady state kinetics in which specificity is quantified in terms of the relative k_{cat}/K_m values for different substrates. In many of these studies one finds that the K_m values among different substrates vary very little, perhaps by a factor of 10-fold or less. A good substrate is distinguished from a bad one in these studies mainly by the effects on k_{cat}. Hence, much of the substrate specificity resides in transition state interactions with the enzyme active site. We shall have more to say about this in subsequent sections of this chapter.

6.2 RATE ENHANCEMENT THROUGH TRANSITION STATE STABILIZATION

In Chapter 2 we said that chemical reactions, such as molecule S (for substrate) going to molecule P (for product), will proceed through formation of a high energy, short-lived (typical half-life ca. 10^{-13} second) state known as the transition state ($S^{\ddagger}$). Let us review the minimal steps involved in catalysis, as illustrated in Figure 6.1. The initial encounter (typically through molecular collisions in solution) between enzyme and substrate leads to the reversible formation of the Michaelis complex, ES. Under typical laboratory conditions this equilibrium favors formation of the complex, with ΔG of binding for a typical ES pair being approximately -3 to -12 kcal/mol. Formation of the ES complex leads to formation of the bound transition state species $ES^{\ddagger}$. As with the uncatalyzed reaction, formation of the transition state species is the main energetic barrier to product formation. Once the transition state barrier has been overcome, the reaction is much more likely to proceed energetically downhill to formation of the product state. In the case of the enzyme-catalyzed reaction, this process involves formation of the bound EP complex, and finally dissociation of the EP complex to liberate free product and free enzyme.

Since the enzyme appears on both the reactant and product side of the equation and is therefore unchanged with respect to the thermodynamics of the complete reaction, it can be ignored (Chapter 2). Hence, the free energy of the reaction here will depend only on the relative concentrations of S and P:

$$\Delta G = -RT \ln \left(\frac{[P]}{[S]} \right) \tag{6.1}$$

This is exactly the same equation of ΔG for the uncatalyzed reaction of S $\rightarrow$ P, and it reflects the *path independence* of the function ΔG. In other words, ΔG depends only on the initial and final states of the reaction, not on the various intermediate states (e.g., ES, $ES^{\ddagger}$, and EP) formed during the reaction (ΔG is thus said to be a *state function*). This leads to the important realization that *enzymes cannot alter the equilibrium between products and substrates.*

What then is the value of using an enzyme to catalyze a chemical reaction? The answer is that enzymes, and in fact all catalysts, speed up the rate at which equilibrium is established in a chemical system: *enzymes accelerate the rate of chemical reactions.* Hence, with an ample supply of substrate, one can form much greater amounts of product *per unit time* in the presence of an enzyme than in its absence. This rate acceleration is a critical feature of enzyme usage in metabolic processes. Without the speed imparted by enzyme catalysis, many metabolic reactions would proceed too slowly in vivo to sustain life. Likewise, the *ex vivo* use of enzymes in chemical processes relies on this rate acceleration, as well as the substrate specificity that enzyme catalysis provides. Thus, the great value of enzymes, both for biological systems and in commercial use, is that they provide a means of making more product at a faster rate than can be achieved without catalysis.

How is it that enzymes achieve this rate acceleration? The answer lies in a consideration of the activation energy of the chemical reaction. The key to enzymatic rate acceleration is that by lowering the energy barrier, by stabilizing the transition state, reactions will proceed faster.

Recall from Chapter 2 that the rate or velocity of substrate utilization, v, is related to the activation energy of the reaction as follows:

$$v = \frac{-d[S]}{dt} = \left(\frac{k_B T}{h}\right)[S]\exp\left(-\frac{E_a}{RT}\right) \tag{6.2}$$

Now, for simplicity, let us fix the reaction temperature at 25°C and fix [S] at a value of 1 in some arbitrary units. At 25°C, $RT = 0.59$ and $k_B T/h = 6.2 \times 10^{12}\,\text{s}^{-1}$. Suppose that the activation energy of a chemical reaction at 25°C is 10 kcal/mol. The velocity of the reaction will thus be:

$$v = 6.2 \times 10^{12}\,\text{s}^{-1} \cdot 1\,\text{unit} \cdot e^{-10/0.59} = 2.7 \times 10^5\,\text{units}\,\text{s}^{-1} \tag{6.3}$$

If we somehow reduce the activation energy to 5 kcal/mol, the velocity now becomes:

$$v = 6.2 \times 10^{12}\,\text{s}^{-1} \cdot 1\,\text{unit} \cdot e^{-5/0.59} = 1.3 \times 10^9\,\text{units}\,\text{s}^{-1} \tag{6.4}$$

Thus by lowering the activation energy by 5 kcal/mol we have achieved an increase in reaction velocity of about 5000! In general a linear decrease in activation energy results in an exponential increase in reaction rate. This is exactly how enzymes function. They accelerate the velocity of chemical reactions by stabilizing the transition state of the reaction, hence lowering the energetic barrier that must be overcome.

Let us look at the energetics of a chemical reaction in the presence and absence of an enzyme. For the enzyme-catalyzed reaction we can estimate the free energies associated with different states from a combination of equilibrium and kinetic measurements. If we normalize the free energy of the free E + S starting point to zero, we can calculate the free energy change associated with $ES^{\ddagger}$ (under experimental conditions of subsaturating substrate) as follows:

$$E_a = \Delta G_{ES^{\ddagger}} = -RT\ln\left(\frac{k_{cat}}{K_s}\right) + RT\ln\left(\frac{k_B T}{h}\right) \tag{6.5}$$

The free energy change associated with formation of the ES complex can, in favorable cases, be determined from measurement of K_s by equilibrium methods (see Chapter 4) or from kinetic measurements (see Chapter 5):

$$\Delta G_{ES} = -RT\ln\left(\frac{1}{K_s}\right) \tag{6.6}$$

Alternatively, from steady state measurements one can calculate the free energy change associated with k_{cat} from the Eyring equation:

$$\Delta G_{k_{cat}} = RT \left(\ln \left(\frac{k_B T}{h} \right) - \ln (k_{cat}) \right) \tag{6.7}$$

If one then subtracts Equation 6.7 from Equation 6.5, the difference is equal to ΔG_{ES}. Thus, we see that the overall activation energy E_a is composed of two terms, ΔG_{ES} and $\Delta G_{k_{cat}}$. The term $\Delta G_{k_{cat}}$ is the amount of energy that must be expended to reach the transition state (i.e., bond-making and bond-breaking steps), while the term ΔG_{ES} is the net energy gain that results from the realization of enzyme–substrate binding energy (Fersht, 1974; So et al., 1998).

The free energy change associated with the EP complex can also be determined from equilibrium measurements or from the inhibitory effects of product on the steady state kinetics of the reaction (see Chapters 8 and 11).

For either the catalyzed or uncatalyzed reaction, the activation energy can also be determined from the temperature dependence of the reaction velocity according to the Arrhenius equation (see also Chapters 2 and 7):

$$k_{cat} = A \exp \left(\frac{-E_a}{RT} \right) \tag{6.8}$$

Note that for the uncatalyzed reaction, k_{cat} is replaced in Equation 6.8 by the first-order rate constant for reaction.

From such measurements one can construct a reaction energy level diagram as illustrated in Figure 6.4. In the absence of enzyme, the reaction proceeds from substrate to product by overcoming the sizable energy barrier required to reach the transition state $S^{\ddagger}$. In the presence of enzyme, on the other hand, the reaction first proceeds through formation of the ES complex. The ES complex represents an intermediate along the reaction pathway that is not available in the uncatalyzed reaction; the binding energy associated with ES complex formation can, in part, be used to drive transition state formation. Once binding has occurred, molecular forces in the bound molecule (as discussed shortly) have the effect of simultaneously destabilizing the ground state configuration of the bound substrate molecule, and energetically favoring the transition state. The complex $ES^{\ddagger}$ thus occurs at a lower energy than the free $S^{\ddagger}$ state, as shown in Figure 6.4.

The reaction next proceeds through formation of another intermediate state, the enzyme–product complex, EP, before final product release to form the free product plus free enzyme state. Again, the initial and final states are energetically identical in the catalyzed and uncatalyzed reactions. However, the overall activation energy barrier has been substantially reduced in the enzyme-catalyzed case. This reduction in activation barrier results in a significant acceleration of reaction velocity in the presence of the enzyme, as we have seen above (Equations 6.2–6.4). This is the common strategy for rate acceleration used by all enzymes:

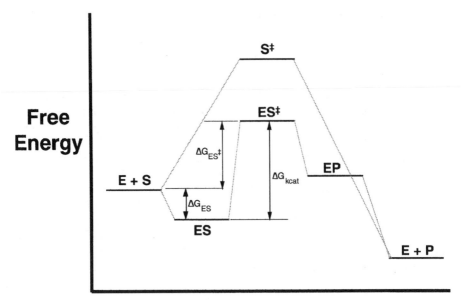

Reaction Coordinate

Figure 6.4 Energy level diagram of an enzyme-catalyzed reaction and the corresponding uncatalyzed chemical reaction. The symbols E, S, S‡, ES, ES‡, EP, and P represent the free enzyme, the free substrate, the free transition state, the enzyme–substrate complex, the enzyme–transition state complex, the enzyme–product complex, and the free product states, respectively. The activation energy, $\Delta G_{ES‡}$ and its components, ΔG_{ES} and $\Delta G_{k_{cat}}$, are as described in the text. The energy levels depicted relate to the situation in which the substrate is present in concentrations greater than the dissociation constant for the ES complex. When [S] is less than K_S, the potential energy of the ES state is actually greater than that of the E + S initial state (see Fersht, 1985, for further details).

Enzymes accelerate the rates of chemical reactions by stabilizing the transition state of the reaction, hence lowering the activation energy barrier to product formation.

6.3   CHEMICAL MECHANISMS FOR TRANSITION STATE STABILIZATION

The transition state stabilization associated with enzyme catalysis is the result of the structure and reactivity of the enzyme active site, and how these structural features interact with the bound substrate molecule. Enzymes use numerous detailed chemical mechanisms to achieve transition state stabilization and the resulting reaction rate acceleration. These can be grouped into five major categories (Jencks, 1969; Cannon and Benkovic, 1998):

1. Approximation (i.e., proximity) of reactants
2. Covalent catalysis
3. General acid–base catalysis
4. Conformational distortion
5. Preorganization of the active site for transition state complementarity

We shall discuss each of these separately. However, the reader should realize that in any catalytic system, several or all of these effects can be utilized in concert to achieve overall rate enhancement. They are thus often interdependent, which means that the line of demarcation between one mechanism and another often is unclear, and to a certain extent arbitrary.

6.3.1 Approximation of Reactants

Several factors associated with simply binding the substrate molecule within the enzyme active site contribute to rate acceleration. One of the more obvious of these is that binding brings into close proximity (hence the term *approximation*), the substrate molecule(s) and the reactive groups within the enzyme active site. Let us consider the example of a bimolecular reaction, involving two substrates, A and B, that react to form a covalent species A–B. For the two molecules to react in solution they must (1) encounter each other through diffusion-limited collisions in the correct mutual orientations for reaction; (2) undergo changes in solvation that allow for molecular orbital interactions; (3) overcome van der Waals repulsive forces; and (4) undergo changes in electronic orbitals to attain the transition state configuration.

In solution, the rate of reaction is determined by the rate of encounters between the two substrates. The rate of collisional encounters can be marginally increased in solution by elevating the temperature, or by increasing the concentrations of the two reactants. In the enzyme-catalyzed reaction, the two substrates bind to the enzyme active site as a prerequisite to reaction. When the substrates are sequestered within the active site of the enzyme, their *effective* concentrations are greatly increased with respect to their concentrations in solution.

A second aspect of approximation effects is that the structure of the enzyme active site is designed to bind the substrates in a specific orientation that is optimal for reaction. In most bimolecular reactions, the two substrates must achieve a specific mutual alignment to proceed to the transition state. In solution, there is a distribution of rotamer populations for each substrate that have the effect of retarding the reaction rate. By locking the two substrates into a specific mutual orientation in the active site, the enzyme overcomes these encumberances to transition state attainment. Of course, these severe steric and orientational restrictions are associated with some entropic cost to reaction. However, such alignment must occur for reaction in solution as well as in the enzymatic reaction. Hence, there is actually a considerable entropic advantage associated with reactant approximation. By having the two substrates bound in the enzyme active site, the entropic cost associated with the solution reaction is largely eliminated; in enzymatic catalysis this energetic cost is compensated for in terms of the binding energy of the ES complex. Together, the concentration and orientation effects associated with substrate binding are referred to as the *proximity effect* or the *propinquity effect*.

Some sense of the effects of proximity on reaction rate can be gleaned from studies comparing the reaction rates of intramolecular reactions with those of comparable intermolecular reactions (see Kirby, 1980, for a comprehensive review of this subject). For example, Fersht and Kirby (1967) compared the reaction rate of aspirin hydrolysis catalyzed by the intramolecular carboxylate group with that for the same reaction catalyzed by acetate ions in solution (Scheme 1).

The intramolecular reaction proceeds with a first-order rate constant of $1.1 \times 10^{-5} \, s^{-1}$. The same reaction catalyzed by acetate ions in solution has a second-order rate constant of $1.27 \times 10^{-6} \, L \, mol^{-1} \, s^{-1}$. In comparing these two reactions we can ask what effective molarity of acetate ions would be required to make the intermolecular reaction go at the same rate as the intramolecular reaction. This is measured as the ratio of the first-order rate constant to the second-order rate constant (k_1/k_2); this ratio has units of molarity, and its value for the present reaction is 8.7 M. However, because acetate is more basic (pK_a 4.76) than the carboxylate of aspirin (pK_a 3.69), one must adjust the value of k_2 to account for this difference. When this is done, the effective molarity is 13 M. Thus, with the pK_a adjustment corrected for, the overall rate of the intramolecular reaction is far greater than that of the intermolecular reaction. Additional examples of such effects have been presented in Jencks (1969) and in Kirby (1980).

A concept related to proximity effects is that of *orbital steering*. The orbital steering hypothesis suggests that the juxtaposition of reactive groups among the substrates and active site residues is not sufficient for catalysis. In addition to this positioning, the enzyme needs to precisely steer the molecular orbitals of the substrate into a suitable orientation. According to this hypothesis, enzyme active site groups have evolved to optimize this steering upon substrate binding. While some degree of orbital steering no doubt occurs in enzyme catalysis, there are two strong arguments against a major role for this effect in transition state stabilization:

1. Thermal vibrations of the substrate molecules should give rise to large changes in the orientation of the reacting atoms within the active site structure. The magnitude of such vibrational motions at physiological temperatures contradicts the idea of rigidly oriented molecular orbitals as required for orbital steering.

2. Recent molecular orbital calculations predict that orbital alignments result in shallow total energy minima (as in a Morse potential curve, such as seen in Chapter 2), whereas the orbital steering hypothesis would require deep,

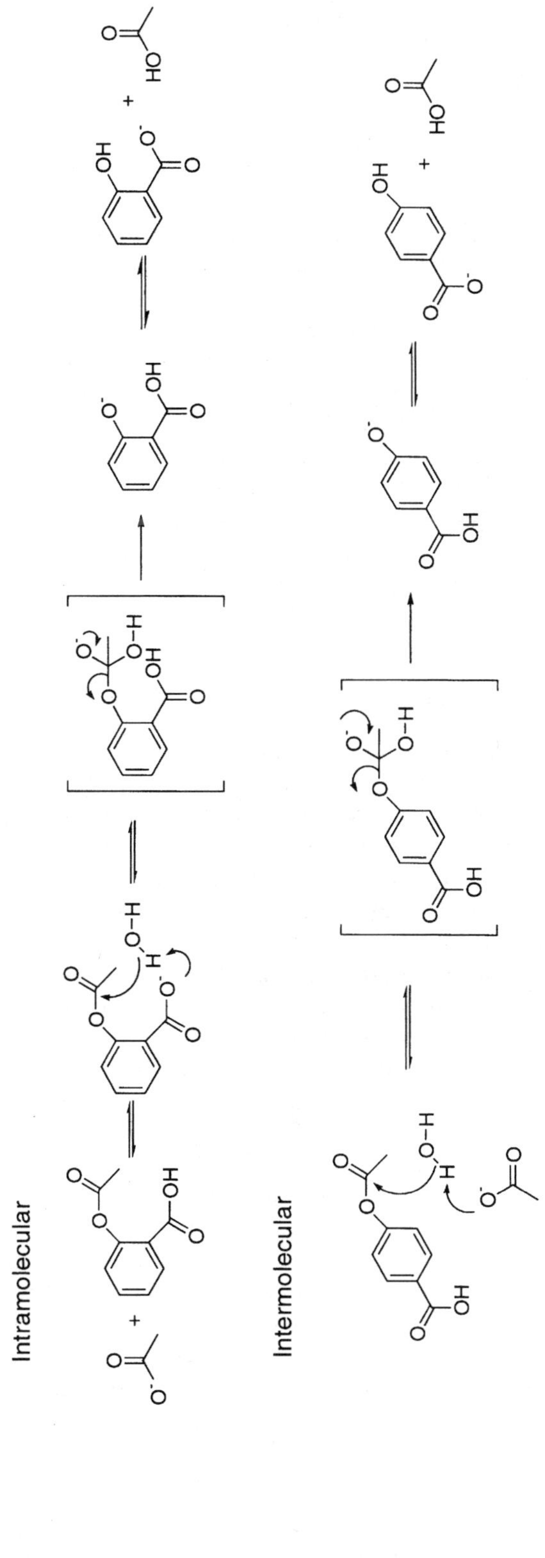

SCHEME 6.1

narrow energy minimal to retain the exact alignment. An expanded discussion of orbital steering and the arguments for and against this hypothesis has been provided by Bender et al. (1984).

Changes in solvation are also required for reaction between two substrates to occur. In solution, desolvation energy can be a large barrier to reaction. In enzymatic reactions the desolvation of reactants occurs during the binding of substrates to the hydrophobic enzyme active site, where they are effectively shielded from bulk solvent. Hence desolvation costs are offset by the binding energy of the complex and do not contribute to the activation barrier in the enzymatic reaction (Cannon and Benkovic, 1998).

Finally, overcoming van der Waals repulsions and changes in electronic overlap are important aspects of intramolecular reactions and enzyme catalysis. These ends are accomplished in part by the orientation effects discussed above, and through induction of strain, as discussed latter in this chapter.

Together these different properties lead to an overall approximation effect that results from the binding of substrates in the enzyme active site. Approximation effects contribute to the overall rate acceleration seen in enzyme catalysis, with the binding forces between the enzyme and substrate providing much of the driving force for these effects.

6.3.2 Covalent Catalysis

There are numerous examples of enzyme-catalyzed reactions that go through the formation of a covalent intermediate between the enzyme and the substrate molecule. Experimental evidence for such intermediates has been obtained from kinetic measurements, from isolation and identification of stable covalent adducts, and more recently from x-ray crystal structures of the intermediate species. Several families of enzymes have been demonstrated to form covalent intermediates, including serine proteases (acyl–serine intermediates), cysteine proteases (acyl–cysteine intermediates), protein kinases and phosphatase (phospho–amino acid intermediates), and pyridoxal phosphate-utilizing enzymes (pyridoxal–amino acid Schiff bases).

For enzymes that proceed through such mechanisms, formation of the covalent adduct is a required step for catalysis. Generation of the covalent intermediate brings the system along the reaction coordinate toward the transition state, thus helping to overcome the activation energy barrier. Enzymes that utilize covalent intermediates have evolved to break this difficult reaction down into two steps — formation and breakdown of the covalent intermediate — rather than catalysis of the single reaction directly. The rate-limiting step in the reactions of these enzymes is often the formation or decomposition of the covalent intermediate. This can be seen, for example, in Figure 6.5, which illustrates the steady state kinetics of p-nitrophenylethyl carbonate hydrolysis by chymotrypsin (Hartley and Kilby, 1954; see also

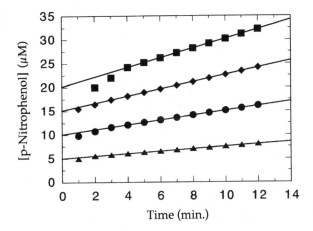

Figure 6.5 Illustration of burst phase kinetics. The data represent production of *p*-nitrophenol from the chymotrypsin-catalyzed hydrolysis of *p*-nitrophenylethyl carbonate. The nominal chymotrypsin concentrations used were 8 (triangles), 16 (circles), 24 (diamonds), and 32 (squares) μM. From the intercept values, the fraction of active enzyme in these samples was estimated to be 0.63. Note the apparent curvature in the early time points at high enzyme concentration, demonstrating a pre–steady state phase (i.e., the burst) in these reactions. [Data approximated and redrawn from Hartley and Kilby (1954).]

Chapter 7). Figure 6.5 shows the steady state progress curves for hydrolysis at several different enzyme concentrations. For an experiment of this type, one would expect the steady state rate to increase with enzyme concentration (i.e., the slopes of the lines should increase with increasing enzyme), but all the curves should converge at zero product concentration at zero time. Instead, one sees in Figure 6.5 that the *y* intercept for each progress curve is nonzero, and the value of the intercept increases with enzyme concentration. In fact, extrapolation of the steady state lines to time zero results in a *y* intercept equal to the concentration of enzyme active sites present in solution. The early "burst" in product formation is the result of a single turnover of the enzyme with substrate, and formation of a stoichiometric amount of acyl–enzyme intermediate and product. Formation of the acyl intermediate in this case is fast, but the subsequent decomposition of the intermediate is rate limiting. Since further product formation cannot proceed without decomposition of the acyl intermediate, a burst of rapid kinetics is observed, followed by a much slower steady state rate of catalysis.

Covalent catalysis in enzymes is facilitated mainly by nucleophilic and electrophilic catalysis, and in more specialized cases by redox catalysis. We turn next to a thorough discussion of nucleophilic and electrophilic catalysis. A detailed treatment of redox reactions in enzyme catalysis can be found in the text by Walsh (1979).

6.3.2.1 Nucleophilic Catalysis.

6.3.2.1 Nucleophilic Catalysis. Nucleophilic reactions involve donation of electrons from the enzyme nucleophile to a substrate with partial formation of a covalent bond between the groups in the transition state of the reaction:

$$\text{Nuc:} + \text{Sub} \rightleftharpoons \overset{\delta+}{\text{Nuc}} \text{---} \overset{\delta+}{\text{Sub}}$$

The reaction rate in nucleophilic catalysis depends both on the strength of the attacking nucleophile and on the susceptibility of the substrate group (electrophile) that is being attacked (i.e., how good a "leaving group" the attacked species has). The electron-donating ability, or *nucleophilicity*, of a group is determined by a number of factors; one of the most important of these factors is the basicity of the group. Basicity is a measure of the tendency of a species to donate an electron pair to a proton, as discussed in Chapter 2 and further in Section 6.3.3. Generally, the rate constant for reaction in nucleophilic catalysis is well correlated with the pK_a of the nucleophile. A plot of the logarithm of the second-order rate constant (k_2) of nucleophilic reaction as a function of nucleophile pK_a yields a straight line within a family of structurally related nucleophiles (Figure 6.6). A graph such as Figure 6.6 is known as a Brønsted plot because it was first used to relate the reaction rate to catalyst pK_a in general acid–base catalysis (see Section 6.3.3).

Note that in Figure 6.6 different structural families of nucleophiles all yield linear Brønsted plots, but with differing slopes, depending on the chemical nature of the nucleophile. Other factors that affect the strength of a nucleophile include oxidation potential, polarizability, ionization potential, electronegativ-

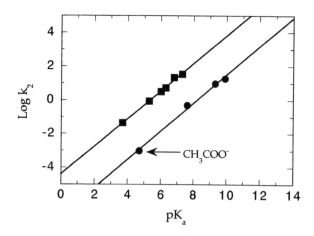

Figure 6.6 Brønsted plots for nucleophilic attack of *p*-nitrophenyl acetate by imidazoles (squares) and phenolates (circles). Unlike general base catalysis, in this illustration N- and O-containing nucleophiles of similar basicity (pK_a) show distinct Brønsted lines. Note that the data points for acetate ion and the phenolate nucleophiles fall on the same Brønsted line [Data from Bruice and Lapinski (1958).]

ity, potential energy of its highest occupied molecular orbital (HOMO); covalent bond strength, and general size of the group. Hence the reaction rate for nucleophilic catalysis depends not just on the pK_a of the nucleophile but also on the chemical nature of the species (a more comprehensive treatment of some of these factors can be found in Jencks, 1969, and Walsh, 1979). This is one property that distinguishes nucleophilic catalysis from general base catalysis. While the Brønsted plot slope depends on the nature of the nucleophilic species in nucleophilic catalysis, in general base catalysis the slope depends solely on catalyst pK_a.

The most distinguishing feature of nucleophilic catalysis, however, is the formation of a stable covalent bond between the nucleophile and substrate along the path to the transition state. Often these covalent intermediates resemble isolable reactive species that are common in small molecule organic chemistry. This and other distinctions between nucleophilic and general base catalysis are presented in Section 6.3.3.

The susceptibility of the electrophile is likewise affected by several factors. Again, the pK_a of the leaving group, hence its state of protonation, appears to be a dominant factor. Studies of the rates of catalysis by a common nucleophile on a series of leaving groups demonstrate a clear correlation between rate of attack and the pK_a of the leaving group; generally, the weaker the base, the better leaving group the species. As with the nucleophile itself, the chemical nature of the leaving group, not its pK_a alone, also affects catalytic rate. Other factors influencing the ability of a group to leave can be found in the texts by Fersht (1985) and Jencks (1969), and in most general physical organic chemistry texts.

In enzymatic nucleophilic catalysis, the nucleophile most often is an amino acid side chain within the enzyme active site. From the preceding discussion, one might expect the most basic amino acids to be the best nucleophiles in enzymes. Enzymes, however, must function within a narrow physiological pH range (around pH 7.4), and this limits the correlation between pK_a and nucleophilicity just described. For example, referring to Table 3.1, we might infer that the guanidine group of arginine would be a good nucleophile (pK_a 12.5). Consideration of the Henderson–Hasselbalch equation (see Chapter 2), however, reveals that at physiological pH (ca. 7.4) this group would exist almost entirely in the protonated conjugate acid form. Hence, arginine side chains do not generally function as nucleophiles in enzyme catalysis.

The amino acids that are capable of acting as nucleophiles are serine, threonine, cysteine, aspartate, glutamate, lysine, histidine, and tyrosine. Examples of some enzymatic nucleophiles and the covalent intermediates they form are given in Table 6.1. A more comprehensive description of nucleophilic catalysis and examples of its role in enzyme mechanisms can be found in the text by Walsh (1979).

6.3.2.2 *Electrophilic Catalysis.* In electrophilic catalysis covalent intermediates are also formed between the cationic electrophile of the enzyme and

Table 6.1 Some examples of enzyme nucleophiles and the covalent intermediates formed in their reactions with substrates

Nucleophilic Group	Example Enzyme	Covalent Intermediate
Serine (—OH)	Serine proteases	Acyl enzyme
Cysteine (—SH)	Thiol proteases	Acyl enzyme
Aspartate (—COO⁻)	ATPases	Phosphoryl enzyme
Lysine (—NH₂)	Pyridoxal-containing enzymes	Schiff bases
Histidine	Phosphoglycerate mutase	Phosphoryl enzyme
Tyrosine (—OH)	Glutamine synthase	Adenyl enzyme

Source: Adapted from Hammes (1982).

an electron-rich portion of the substrate molecule. The amino acid side chains do not provide very effective electrophiles. Hence, enzyme electrophilic catalysis most often require electron-deficient organic cofactors or metal ions.

There are numerous examples of enzymatic reactions involving active site metal ions in electrophilic catalysis. The metal can play a number of possible roles in these reactions: it can shield negative charges on substrate groups that would otherwise repel attack of an electron pair from a nucleophile; it can act to increase the reactivity of a group by electron withdrawal; and it can act to bridge a substrate and nucleophilic group; they can alter the pK_a and reactivity of nearby nucleophiles.

Metal ions are also used in enzyme catalysis as binding centers for substrate molecules. For example, in many of the cytochromes, the heme iron ligates the substrate at one of its six coordination sites and facilitates electron transfer to the substrate. Metal ions bound to substrates can also affect the substrate conformation to enhance catalysis; that is, they can change the geometry of a substrate molecule in such a way as to facilitate reactivity. In the ATP-dependent protein kinases (enzymes that transfer a γ-phosphate group from ATP to a protein substrate), for example, the substrate of the enzyme is not ATP itself, but rather an ATP–Mg^{2+} coordination complex. The Mg^{2+} binds at the terminal phosphates, positioning these groups to greatly facilitate nucleophilic attack on the γ-phosphate.

The most common mechanism of electrophilic catalysis in enzyme reactions is one in which the substrate and the catalytic group combine to generate, in situ, an electrophile containing a cationic nitrogen atom. Nitrogen itself is not a particularly strong electrophile, but it can act as an effective electron sink in such reactions because of its ease of protonation and because it can form cationic unsaturated adducts with ease. A good example of this is the family of electrophilic reactions involving the pyridoxal phosphate cofactor.

Pyridoxal phosphate (see Chapter 3) is a required cofactor for the majority of enzymes catalyzing chemical reactions at the alpha, beta, and gamma carbons of the α-amino acids (Chapter 3).

The first step in reactions between the amino acids and the cofactor is the formation of a cationic imine (Schiff base), which plays a key role in lowering the activation energy.

Pyridoxal phosphate	Amine	Schiff base

The function of the pyridoxal phosphate here is to act as an electron sink, stabilizing the carbanion intermediate that forms during catalysis. Electron withdrawal from the alpha-carbon of the attached amino acid toward the cationic nitrogen activates all three substituents for reaction; hence any one of these can be cleaved to form an anionic center. Because the cationic imine is conjugated to the heteroaromatic pyridine ring, significant charge delocalization is provided, thus making the pyridoxal phosphate group a very efficient catalyst for electrophilic reactions.

All pyridoxal-containing enzymes proceed through three basic steps: formation of the cationic imine, chemical changes through the carbanion intermediate, and hydrolysis of the product imine. A common reaction of pyridoxal phosphate with α-amino acids is removal of the α-hydrogen (Figure 6.7) to give a key intermediate in a variety of amino acid reactions, including transamination, racemization, decarboxylation, and side chain interconversion (Fersht, 1985). In transamination (Figure 6.7), for example, removal of the α-hydrogen is followed by proton donation to the pyridoxal phosphate carbonyl carbon, leading to formation of an α-keto acid–pyridoxamine Schiff base. Subsequent hydrolysis of this species yields the free α-keto acid and the pyridoxamine group. An imine is then formed between the keto acid and pyridoxamine, and reversed proton transfer occurs to generate a new amino acid and to regenerate the pyridoxal, thus completing the catalytic cycle.

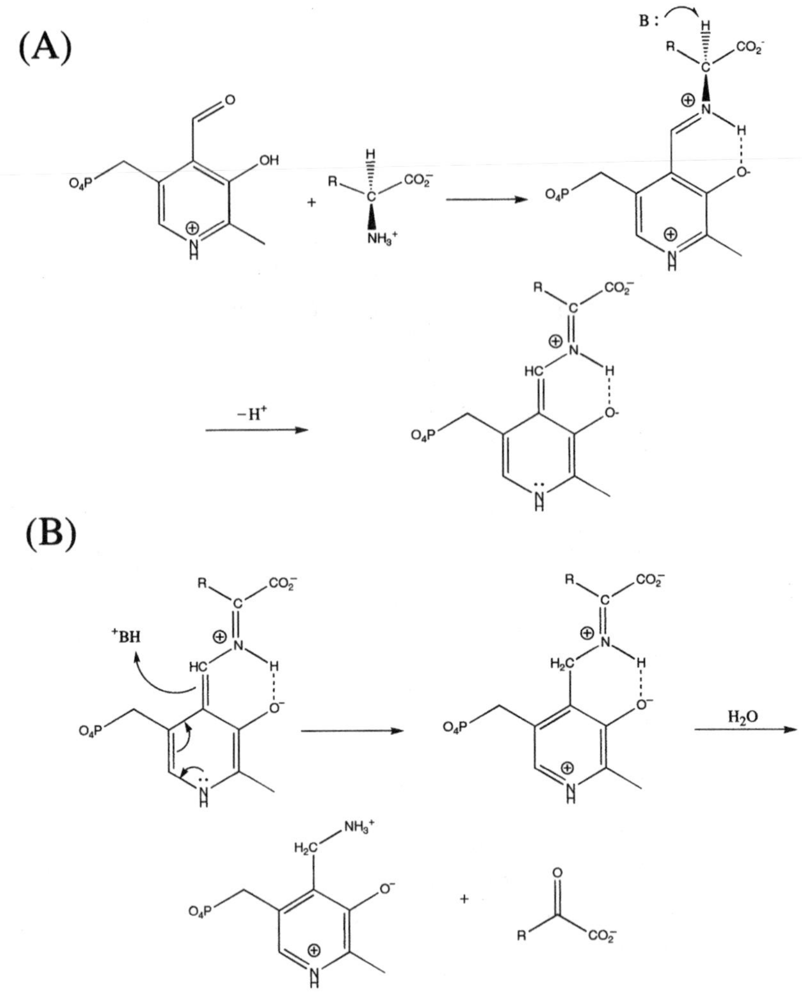

(A)

(B)

Figure 6.7 Examples of reactions of amino acids facilitated by electrophilic catalysis by pyridoxal phosphate: (A) α-hydrogen removal and (B) subsequent transamination. See text for further details.

Examples of some enzymatic reactions involving electrophilic catalysis are provided in Table 6.2. A more comprehensive treatment of these reactions can be found in the texts by Jencks (1969), Bender et al. (1984), and Walsh (1979).

6.3.3 General Acid/Base Catalysis

Just about every enzymatic reaction involves some type of proton transfer that requires acid and/or base group participation. In small molecule catalysis, and in some enzyme examples, protons (from hydronium ion H_3^+O) and hydroxide

Table 6.2   Some examples of electrophilic catalysis in enzymatic reactions

Enzyme	Electrophile
Acetoacetate decarboxylase	Lysine–substrate Schiff base
Aldolase	Lysine–substrate Schiff base
Aspartate aminotransferase	Pyridoxal phosphate
Carbonic anhydrase	Zn^{2+}
L-Malic enzyme	Mn^{2+}
Pyruvate decarboxylase	Thiamine pyrophosphate

ions (OH^-) act directly as the acid and base groups in activities referred to as *specific acid* and *specific base catalysis*. Most often, however, in enzymatic reactions organic substrates, cofactors, or amino acid side chains from the enzyme fulfill this role by acting as Brønsted–Lowry acids and bases in what is referred to as *general acid* and *general base catalysis*.

For catalysis by small molecules (nonenzymatic reactions), general acid/base catalysis can be distinguished from specific acid/base catalysis on the basis of the effects of acid or base concentration on reaction rate. In general acid/base catalysis, the reaction rate is dependent on the concentration of the general acid or base catalyst. Specific acid/base catalysis, in contrast, is independent of the concentrations of these species (Figure 6.8). Although most enzymatic reactions rely on general acid/base catalysis, it is difficult to define the extent of this reliance by changing acid/base group concentration as in Figure 6.8, since the acid and base groups reside within the enzyme molecule

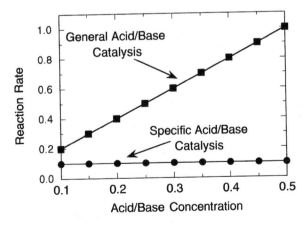

Figure 6.8   Effect of general acid or base concentration on the rate of reaction for general and specific acid- or base-catalyzed reactions.

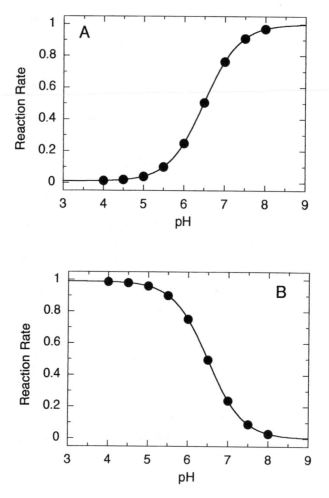

Figure 6.9 Effect of pH on the rate of reaction for (A) a general base-catalyzed reaction and (B) a general acid-catalyzed reaction.

itself. Identification of the group(s) participating in general acid/base catalysis in enzymes has generally come from studies of reaction rate pH profiles (see below), amino acid–specific chemical modification (see Chapter 10), site-directed mutagenesis, and x-ray crystal structures.

General acids and bases will be functional only below or above their pK_a values, respectively. Hence a plot of reaction rate as a function of pH will display the type of sigmoidal curve we are used to seeing for acid–base titrations (Figure 6.9). If we plot the same data as log (reaction rate constant) as a function of pH over a finite pH range (pH 5–7 in Figure 6.9), we find a linear relationship between log (k) and pH. This empirical relationship is defined by the Brønsted equations for general acid and general base catalysis (Equations 6.9 and 6.10, respectively):

$$\log(k) = A - \alpha p K_a \qquad (6.9)$$

$$\log(k) = A + \beta p K_a \qquad (6.10)$$

where A is a constant associated with a specific reaction, and α and β are measures of the sensitivity of reaction to the acid or conjugate base, respectively. The Brønsted equation for general base catalysis is similar to that described for nucleophilic catalysis. In contrast to nucleophilic catalysis, however, general base catalysis depends only on the pK_a of the catalyst and is essentially independent of the nature of the catalytic group.

Generally, these Brønsted equations indicate that the stronger the acid, the better a general acid catalyst it will be, and likewise, the stronger the base, the better a general base catalyst it will be. It is important to reemphasize, however, that the efficiency of general acid or base catalysis depends on the effective concentration of acid or base species present. The concentrations of these species depend on the pK_a of the catalyst in relation to the solution pH at which the reaction is run. For example, upon consideration of the Brønsted equations, one would say that an acid of pK_a 5 would be a better general acid than one of pK_a 7. However, if the reaction is run near pH 7, only about 1% of the stronger acid is in its acid form, while 99% of it is present as the conjugate base form. On the other hand, at the same pH, 50% of the weaker acid (pK_a 7) is present in its acid form. Thus because of the effective concentrations of the catalytically relevant forms of the two species, the weaker catalyst may be more effective at pH 7. For this reason, one finds that the reaction rates for general acid/base catalysis are maximal when the solution pH is close to the pK_a of the catalytic group. Hence, in enzymatic reactions, the general acid/base functionalities that are utilized are those with pK_a values near physiological pH (pH 7.4). Generally, this means that enzymes are restricted to using amino acid side chains with pK_a values between 4 and 10 as general acids and bases. Surveying the pK_a values of the amino acid side chains (Chapter 2), one finds that the side chains of apartate, glutamate, histidine, cysteine, tyrosine, and lysine, along with the free N- and C-termini of the protein, are the most likely candidates for general acid/base catalysts. However, it is important to realize that the pK_a value of an amino acid side chain can be greatly perturbed by the local protein environment in which it is found. Two extreme examples of this are histidine 159 of papain, which has a side chain pK_a of 3.4, and glutamic acid 35 of lysozyme, with a greatly elevated side chain pK_a of 6.5.

The fundamental feature of general acid/base catalysis is that the catalytic group participates in proton tranfers that stabilize the transition state of the chemical reaction. A good example of this comes from the hydrolysis of ester bonds in water, a reaction carried out by many hydrolytic enzymes. The mechanism of ester hydrolysis requires formation of a transition state involving

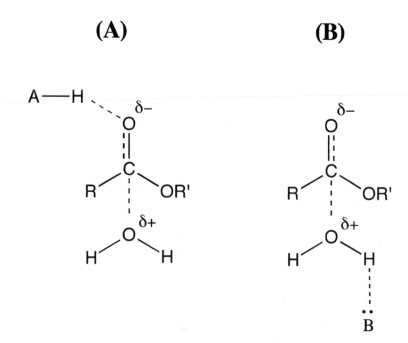

Figure 6.10 Transition state stabilization by a general acid (A) or general base (B) in ester bond hydrolysis by water.

partial charge transfer between the ester and a water molecule, as shown in Figure 6.10. This transition state can be stabilized by a basic group (B:) acting as a partial proton acceptor from the water molecule, thus enhancing the stability of the partial positive charge on the water molecule (Figure 6.10B). Alternatively, the transition state can be stabilized by an acidic group that acts as a partial proton donor to the carbonyl oxygen of the ester (Figure 6.10A). This type of acid/base group transition state stabilization is very common in enzyme catalysis. The active site histidine residue of the serine proteases, for example, acts in this capacity to stabilize the transition state during peptide hydrolysis, as described later.

As we have just stated, general acid/base catalysis is a common mechanism of transition state stabilization in enzymatic catalysis. So too is nucleophilic catalysis, as described in Section 6.3.2. Ambiguity can arise in distinguishing nucleophilic catalysis from general base catalysis because these two mechanisms share some common kinetic features. Bender et al. (1984) suggest the following criteria for distinguishing betweeen these mechanisms:

1. If the leaving group of a reaction can itself catalyze the same reaction, this is evidence of general base rather than nucleophilic catalysis. Consider the reaction:

If this reaction went with nucleophilic attack of B on the carbonyl, the reaction product would be the same as the starting material. Hence, catalysis by B indicates this species must be functioning as a general base.

2. Catalysis by a second equivalent of an attacking species is evidence that general base catalysis is occurring. The covalent intermediate of nucleophilic catalysis forms stoichiometrically with substrate. Hence catalysis by a second equivalent of the attacking species cannot occur via nucleophilic catalysis,

3. As stated above, observation of a transient intermediate that can be identified as a covalent adduct constitutes proof of nucleophilic catalysis, rather than the general base version.

4. As stated above, the Brønsted plot for general base catalysis fits a single straight line for bases of identical basicity, regardless of their structural type. Conversely, nucleophilic catalysis is characterized by Brønsted plots with large differences in rates when one is considering, for example, nitrogen and oxygen pairs of catalysts of equal basicity.

5. Related to point 4, steric hinderance is not an important determinant in general base catalysis but can have significant effects on nucleophilic catalysis. This follows because general base catalysis involves proton abstraction, while nucleophilic catalysis involves attack at a carbon center; the steric requirements for the latter reaction are far greater than that of the former.

6. Deceleration of reaction rate by addition of a common ion is an indication of a reaction mechanism involving a reversible intermediate, hence nucleophilic catalysis. For example, the pyridine-catalyzed hydrolysis of acetic anhydride proceeds with formation of acetate ion:

Addition of acetate ion significantly decreases reaction rate, indicating nucleophilic catalysis by pyridine.

Additional criteria that are more relevant to small molecule catalysis are provided by Bender et al. (1984). The interested reader should consult this text and the references therein.

6.3.4 Conformational Distortion

Two experimental observations lead to the hypothesis that conformational adjustments of the enzyme, the substrate, or both are important aspects of enzymatic catalysis.

First, quantitative studies of enzymatic reactions often lead to the conclusion that the observed rate enhancement cannot be accounted for fully by consideration of approximation, covalent catalysis, and acid/base catalysis alone. Hence, an additional mechanism for rate enhancement must be invoked. It is well established that many proteins, including enzymes, are conformation-

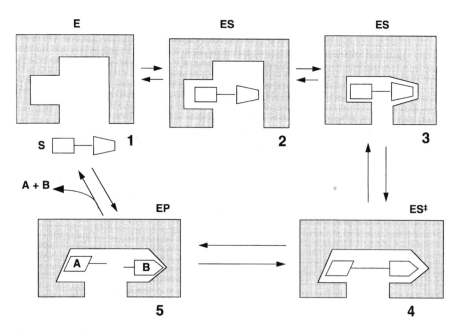

Figure 6.11 The rack model of enzyme–substrate interaction and catalysis: 1, the free enzyme is in solution with the free substrate; 2, initial binding occurs to form the ES complex without structural distortion of the substrate; 3, the enzyme undergoes a conformational change to induce better binding with the substrate (induced fit); 4, the enzyme undergoes further conformational adjustments that induce strain into the substrate molecule, distorting it from its ground state structure toward its transition state structure; 5, the strain induced in 4 leads to bond rupture and release of the two hydrolytic products, A and B.

ally dynamic. Therefore, one hypothesis that has emerged is that conformational adjustments are used to distort the substrate, the enzyme, or both to bring the system toward the transition state structure. Figure 6.11 illustrates this concept in schematic form for a bond cleavage reaction. In this example the enzyme active site undergoes a series of conformational adjustments that distort the substrate toward its transition state structure, thus facilitating catalytic bond rupture. In this illustration, the distortion of the substrate by the enzyme active site is akin to the medieval torture in which a prisioner was stretched on a device known as a rack. For this reason, the term *rack mechanism* is sometimes used to describe the induced-strain hypothesis (Segel, 1975).

The second experimental observation that suggests a need to invoke conformational distortions is the manifestation of substrate specificity in k_{cat}, rather than in K_m. We saw in Chapter 5 that the best measure of substrate specificity is the second-order rate constant obtained by dividing k_{cat} by K_m. If substrate specificity merely involved discrimination in binding of substrates in their ground states, one would expect that differences in K_m would be the dominant factor in distinguishing a good from a bad substrate. Experimentally, however, one often observes that substrate specificity is manifested at high, saturating substrate concentration, rather than at low substrate concentration. In other words, specificity is largely dictated by differences in k_{cat} (i.e., V_{max}), rather than by differences in K_m. A good example comes from studies of the hydrolysis of synthetic peptides by the enzyme pepsin (Bender et al., 1984); the results of these studies are summarized in Table 6.3, which indicates that for a series of peptide substrates, the catalytic efficiency (measured as k_{cat}/K_m) of pepsin varies over 1000-fold. Looking at the individual kinetic constants, we see that k_{cat} likewise varies over about 1000-fold. In contrast, the values for K_m of the various substrates are quite similar, differing only about four-fold at

Table 6.3   Steady state kinetics of synthetic peptide hydrolysis by pepsin

Peptide[a]	K_m (mM)	k_{cat} (s^{-1})	k_{cat}/K_m (mM$^{-1} \cdot$ s^{-1})
Cbz-G-H-F-F-OEt	0.8	2.4300	3.04000
Cbz-H-F-W-OEt	0.2	0.5100	2.55000
Cbz-H-F-F-OEt	0.2	0.3100	1.55000
Cbz-H-F-Y-OEt	0.2	0.1600	0.80000
Cbz-H-Y-F-OEt	0.7	0.0130	0.01860
Cbz-H-Y-Y-OEt	0.2	0.0094	0.04700
Cbz-H-F-L-OMe	0.6	0.0025	0.00417

[a]Cbz, carbobenzyloxy; OEt, ethyl ester of carboxy terminus; OMe, methyl ester of carboxy terminus.

Source: Bender et al. (1984).

most. In fact, the K_m value for the worst substrate is actually lower than that of the best substrate, clearly indicating that ground state substrate binding is not a major determinant of substrate specificity for this enzyme. Instead, the experimental data for pepsin, and for a large number of other enzymes, suggest that substrate specificity is determined mainly by the transition states structure, rather than by that of the ground state.

A similar conclusion is often reached by the study of the effects of point mutations within the active site of enzymes. For example, Leatherbarrow et al. (1985) studied the enzyme tyrosyl–tRNA synthetase, an enzyme that covalently links the amino acid tyrosine to its tRNA. The enzyme requires ATP for catalysis and proceeds through formation of an enzyme-bound tyrosyl–ATP transition state. The crystal structure of the enzyme with the covalent intermediate tyrosyl–AMP bond was known, and Leatherbarrow et al. were able to use this to model the transition state complex. From this modeling the workers identified two amino acid residues within the enzyme active site that should be critical for catalysis: Thr 40 and His 45. They then set about making three mutant enzymes: a Thr 40-Ala mutant, a His 45-Gly mutant, and the double mutant Thr 40-Ala/His 45-Gly. When they compared the kinetic parameters for these three mutants to the wild-type enzyme, they found that the values of K_s for either tyrosine or ATP were hardly affected; the largest change in K_s seen was about four-fold relative to the wild-type enzyme. In contrast, however, the values of k_{cat} were dramatically changed. The value of k_{cat} went from $38 \, s^{-1}$ for the wild-type enzyme to $0.16 \, s^{-1}$ for the His 45-Gly mutant, to $0.0055 \, s^{-1}$ for the Thr 40-Ala mutant, and to $0.00012 \, s^{-1}$ for the double mutant. Thus Leatherbarrow et al. concluded that the main function of these residues was to provide favorable binding interactions with the tyrosyl–ATP transition state structure, and in contrast provided no binding interactions with the ground state substrate.

Jencks (1969) has argued that for reversible enzymatic reactions, distortion of the ES complex is a necessary component of catalysis. Consider a reversible enzymatic reaction that proceeds through nucleophilic attack of a group on the substrate molecule. We could imagine that the enzyme active site and the substrate ground state are both conformationally rigid. The substrate binds to the active site because its ground state configuration fits perfectly into this pocket. In this case, the interatomic distance between the enzymatic nucleophile and the substrate carbon atom that is attacked would be determined by the van der Waals radii of the interacting atoms. If this were so, the rigid product molecule could not have a perfect fit to the same rigid active site structure. Thus the enzyme–product complex would be destabilized relative to the enzyme–substrate complex as a result of steric hindrance (hence loss of binding energy). The enzyme could overcome this by distorting the bound product in such a way as to more closely resemble the substrate molecule. This would facilitate reaction in the reverse direction. Likewise, if the active site were rigid and fit the product molecule perfectly, the imperfect fit of the substrate molecule would now impede reaction in the forward direction. The need for enzymatic rate acceleration in both the forward and reverse reactions is most effectively accomplished by having the active site structure best matched to a structure intermediate between the substrate and product states—that is, by having the active site designed to best match the transition state structure.

However, the foregoing arguments and experimental data are not adequately accounted for by models in which the ground state substrate and the enzyme active site are conformationally rigid, as in the original lock and key model of Fischer. Hence, models are needed that take into account the need for conformational flexibility and optimization of active site–transition state complementarity. Three major models have been put forth to fill these needs: (1) the induced-fit model, (2) the nonproductive binding model, and (3) the induced-strain model.

The induced-fit model, first proposed by Koshland (1958), suggests that the enzyme active site is conformationally fluid. In the absence of substrate, the active site is in a conformation that does not support catalysis. When a "good" substrate binds to the active site, the binding forces between the enzyme and the substrate are used to drive the enzyme into an energetically less favorable, but catalytically active conformation (the rack model illustrated in Figure 6.11 is one interpretation of this induced fit model). In this model a "poor" substrate lacks the necessary structural features to induce the conformational change required for catalytic activity, and thus does not undergo reaction. The expected results from the induced-fit model is that V_{max} is optimal for specific substrates that can best utilize binding interactions to induce the necessary conformational activation of the enzyme molecule.

In the nonproductive binding model the enzyme active site is considered to be rigid; and a "good" substrate would be one that has several structural features, each one being complementary to a specific subsite within the active site structure. Because of the presence of multiple complementary subsite interactions, there would be only one active conformation and orientation of the substrate in the active site of the enzyme; other structures or orientations would lead to binding of an inactive substrate species that would be nonproductive with respect to catalysis. A "poor" substrate in this model might lack one or more key functional group for correct binding. Alternatively, the functional group might be present in a "poor" substrate, but arranged in a fashion that is inappropriate for correct binding in the enzyme active site; these concepts are illustrated in Figure 6.12. The expected result from this model is that V_{max} for a "poor" substrate would be greatly diminished because only a fraction of the "poor" substrate—that happened to bind in a productive mode—would lead to catalytic activity. As discussed in Chapter 5, the nonproductive binding model is often invoked to explain the phenomenon of substrate inhibition at high substrate concentrations.

The third major model for conformational distortion is referred to as the induced-strain model. In this model the binding forces between enzyme and substrate are directly used to induce strain in the substrate molecule, distorting it toward the transition state structure to facilitate reaction. The enzyme active

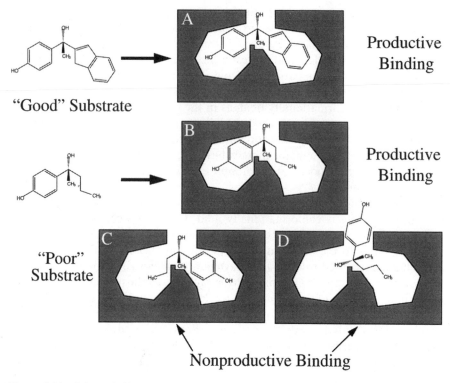

"Good" Substrate

A Productive Binding

B Productive Binding

"Poor" Substrate

C

D

Nonproductive Binding

Figure 6.12   Schematic illustration of the nonproductive binding model. In this model a "good" substrate has specific groups that engage subsites in the enzyme active site, resulting in a unique, productive binding orientation that facilitates catalysis. A "poor" substrate is one that lacks a keep functionality to help orient the substrate properly in the active site. The poor substrate therefore may bind in a number of orientations, only a fraction of which lead to efficient catalysis. The other, nonproductive, binding orientations are inhibitory to catalysis.

site is considered to be flexible in this model. The most stable (i.e., lowest energy) conformation of the active site is one that does not optimally fit the ground state substrate conformation, but instead is complementary to the transition state of the reaction. For the ground state substrate to bind, the enzyme must undergo a conformational deformation that is energetically unfavorable. Hence, in the ES complex there will be a driving force for the enzyme molecule to return to its original lower energy conformation, and this will be accompanied by distortions of the substrate molecule to bring it into the transition state structure that is complementary to the lowest energy conformation of the active site (Figure 6.13). At first glance, this model seems similar to the induced-fit model, in which binding energy is used to bring reactive groups within the active site into proper register with respect to a "good" substrate to provide substrate specificity. In contrast, the induced-

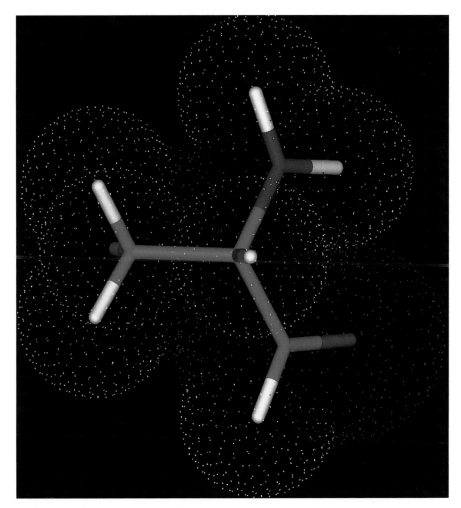

Figure 2.14 Van der Waals radii for the atoms of the amino acid alanine. The "tubes" represent the bonds between atoms. Oxygen is colored red, nitrogen is blue, carbon is green, and hydrogens are gray. The white dimpled spheres around each atom represent the van der Waals radii. (Diagram courtesy of Karen Rossi, Department of Computer Aided Drug Design, The DuPont Pharmaceuticals Company.)

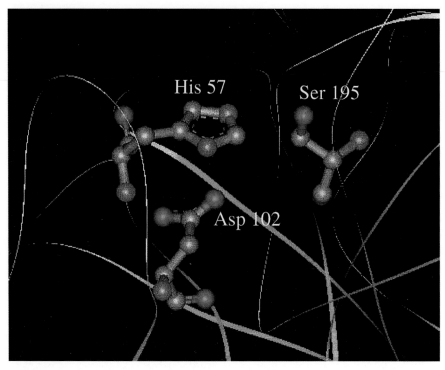

Figure 3.14 The active site triad of the serine protease α-chymotrypsin. [Adapted from the crystal structure reported by Frigerio et al. (1992) *J. Mol. Biol.* **225**, 107.]

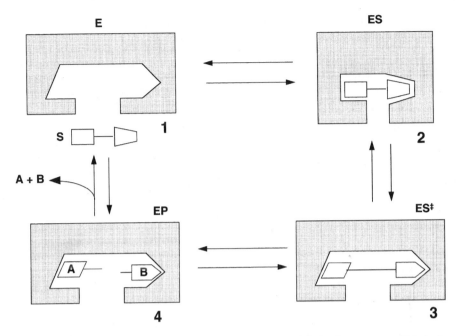

Figure 6.13 Schematic illustration of the induced-strain model: the free enzyme (1) is in its lowest energy conformation that is most complementary to the transition state structure of the substrate. Upon substrate binding (2), the enzyme undergoes a conformational transition to engage the substrate. The conformational strain induced in (2) is relieved by formation of the bound transition state in 3. Catalysis then leads to formation of the enzyme–product complex (4), before product release to re-form the free enzyme.

strain model utilizes the binding energy to directly decrease the free energy of activation for the reaction by coupling substrate distortion to an energetically favorable conformational relaxation of the enzyme molecule.

Steric effects are not the only mechanism for inducing strain in a bound substrate molecule. Electrostatic effects can also provide for induction of strain in the ground state, which is subsequently relieved in the transition state structure. For example, a negatively charged amino acid side chain in the hydrophobic enzyme active site would have a strong polarizing effect that would be unfavorable to binding of an uncharged ground state substrate. If, however, the transition state involved formation of a cationic center, the negative counterion would now provide a favorable interaction. This relief of ground state destabilization can be effectively used as a driving force for transition state formation. Likewise, hydrogen bonding, hydrophobic interactions, and other noncovalent forces can provide unfavorable interactions with the ground state substrate that drive distortion to the transition state structure, where these interactions are energetically favorable.

6.3.5 Preorganized Active Site Complementarity to the Transition State

An alternative to mechanisms utilizing conformational distortion is one in which the enzyme active site is, relatively speaking, conformationally rigid and preorganized to optimally fit the substrate in its transition state conformation. This is somewhat reminiscent of the lock and key model of Fischer, but here the complementarity is with the substrate transition state, rather than the ground state. In a recent review of this mechanism, Cannon and Benkovic (1998) use the following thermodynamic cycle to define an equilibrium dissociation constant for the $ES^{\ddagger}$ binary complex, K_{TS}, which is equivalent to the ratio $(k_{cat}/K_m)/k_{non}$, where k_{non} is the rate constant for formation of the substrate transition state in the absence of enzyme catalysis.

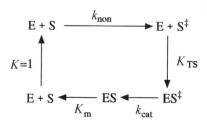

From our earlier discussion, in which we learned that enzymes bind the transition state better than the ground state substrate, it is clear that K_{TS} is typically much smaller than K_s. Hence there is usually a large, favorable free energy of binding for the $ES^{\ddagger}$ complex. Cannon and Benkovic suggest two potential ways to explain the difference in free energy of binding between $S^{\ddagger}$ and S. First, there may be significantly stronger binding interactions between the enzyme and the transition state conformation of the substrate. This possibility was discussed above, and, at least for some enzymes there is experimental evidence for stronger interaction between the enzyme and transition state analogues, than between the enzyme and the ground state substrate. The second possibility is that the free energy of interaction between the solvent and the transition state is very much less favorable than that for the solvent and the ground state substrate. Hence, in solution the attainment of the transition state must overcome a significant free energy change due to solvation effects. In other words, in this case significantly better interactions between the enzyme and $S^{\ddagger}$ relative to S are not the main driving forces for transition state stabilization. Instead, by removing S from solvent, the enzyme avoids much of the solvation-related barrier to formation of $S^{\ddagger}$. In this model the enzyme does not so much stabilize the transition state as avoid the destabilizing effect of the solvent by sequestering the substrate. Cannon and Benkovic suggest that both possibilities occur in enzymatic catalysis, but that the latter is a more dominant effect.

As evidence for this mechanism, Cannon and Benkovic have plotted K_{TS} against k_{cat} and k_{non} for a series of first-order, or pseudo-first-order, enzymatic reactions and appropriate model reactions in solution. If the main function of enzymes were to bind $S^{\ddagger}$ much more tightly than S, one would expect that the catalytically most powerful enzymes (i.e., those with the lowest values of K_{TS}) would display the fastest turnover rates (k_{cat}). On the other hand, if rate acceleration by enzymes is due mainly to relief of the retardation of solution reactions due to solvation effects, the enzymes displaying small K_{TS} values would be those catalyzing reactions that proceed very slowly in solution (i.e., with small values of k_{non}). The data plot provided by the authors shows a significant, negative correlation between the values of K_{TS} and k_{non}, but a general lack of correlation between K_{TS} and k_{cat}. Cannon and Benkovic conclude from this plot that the kinetics of reactions in solution are the dominant determinants of K_{TS}.

The physical explanation for these observations is that the effect of solvent on the solution reactions is counterproductive. Significant solvent reorganization is required for the solution reaction to proceed to the transition state, and this reorganization has a retarding effect on the rate of reaction. The enzyme thus functions as a mechanism for solvent substitution for the reactants. Cannon and Benkovic point out that the solvent effect on reaction will depend on both the dielectric response and the polarity of the medium. Because enzyme active sites are largely hydrophobic, they will generally have low dielectric constants; but they can be highly polar (as a result of the effect of placing a charge within a low dielectric medium), thus producing very intense electric fields. By judicious placement of charged groups within the active site, the enzyme can achieve electrostatic complementarity with the transition state structure of the substrate, thus eliminating the solvent retardation effects.

The foregoing argument suggests that the enzyme active site is preorganized to be complementary to the transition state of the substrate, thus minimizing the energetic cost of reorganizations such as those occurring in solution. This mechanism would disfavor conformationally induced distortions of the substrate, which would only add to the reorganizational cost to catalysis. Cannon and Benkovic point out further that the time scale of most protein conformational changes is inconsistent with the rates of catalysis for many enzymes. It is important to note, however, that vibrational bond motions occur on a time scale (ca. 10^{-14}–10^{-13} s^{-1}) that is consistent with enzymatic catalysis. Hence, small vibrational adjustments of the enzyme–substrate complex cannot be ruled out by the foregoing argument.

Thus, the mechanism proposed by Cannon and Benkovic relies on the preorganization of the enzyme active site in a configuration that is complementary to the transition state of the substrate (but does not necessarily bind the transition state extremely tightly). This preorganized active site features strategically located reactive groups (general acids/bases; active site nucleophiles, hydrogen-bonding partners, etc.) in a low dielectric medium that greatly relieves the destabilization of the transition state associated with reaction in solution. Thus, in this mechanism, enzymes primarily accelerate reaction rates not by stabilizing the transition state through tight binding interactions per se, but instead by avoiding the energetic penalties that accompany transition state formation in bulk solution. The authors point out that this mechanism suggests that one could capture most of the catalytic efficiency of enzymes by providing a properly preorganized transition state binding pocket within an engineered protein. For example, the immune system normally produces antibodies that recognize and bind tightly to proteins and peptides from an infecting organism. Antibodies can also be produced that recognize and bind small molecules, referred to as haptens. One might therefore expect antibodies raised against transition state analogues of specific reactions to display some ability to act as catalysis of the reaction. This approach has been experimentally verified (Lerner et al., 1991), although the resulting *catalytic antibodies* have thus far showed only modest catalytic activity relative to natural enzymes.

In this section we have discussed a variety of strategies by which enzymes can effect transition state stabilization. Which of these play significant roles in enzyme catalysis? The most likely answer is that each of these strategies is used to varying degrees by different enzymes to achieve transition state stabilization. The essential point to take away from this discussion is that enzyme active sites have evolved to bind preferentially the transition state of the substrate. Through preferential binding and stabilization of the transition state, the enzyme provides a reaction pathway that is energetically much more favorable than any pathway that can be achieved in its absence.

6.4 THE SERINE PROTEASES: AN ILLUSTRATIVE EXAMPLE

The serine proteases are a family of enzymes that catalyze the cleavage of specific peptide bonds in proteins and peptides. As we briefly mentioned earlier (Chapter 3), the serine proteases have a common mechanism of catalysis that requires a triad of amino acids at the active sites of these enzymes; a serine residue (hence the family name) acts as the primary nucleophile for attack of the peptide bond, and the nucleophilicity of this group is enhanced by specific interactions with a histidine side chain (a general acid/base catalyst), which in turn interacts with an aspartate side chain. The catalytic importance of the active site serine and histidine residues has been demonstrated recently by site-directed mutagenesis studies, in which replacement of either the serine or the histidine or both reduced the rate of reaction by the enzyme by as much as 10^6 fold (Carter and Wells, 1988).

Because of their ease of isolation and availability in large quantities from the gastric juices of large animals, the serine proteases were among the first enzymes studied. While the members of this family studied initially were all digestive enzymes, we now know that the serine proteases perform a wide variety of catalytic functions in most organisms from bacteria to higher mammals. In man, for example, serine proteases take part not only in digestive processes, but also in the blood clotting cascade, inflammation, wound healing, general immune response, and other physiologically important events. These enzymes are among the most well-studied proteins in biochemistry. A great deal of structural and mechanistic information on this class of enzymes is available from crystallographic, classical biochemical, mechanistic, and mutagenesis studies (see Perona and Craik, 1995, for a recent review). Because of this wealth of information, the serine proteases provide a model for discussing in concrete chemical and structural terms some of the concepts of substrate binding specificity and transition state stabilization covered thus far in this chapter.

To cleave a peptide bond within a polypeptide or protein, a protease must recognize and bind a region of the polypeptide chain that brackets the scissile peptide bond (i.e., the bond that is to be cleaved). Proteases vary in the length of polypeptide that forms their respective recognition sequences, but most bind several amino acid residues in their active sites. A nomenclature system has been proposed by Schechter and Berger (1967) to keep track of the substrate amino acid residues involved in binding and catalysis, and the corresponding sites in the enzyme active site where these residues make contact. In this system, the bond that is to be hydrolyzed is formed between residue P1 and P1' of the substrate; P1 is the residue that is on the N-terminal side of the scissile bond, and P1' is the C-terminal hydrolyzed residue (the "P" stands for "peptide" to designate these residues as belonging to the substrate of the reaction). The residue adjacent to P1 on the N-terminal side of the scissile bond is designated P2, and the residue adjacent to the P1' residue on the C-terminal side is P2'. The "subsite" within the enzyme active site that residue P1 fits into is designated S1, and the "subsite" into which residue P1' fits is designated S1'. The numbering continues in this manner, as illustrated in Figure 6.14 for a six-residue peptide substrate. We shall use this nomenclature system from now on when discussing proteolytic enzymes.

On the basis of their structural properties, the serine proteases have been divided into three classes, called the chymotrypsin-like, the subtilisin-like, and the serine carboxypeptidase II–like families. The secondary and tertiary structures of the proteins vary considerably from one family to another, yet in all three families the active site serine, histidine, and aspartate are conserved and a common mechanism of catalysis is used. All these enzymes catalyze the hydrolysis of ester and peptide bonds through the same acyl transfer mechanism (Figure 6.15), with a rate acceleration of 10^9 or more relative to the uncatalyzed reaction. After formation of the ES complex, the carbonyl carbon of the scissile peptide bond (i.e., that on P1) is attacked by the active site serine, forming a tetrahedral intermediate with an oxyanionic center on the carbonyl carbon that is highly reminiscent of the transition state of the reaction. This transition state is stabilized by specific hydrogen-bonding interactions between residues in the active site pocket and the oxyanion center of the substrate. In subtilisin this hydrogen bonding is provided by the backbone nitrogen of Ser

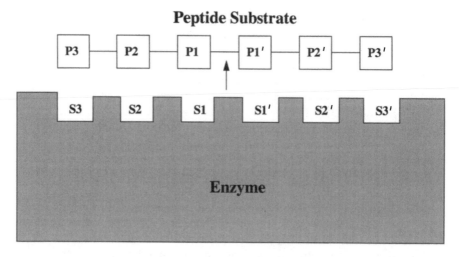

Figure 6.14 The protease subsite nomenclature of Schechter and Berger (1967): residues on the peptide substrate are labeled P1–Pn on the N-terminal side of the scissile bond, and P1′–Pn′ on the C-terminal side of this bond; the scissile bond is thus between residues P1 and P1′. The corresponding subsites into which these residues fit in the enzyme active site are labeled S1–Sn and S1′–Sn′.

195 (the active site nucleophile) and the side chain of Asn 155. In the chymotrypsin-like enzymes, these H bonds are provided by the backbone nitrogens of the nucleophilic serine residue and an active site glycine, while in the serine carboxypeptidases these bonds are formed by the backbone nitrogens of a tyrosine and a glycine in the active site. Crystallographic data from studies of subtilisin indicate that a weak hydrogen bond exists between Asn 155 and the substrate in the ES complex, and that this H-bonding is significantly strengthened in the transition state (an example of conformational alterations in the enzyme active site that facilitate transition state stabilization). Mutation of Asn 155 to any of a variety of non-hydrogen-bonding amino acids significantly decreases the rate of the enzymatic reaction, as would be expected from our discussion of enzymatic rate enhancement by transition state stabilization.

The transition state then decays as a proton is donated from the active site histidine to the amine group of P1′, followed by dissociation of the first product of the reaction, the peptide starting at P1′, and simultaneous formation of a covalent intermediate with an acyl group (from P1) bound to the active site serine. The enzyme is then deacylated by nucleophilic attack by a water molecule that enters the enzyme active site from the cavity resulting from the departure of the first product peptide. The deacylation reaction proceeds with formation of another tetrahedral transition state, very similar to that formed during the acylation reaction, and engaging the same stabilizing H bonds with

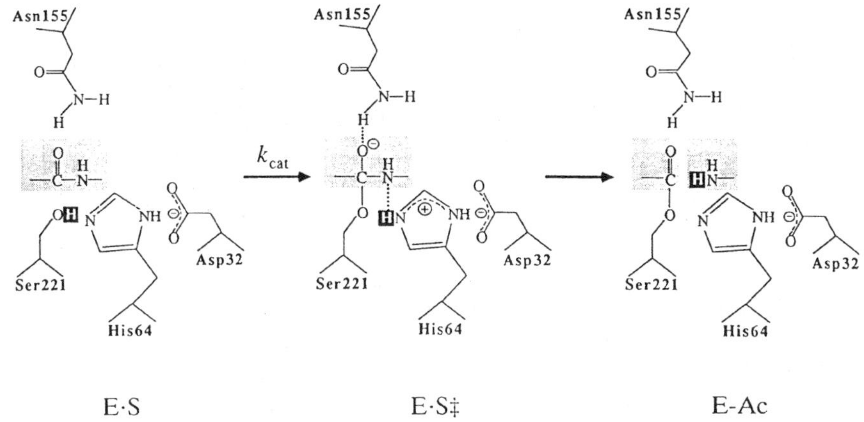

E·S   E·S‡   E–Ac

Figure 6.15 Schematic representation of the general acyl transfer mechanism of serine proteases. [Reprinted with permission from *Nature* (Carter and Wells (1988) **332**, 564), copyright (1988) Macmillan Magazines Limited.]

the enzyme. This transition state decays with proton transfer to the active site histidine and release of the second peptide product having the P1′ group at its amino terminus.

The active site aspartate residue is a common feature of the serine proteases and has been shown, through mutagenesis studies, to be critical for catalysis. The role of this residue in catalysis is not completely clear. Early studies suggested that together with the serine and histidine residues, this aspartate formed a catalytic "triad" that acts as a proton shuttle. Such a specific interaction requires a precise geometric relationship between the side chains of the aspartate and histidine residues to ensure strong H-bonding. However, the orientation of the aspartate side chain relative to the Ser-His active site residues varies considerably among the three classes of serine protease, making it unlikely that direct proton transfer occurs between the histidine and aspartate side chains in all these enzymes. In fact, some workers have suggested that the catalytic machinery of the serine proteases is most correctly viewed as two distinct catalytic dyads — one comprising the serine and histidine residue and the other comprising the histidine and aspartate residues — rather than as a single catalytic triad (Liao et al., 1992). Regardless of the molecular details, experimental data demonstrate that the presence of the carboxylate anion of the aspartate influences the reactivity of the histidine residues in a way that is critical for catalysis.

The foregoing discussion provides a good example of the interplay between substrate and enzyme active site that must accompany transition state stabilization and reaction rate enhancement. Substrate binding, however, must precede these catalytic steps, and formation of the enzyme–substrate complex is also governed by the stereochemical relationships between groups on the substrate and their counterpart subsites in the enzyme active site. For example, the differences in substrate specificity within the chymotrypsin-like and subtilisin-like serine proteases can be explained on the basis of their active site structures. In all these enzymes, substrate binding is facilitated by H-bonding to form β-pleated sheet structures between residues in the enzyme active site and the P1–P4 residues of the substrate (Figure 6.16). These interactions provide binding affinity for the substrates but do not significantly differentiate one peptide substrate from another. The P1–S1 site interactions appear to play a major role in defining substrate specificity in these enzymes. Subtilisins generally show broad substrate specificity, with a preference for large, hydrophobic groups at the substrate P1 position. The order of preference is approximately Tyr, Phe > Leu, Met, Lys > His, Ala, Gln, Ser ≫ Glu, Gly (Perona and Craik, 1995). The S1 site of the subtilisins exists as a broad, shallow cleft that is formed by two strands of β-sheet structure and a loop region of variable size (Figure 6.16B). In a subclass of these enzymes, in which a P1 Lys residue is accommodated, the loop contains a glutamate residue positioned to form a salt bridge with the substrate lysine residue, to neutralize the charge upon substrate binding.

In the chymotrypsin-like enzymes, the S1 pocket consists of a deep cleft into which the substrate P1 residue must fit (Figure 6.16A). The identity of the amino acid residues within this cleft will influence the types of substrate residue that are tolerated at this site. The chymotrypsin-like enzymes can be further subdivided into three subclasses on this basis. Enzymes within the trypsinlike subclass have a conserved aspartate residue at position 189, located at the bottom of the S1 well; this explains the high preference of these enzymes for substrates with arginine or lysine residues at P1. This aspartate is replaced by a serine or small hydrophobic residue in the chymotrypsin and elastase subclasses; hence both subclasses show specificity for nonpolar P1 residues. The other amino acid residues lining the S1 pocket further influence substrate specificity. The elastase subclass contains in this pocket large nonpolar groups that tend to exclude bulky substrate residues. Hence, the elastase subclass favors substrates with small hydrophobic residues at P1. In contrast, the chymotrypsin subclass has small residues in these positions (e.g., glycines), and these enzymes thus favor larger hydrophobic residues, such as tyrosine and phenylalanine at the P1 position of their substrates.

Within this overview of the serine proteases we have observed specific examples of how the active site structure of enzymes (1) engages the substrate and binds it in an appropriate orientation for catalysis (e.g., the H-bonding network developed between the P1–P4 residues of the substrate and the active site residues), (2) stabilizes the transition state to accelerate the reaction rate (e.g., the stabilization of the tetrahedral oxyanionic intermediate through H-bonding interactions), and (3) differentiates between potential substrates on the basis of stereochemical relationships between the substrate and active site subsites. While the molecular details differ from one enzyme to another, the

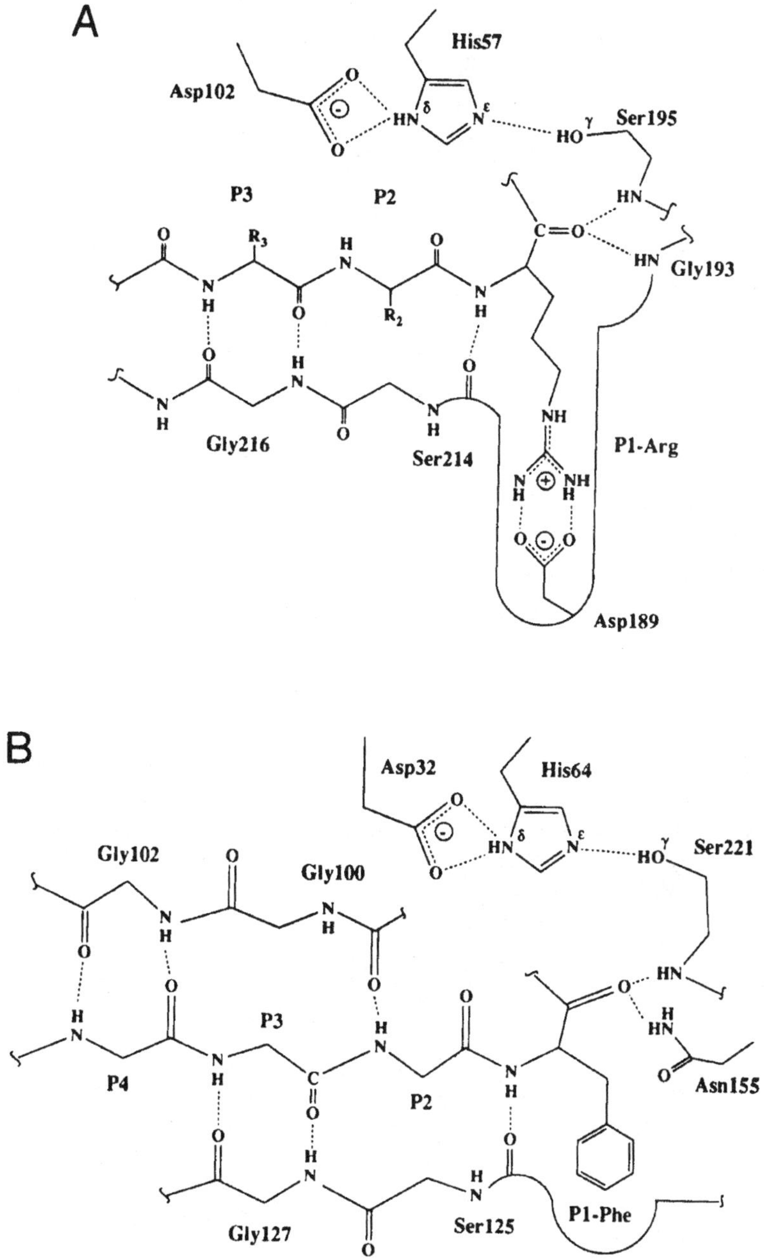

Figure 6.16 Substrate–active site interactions in the serine proteases. (A) Interactions within the trypsinlike class of serine proteases. (B) Interactions within the subtilisin-like class of serine proteases. [Reprinted with the permission of Cambridge University Press from Perona and Craik (1995).]

general types of interaction illustrated with the serine proteases also govern substrate binding and chemical transformations in all the enzymes nature has devised.

6.5 ENZYMATIC REACTION NOMENCLATURE

The hydrolytic activity illustrated by the serine proteases is but one of a wide variety of bond cleavage and bond formation reactions catalyzed by enzymes. From the earliest studies of these proteins, scientists have attempted to categorize them by the nature of the reactions they provide. Group names have been assigned to enzymes that share common reactivities. For example, "protease" and "proteinase" are used to collectively refer to enzymes that hydrolyze peptide bonds. Common names for particular enzymes are not always universally used, however, and their application in individual cases can lead to confusion. For example, there is a metalloproteinase known by the common names stromelysin, MMP-3 (for matrix metalloproteinase number 3), transin, and proteoglycanase. Some workers refer to this enzyme as stromelysin, others call it MMP-3, and still others call it transin or proteoglycanase. A newcomer to the metalloproteinase field could be quite frustrated by this confusing nomenclature. For this reason, the International Union of Pure and Applied Chemistry (IUPAC) formed the Enzyme Commission (EC) to develop a systematic numerical nomenclature for enzymes. While most workers still use common names for the enzymes they are working with, literature references should always include the IUPAC EC designations, which have been universally accepted, to let the reader know precisely what enzymes are being discussed. The EC classifications are based on the reactions that enzymes catalyze. Six general categories have been defined, as summarized in Table 6.4. Within each of these broad categories, the enzymes are further differentiated by a second number that more specifically defines the substrates on which they act. For example, 11 types of hydrolase (category 3) can be defined, as summarized in Table 6.5.

Table 6.4   The IUPAC EC classification of enzymes into six general categories according to the reactions they catalyze

First EC Number	Enzyme Class	Reaction
1	Oxidoreductases	Oxidation–reduction
2	Transferases	Chemical group transfers
3	Hydrolases	Hydrolytic bond cleavages
4	Lyases	Nonhydrolytic bond cleavages
5	Isomerases	Changes in arrangements of atoms in molecules
6	Ligases	Joining together of two or more molecules

Table 6.5 The IUPAC EC subclassifications of the hydrolases

First Two EC Numbers	Substrates[a]
3.1	Esters, —C(O)—O---R, or with S or P replacing C, or —C(O)—S---R
3.2	Glycosyl, sugar—C—O---R, or with N or S replacing O
3.3	Ether, R—O---R, or with S replacing O
3.4	Peptides, C---N
3.5	Nonpeptides C---N
3.6	Acid anhydrides, R—C(O)—O---C(O)—R′
3.7	C---C
3.8	Halides (X), C---X, or with P replacing C
3.9	P---N
3.10	S---N
3.11	C---P

[a]Hydrolyzed bonds shown as dashed lines.

Individual enzymes in each subclass are further defined by a third and a fourth number. In this way any particular enzyme can be uniquely identified. Examples of the common names for some enzymes, and their EC designations are given in Table 6.6. (The enzymes selected have served as illustrative examples in my laboratory.)

The detailed rules for assigning an EC number to a newly discovered enzyme were set forth in Volume 13 of the series *Comprehensive Biochemistry* (Florkin and Stotz, 1973); an updated version of the nomenclature system was published nearly 20 years later (Webb, 1992). Most of the enzymes the reader is likely to encounter or work with already have EC numbers. One can often obtain the EC designation directly from the literature pertaining to the enzyme of interest. Another useful source for this information is the Medical Subject Headings Supplementary Chemical Records, published by the National Library of Medicine (U.S. Department of Health and Human Services, Bethesda,

Table 6.6 Some examples of enzyme common names and their EC designations

Common Name(s)	EC Designation
Cytochrome oxidases (cytochrome *c* oxidase)	EC 1.9.3.1
Prostaglandin G/H synthase (cyclooxygenase)	EC 1.14.99.1
Stromelysin (MMP-3, proteoglycanase)	EC 3.4.24.17
Dihydroorotate dehydrogenase	ED 1.3.99.11
Rhodopsin kinase	EC 2.7.1.125

MD). This volume lists the common names of chemicals and reagents (including enzymes) that are referred to in the medical literature covered by the *Index Medicus* (a source book for literature searching of medically related subjects). Enzymes are listed here under their common names (with cross-references for enzymes having more than one common name) and the EC designation is provided for each. Most college and university libraries carry the *Index Medicus* and will have this supplement available, or one can purchase the supplement directly from the National Library of Medicine. Yet another resource for determining the EC designation of an enzyme is the Enzyme Data Bank, which can be accessed on the Internet.* This data bank provides EC numbers, recommended names, alternative names, catalytic activities, information on cofactor utilization, and associated diseases for a very large collection of enzymes. A complete description of the data bank and its uses can be found in Bairoch (1993).

6.6 SUMMARY

In this chapter we have explored the chemical nature of enzyme catalysis. We have seen that enzymes function by enhancing the rates of chemical reaction by lowering the energy barrier to attainment of the reaction transition state. The active site of enzymes provides the structural basis for this transition state stabilization through a number of chemical mechanisms, including approximation effects, covalent catalysis, general acid/base catalysis, induced strain, and solvent replacement effects. The structural architecture of the enzyme active site further dictates the substrate specificity for reaction. A structural complementarity exists between the enzyme active site and the substrate in its transition state configuration. Several models have been presented to describe this structural complementarity and the interplay of structural forces that dictate enzyme specificity and catalytic efficiency.

REFERENCES AND FURTHER READING

Bairoch, A. (1993) *Nucl. Acid. Res.* **21**, 3155.

Bender, M. L., Bergeron, R. J., and Komiyama, M. (1984) *The Bioorganic Chemistry of Enzymatic Catalysis*, Wiley, New York.

Bruice, T. C., and Lapinski, R. (1958) *J. Am. Chem. Soc.* **80**, 2265.

Cannon, W. R., and Benkovic, S. J. (1998) *J. Biol. Chem.* **273**, 26257.

Carter, P., and Wells, J. A. (1988) *Nature*, **332**, 564.

Fersht, A. R. (1974) *Proc. R. Soc. London B*, **187**, 397.

Fersht, A. (1985) *Enzyme Structure and Mechanism*, Freeman, New York.

Fersht, A. R., and Kirby, A. J. (1967) *J. Am. Chem. Soc.* **89**, 4853, 4857.

*http://192.239.77.6/Dan/proteins/ec-enzyme.html.

Fischer, E. (1894) *Berichte*, **27**, 2985.

Florkin, M., and Stotz, E. H. (1973) *Comprehensive Biochemistry*, Vol. 13, Elsevier, New York.

Goldsmith, J. O., and Kuo, L. C. (1993) *J. Biol. Chem.* **268**, 18481.

Hammes, G. G. (1982) *Enzyme Catalysis and Regulation*, Academic Press, New York.

Hartley, B. S., and Kilby, B. A. (1954) *Biochem. J.* **56**, 288.

Jencks, W. P. (1969) *Catalysis in Chemistry and Enzymology*, McGraw-Hill, New York.

Jencks, W. P. (1975) *Adv. Enzymol.* **43**, 219.

Kirby, A. J. (1980) *Effective molarity for intramolecular reactions*, in *Advances in Physical Organic Chemistry*, Vol. 17, V. Gold and D. Bethel, Eds., Academic Press, New York, pp. 183 ff.

Koshland, D. E. (1958) *Proc. Natl. Acad. Sci. USA* **44**, 98.

Leatherbarrow, R. J., Fersht, A. R., and Winter, G. (1985) *Proc. Natl. Acad. Sci. USA*, **82**, 7840.

Lerner, R. A., Benkovic, S. J., and Schultz, P. G. (1991) *Science*, **252**, 659.

Liao, D., Breddam, K., Sweet, R. M., Bullock, T., and Remington, S. J. (1992) *Biochemistry*, **31**, 9796.

Loewus, F., Westheimer, F., and Vennesland, B. (1953) *J. Am. Chem. Soc.* **75**, 5018.

Menger, F. M. (1992) *Biochemistry*, **31**, 5368.

Murphy, D. J., and Benkovic, S. J. (1989) *Biochemistry*, **28**, 3025.

Pauling, L. (1948) *Nature*, **161**, 707.

Perona, J. J., and Craik, C. S. (1995) *Protein Sci.* **4**, 337.

Schechter, I., and Berger, A. (1967) *Biochem. Biophys. Res. Commun.* **27**, 157.

Schowen, R. L. (1978) in *Transition States of Biochemical Processes* (Grandous, R. D. and Schowen, R. L., Eds.), Chapter 2, Plenum, New York.

Segal, I. H. (1975) *Enzyme Kinetics*, Wiley, New York.

So, O.-Y., Scarafia, L. E., Mak, A. Y., Callan, O. H., and Swinney, D. C. (1998) *J. Biol. Chem.* **273**, 5801.

Storm, D. R., and Koshland, D. E. (1970) *Proc. Natl. Acad. Sci. USA* **66**, 445.

Walsh, C. (1979) *Enzyme Reaction Mechanisms*, Freeman, New York.

Webb, E. C. (1992) *Enzyme Nomenclature*, Academic Press, San Diego, CA.

Wison, C., and Agard, D. A. (1991) *Curr. Opin. Struct. Biol.* **1**, 617.

Wolfenden, R. (1999) *Bioorg. Med. Chem.* **7**, 647.

Yagisawa, S. (1995) *Biochem. J.* **308**, 305.

7

EXPERIMENTAL
MEASURES OF
ENZYME ACTIVITY

Enzyme kinetics offers a wealth of information on the mechanisms of enzyme catalysis and on the interactions of enzymes with ligands, such as substrates and inhibitors. Chapter 5 provided the basis for determining the kinetic constants k_{cat} and K_m from initial velocity measurements taken at varying substrate concentrations during steady state catalysis. The determination of these kinetic constants rests on the ability to measure accurately the initial velocity of an enzymatic reaction under well-controlled conditions.

In this chapter we describe some of the experimental methods used to determine reaction velocities. We shall see that numerous strategies have been developed for following over time the loss of substrate or the appearance of products that result from enzyme turnover. The velocity of an enzymatic reaction is sensitive to many solution conditions, such as pH, temperature, and solvent isotopic composition; these conditions must be well controlled if meaningful data are to be obtained. Controlled changes in these solution conditions and measurement of their effects on the reaction velocity can provide useful information about the mechanism of catalysis as well. Like all proteins, enzymes are sensitive to storage conditions and can be denatured easily by mishandling. Therefore we also discuss methods for the proper handling of enzymes to ensure their maximum catalytic activity and stability.

7.1 INITIAL VELOCITY MEASUREMENTS

7.1.1 Direct, Indirect, and Coupled Assays

To measure the velocity of a reaction, it is necessary to follow a signal that reports product formation or substrate depletion over time. The type of signal that is followed varies from assay to assay but usually relies on some unique physicochemical property of the substrate or product, and/or the analyst's ability to separate the substrate from the product. Generally, most enzyme assays rely on one or more of the following broad classes of detection and separation methods to follow the course of the reaction:

Spectroscopy
Polarography
Radioactive decay
Electrophoretic separation
Chromatographic separation
Immunological reactivity

These methods can be used in *direct assay*: the direct measurement of the substrate or product concentration as a function of time. For example, the enzyme cytochrome c oxidase catalyzes the oxidation of the heme-containing protein cytochrome c. In its reduced (ferrous iron) form, cytochrome c displays a strong absorption band at 550 nm, which is significantly diminished in intensity when the heme iron is oxidized (ferric form) by the oxidase. One can thus measure the change in light absorption at 550 nm for a solution of ferrous cytochrome c as a function of time after addition of cytochrome c oxidase; the diminution of absorption at 550 nm that is observed is a direct measure of the loss of substrate (ferrous cytochrome c) concentration (Figure 7.1).

In some cases the substrate and product of an enzymatic reaction do not provide a distinct signal for convenient measurement of their concentrations. Often, however, product generation can be coupled to another, nonenzymatic,

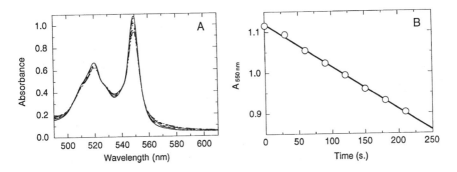

Figure 7.1 (A) Absorption of ferrocytochrome c as a function of time after addition of the enzyme cytochrome c oxidase. As the cytochrome c iron is oxidized by the enzyme, the absorption feature at 550 nm decreases. (B) Plot of the absorption at 550 nm for the spectra in (A), as a function of time. Note that in this early stage of the reaction ($\leqslant 10\%$ of the substrate has been converted), the plot yields a linear relationship between absorption and time. The reaction velocity can thus be determined from the slope of this linear function.

reaction that does produce a convenient signal; such a strategy is referred to as an *indirect* assay. Dihydroorotate dehydrogenase (DHODase) provides an example of the use of an indirect assays. This enzyme catalyzes the conversion of dihydroorotate to orotic acid in the presence of the exogenous cofactor ubiquinone. During enzyme turnover, electrons generated by the conversion of dihydroorotate to orotic acid are transferred by the enzyme to a ubiquinone cofactor to form ubiquinol. It is difficult to measure this reaction directly, but the reduction of ubiquinone can be coupled to other nonenzymatic redox reactions.

Several redox active dyes are known to change color upon oxidation or reduction. Among these, 2,6-dichlorophenolindophenol (DCIP) is a convenient dye with which to follow the DHODase reaction. In its oxidized form DCIP is bright blue, absorbing light strongly at 610 nm. Upon reduction, however, this absorption band is completely lost. DCIP is reduced stoichiometrically by ubiquinol, which is formed during DHODase turnover. Hence, it is possible to measure enzymatic turnover by having an excess of DCIP present in a solution of substrate (dihydroorotate) and cofactor (ubiquinone), then following the loss of 610 nm absorption with time after addition of enzyme to initiate the reaction.

A third way of following the course of an enzyme-catalyzed reaction is referred to as the *coupled* assays method. Here the enzymatic reaction of interest is paired with a second enzymatic reaction, which can be conveniently measured. In a typical coupled assay, the product of the enzyme reaction of interest is the substrate for the enzyme reaction to which it is coupled for convenient measurement. An example of this strategy is the measurement of activity for hexokinase, the enzyme that catalyzes the formation of glucose 6-phosphate and ADP from glucose and ATP. None of these products or substrates provide a particularly convenient means of measuring enzymatic activity. However, the product glucose 6-phosphate is the substrate for the enzyme glucose 6-phosphate dehydrogenase, which, in the presence of $NADP^+$, converts this molecule to 6-phosphogluconolactone. In the course of the second enzymatic reaction, $NADP^+$ is reduced to NADPH, and this cofactor reduction can be monitored easily by light absorption at 340 nm.

This example can be generalized to the following scheme:

$$A \xrightarrow{v_1} B \xrightarrow{v_2} C$$

where A is the substrate for the reaction of interest, v_1 is the velocity for this reaction, B is the product of the reaction of interest and also the substrate for the coupling reaction, v_2 is the velocity for the coupling reaction, and C is the product of the coupling reaction being measured. Although we are measuring C in this scheme, it is the steady state velocity v_1 that we wish to study. To accomplish this we must achieve a situation in which v_1 is rate limiting (i.e., $v_1 \ll v_2$) and B has reached a steady state concentration. Under these conditions B is converted to C almost instantaneously, and the rate of C production

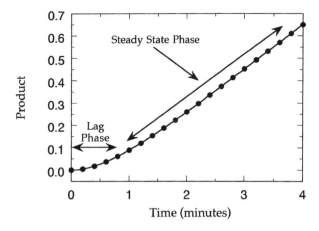

Figure 7.2 Typical data for a coupled enzyme reaction illustrating the lag phase that precedes the steady state phase of the time course.

is a reflection of v_1. The measured rate will be less than the steady state rate v_1, however, until [B] builds up to its steady state level. Hence, in any coupled assay there will be a lag phase prior to steady state production of C (Figure 7.2), which can interfere with the measurement of the initial velocity. Thus to measure the true initial velocity of the reaction of interest, conditions must be sought to minimize the lag phase that precedes steady state product formation, and care must be taken to ensure that the velocity is measured during the steady state phase.

The velocity of the coupled reaction, v_2, follows simple Michaelis–Menten kinetics as follows:

$$v_2 = \frac{V_2[B]}{K_m^2[B]} \tag{7.1}$$

where K_m^2 refers to the Michaelis constant for enzyme E_2, not the square of the K_m. Early in the reaction, v_1 is constant for a fixed concentration of E_1. Hence the rate of B formation is given by:

$$\frac{d[B]}{dt} = v_1 - v_2 = v_1 - \frac{V_2[B]}{K_m^2 + [B]} \tag{7.2}$$

Equation 7.2 was evaluated by integration by Storer and Cornish-Bowden (1974), who showed that the time required for [B] to reach some percentage of its steady state level $[B]_{ss}$ can be defined by the following equation:

$$t_{99\%} = \frac{\phi K_m^2}{v_1} \tag{7.3}$$

Table 7.1 Values of ϕ, for [B] = 99% [B]$_{ss}$, useful in designing coupled assays

v_1/V_2	ϕ
0.1	0.54
0.2	1.31
0.3	2.42
0.4	4.12
0.5	6.86
0.6	11.70
0.7	21.40
0.8	45.50
0.9	141.00

Source: Adapted from Storer and Cornish-Bowden (1974).

where $t_{99\%}$ is the time required for [B] to reach 99% [B]$_{ss}$ and ϕ is a dimensionless value that depends on the ratio v_1/V_2 and v_2/v_1. Recall from Chapter 5 that the maximal velocity V_2 is the product of k_{cat} for the coupling enzyme and the concentration of coupling enzyme [E$_2$]. The values of k_{cat} and K_m for the coupling enzyme are constants that cannot be experimentally adjusted without changes in reaction conditions. The maximal velocity V_2, however, can be controlled by the researcher by adjusting the concentration [E$_2$]. Thus by varying [E$_2$] one can adjust V_2, hence the ratio v_1/V_2, hence the lag time for the coupled reaction.

Let us say that we can measure the true steady state velocity v_1 after [B] has reached 99% of [B]$_{ss}$. How much time is required to achieve this level of [B]$_{ss}$? We can calculate this from Equation 7.3 if we know the values of v_1 and ϕ. Storer and Cornish-Bowden tabulated the ratios v_1/V_2 that yield different values of ϕ for reaching different percentages of [B]$_{ss}$. Table 7.1 lists the values for [B] = 99% [B]$_{ss}$. This percentage is usually considered to be optimal for measuring v_1 in a coupled assay. In certain cases this requirement can be relaxed. For example, [B] = 90% [B]$_{ss}$ would be adequate for use of a coupled assay to screen column fractions for the presence of the enzyme of interest. In this situation we are not attempting to define kinetic parameters, but merely wish a relative measure of primary enzyme concentration among different samples. The reader should consult the original paper by Storer and Cornish-Bowden (1974) for additional tables of ϕ for different percentages of [B]$_{ss}$.

Let us work through an example to illustrate how the values in Table 7.1 might be used to design a coupled assay. Suppose that we adjust the concentration of our enzyme of interest so that its steady state velocity v_1 is 0.1 mM/min, and the value of K_m for our coupling enzyme is 0.2 mM. Let us say that we wish to measure velocity over a 5-minute time period. We wish the lag phase to be a small portion of our overall measurement time, say $\leqslant 0.5$ minute. What value of V_2 would we need to reach these assay conditions? From

Equation 7.3 we have:

$$0.5 = \phi \frac{0.2}{0.1} \tag{7.4}$$

Rearranging this we find that $\phi = 0.25$. From Table 7.1 this value of ϕ would require that $v_1/V_2 < 0.1$. Since we know that v_1 is 0.1 mM/min, V_2 must be > 1.0 mM/min. If we had taken the time to determine k_{cat} and K_m for the coupling enzyme, we could then calculate the concentration of $[E_2]$ required to reach the desired value of V_2.

Easterby (1973) and Cleland (1979) have presented a slightly different method for determining the duration of the lag phase for a coupled reaction. From their treatments we find that as long as the coupling enzyme(s) operate under first-order conditions (i.e., $[B]_{ss} \ll K_m^2$), we can write:

$$[B]_{ss} = v_1 \left(\frac{K_m^2}{V_2} \right) \tag{7.5}$$

and

$$\tau = \frac{K_m^2}{V_2} \tag{7.6}$$

where τ is the lag time. The time required for $[B]$ to approach $[B]_{ss}$ is exponentially related to τ so that $[B]$ is 92% $[B]_{ss}$ at 2.5τ, 95% $[B]_{ss}$ at 3τ, and 99% $[B]_{ss}$ at 4.6τ (Easterby, 1973). Product (C) formation as a function of time (t) is dependent on the initial velocity v_1 and the lag time (τ) as follows:

$$[C] = v_1(t + \tau e^{-t/\tau} - \tau) \tag{7.7}$$

To illustrate the use of Equation 7.6, let us consider the following example. Suppose our coupling enzyme has a K_m for substrate B of 10 μM and a k_{cat} of 100 s^{-1}. Let us say that we wish to set up our assay so as to reach 99% of $[B]_{ss}$ within the first 20 seconds of the reaction time course. To reach 0.99 $[B]_{ss}$ requires 4.6τ (Easterby, 1973). Thus $\tau = 20$ s/4.6 = 4.3 seconds. From rearrangement of Equation 7.6 we can calculate that V_2 needed to achieve this desired lagtime would be 2.30 μM/s. Dividing this by k_{cat}(100 s^{-1}), we find that the concentration of coupling enzyme required would be 0.023 μM or 23 nM.

If more than one enzyme is used in the coupling steps, the overall lag time can be calculated as $\Sigma(K_m^i/V_i)$. For example, if one uses two consecutive coupling enzymes, A and B to follow the reaction of the primary enzyme of interest, the overall lag time would be given by:

$$\tau = \frac{K_m^A}{V_A} + \frac{K_m^B}{V_B} \tag{7.8}$$

Because coupled reactions entail multiple enzymes, these assays present a number of potential problems that are not encountered with direct or indirect assays. For example, to obtain meaningful data on the enzyme of interest in coupled assays, it is imperative that the reaction of interest remain rate limiting under all reaction conditions. Otherwise, any velocity changes that accompany changes in reaction conditions may not reflect accurately effects on the target enzyme. For example, to use a coupled reaction scheme to determine k_{cat} and K_m for the primary enzyme of interest, it is necessary to ensure that v_1 is still rate limiting at the highest values of [A] (i.e., substrate for the primary enzyme of interest).

Use of a coupled assay to study inhibition of the primary enzyme might also seem problematic. The presence of multiple enzymes could introduce ambiguities in interpreting the results of such experiments: for example, which enzyme(s) are really being inhibited? Easterby (1973) points out, however, that using coupled assays to screen for inhibitors makes it relatively easy to distinguish between inhibitors of the primary enzyme and the coupling enzyme(s). Inhibitors of the primary enzyme would have the effect of diminishing the steady state velocity v_1 (Figure 7.3A), while inhibitors of the coupling enzyme(s) would extend the lag phase without affecting v_1 (Figure 7.3B). In practice these distinctions are clear-cut only when one measures product formation over a range of time points covering significant portions of both the lag phase and the steady state phase of the progress curves (Figure 7.3). Rudolph et al. (1979), Cleland (1979), and Tipton (1992) provide more detailed discussions of coupled enzyme assays.

7.1.2   Analysis of Progress Curves: Measuring True Steady State Velocity

In Chapter 5 we introduced the progress curves for substrate loss and product formation during enzyme catalysis (Figures 5.1 and 5.2). We showed that both substrate loss and product formation follow pseudo-first-order kinetics. The full progress curve of an enzymatic reaction is rich in kinetic information. Throughout the progress curve, the velocity is changing as the substrate concentration available to the enzyme continues to diminish. Hence, throughout the progress curve the instantaneous velocity approximates the initial velocity for the instantaneous substrate concentration at a particular time point. Modern computer graphing programs often provide a means of computing the derivative of the data points from a plot of y versus x, hence one could determine the instantaneous velocity ($v = d[P]/dt = -d[S]/dt$) from derivitization of the enzyme progress curve. At each time point for which the instantaneous velocity is determined, the instantaneous value of [S] can be determined, and so the instantaneous velocity can be replotted as a function of [S]. This replot is hyperbolic and can be fit to the Michaelis–Menten equation to extract estimates of k_{cat} and K_m from a single progress curve. This approach and its limitations have been described by several authors (Cornish-Bowden,

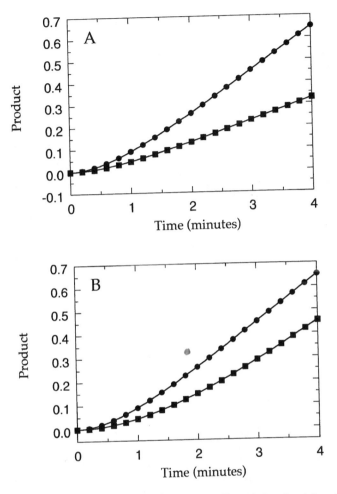

Figure 7.3 Effects of inhibitors on a coupled enzyme reaction: circles, the data points for the enzyme in the absence of inhibitor; squares, the data points when inhibitor is present as some fixed concentration. (A) Effect of an inhibitor of the primary enzyme of interest: the lag phase is unaffected, but the steady state rate (slope) is diminished. (B) Effect of an inhibitor of the coupling enzyme: the lag phase is extended, but the steady state rate is unaffected.

1972; Duggleby and Morrison, 1977; Waley, 1982; Kellershohn and Laurent, 1985; Duggleby, 1985, 1994). The use of full progress curve analysis has not become popular because derivatization of the progress curves was not straight-forward until the use of computers became widespread, and because of some of the associated problems with this method as described in the above-referenced literature.

As we described in Chapter 5, early in the progress curve, the formation of product and the loss of substrate track linearly with time, so that the velocity can be determined from the slope of a linear fit of these early time point data. For the remainder of this chapter our discussions are restricted to measurement at these early times, so that we are evaluating the initial velocity. Nevertheless, it is useful to determine the full progress curve for the enzymatic reaction at least at a few combinations of enzyme and substrate concentrations, to ensure that the system is well behaved, that the reaction goes to completion (i.e., $[P]_\infty = [S]_0$), and that measurements truly are being made during the linear steady state phase of the reaction.

The progress curve for a well-behaved enzymatic reaction should appear as that seen in Figure 5.1. There should be a reasonable time period, early in the reaction, over which substrate loss and product formation are linear with time. As the reaction progresses, one should see curvature and an eventual plateau as the substrate supply is exhausted.

In some situations a lag phase may appear prior to the linear initial velocity phase of the progress curve. This occurs in coupled enzyme assays, as discussed above. Lag phases can be observed for a number of other reasons as well. If the enzyme is stored with a reversible inhibitor present, some time may be required for complete dissociation of the inhibitor after dilution into the assay mixture; hence a lag phase will be observed prior to the steady state (see Chapter 10). Likewise, if the enzyme is stored at a concentration that leads to oligomerization, but only the monomeric enzyme form is active, a lag phase will be observed in the progress curve that reflects the rate of dissociation of the oligomers to monomeric enzyme. Lag phases are also observed when reagent temperatures are not well controlled, as will be discussed in Section 7.4.3.

The linear steady state phase of the reaction may also be preceded by an initial burst of rapid reaction, as we encountered in Chapter 6 for chymotrypsin-catalyzed hydrolysis of p-nitrophenylethyl carbonate (Figure 6.5). *Burst phase kinetics* are observed with some enzymes for several reasons. First, severe product inhibition may occur, so that after a few turnovers, the concentration of product formed by the reaction is high enough to form a ternary ESP complex, which undergoes catalysis at a lower rate than the binary ES complex. Hence, at very early times the rate of product formation corresponds to the uninhibited velocity of the enzymatic reaction, but after a short time the velocity changes to that reflective of the ESP complex. A second cause of burst kinetics is a time-dependent conformational change of the enzyme structure caused by substrate binding. Here the enzyme is present in a highly active form but converts to a less active conformation upon formation of the ES complex. Third, the overall reaction rate may be limited by a slow release of the product from the EP complex. The final, and perhaps the most common, cause for burst kinetics is rapid reaction of the enzyme with substrate to form a covalent intermediate, which undergoes slower steady state decomposition to products.

Enzymes like the serine proteases, for which the reaction mechanism goes through formation of a covalent intermediate, often show burst kinetics because the overall reaction is rate-limited by decomposition of the intermediate species. When this occurs, the intermediate builds up rapidly (i.e., before the steady state velocity is realized) to a concentration equal to that of active enzyme molecules in the sample. For enzymes that demonstrate burst phase kinetics due to covalent intermediate formation, the concentration of active enyzme in a sample can be determined accurately from the intercept value of a linear fit of the data in the steady state portion of the progress curve. For example, referring back to Figure 6.5, we see that the reaction was performed at nominal chymotrypsin concentrations of 8, 16, 24, and 32 μM. Linear fitting of the steady state data for these reactions gave y-intercept values of 5, 10, 15, and 20 μM, respectively. Thus the ratio of the y-intercept value to the nominal concentration of enzyme added is a constant of 0.63, suggesting that about 63% of the enzyme molecules in these samples are in a form that supports catalysis. This method, referred to as *active site titration,* can be a powerful means of determining accurately the active enzyme concentration.

There are several points to consider, however, in the use of the active site titration method. First, one must be sure that the burst phase observed is due to covalent intermediate formation (see Chapter 6 and Fersht, 1985, for methods to establish that the enzyme reaction goes through a covalent intermediate). Second, the y intercepts must be appreciably greater than zero so that they can be measured accurately. This means that the concentration of enzyme required for these assays is much higher than is commonly used for most enzymatic assays. Colorimetric or fluorometric assays typically require micromolar amounts of enzyme to observe a significant burst phase. The amount of enzyme needed can be reduced by use of radiometric detection methods, but even here one typically requires high nanomolar or low microm-olar enzyme concentrations. Finally, the amplitude of the burst does not give a precise measure of the active enzyme concentration if the rates for the burst and the steady state phases are not significantly different. A minimalist scheme for an enzyme that goes through a covalent intermediate is as follows:

$$E + S \underset{k_{-1}}{\overset{k_1}{\rightleftharpoons}} ES \underset{P_1}{\overset{k_2}{\rightarrow}} E - I \overset{k_3}{\rightarrow} E + P_2$$

Fersht (1985) has shown that the burst amplitude (i.e., the y-intercept value from linear fitting of the steady state data: Figure 7.4) can be described as follows:

$$\pi = [E]\left(\frac{k_2}{k_2 + k_3}\right)^2 \tag{7.9}$$

where π is the burst amplitude. From Equation 7.9 we see that if $k_3 \ll k_2$, the rate constant ratio reduces to 1, and so $\pi = [E]$. If however, k_3 is not insignificant in comparison to k_2, we will underestimate the value of $[E]$ from measurement of π. Hence, the active enzyme concentrations determined from

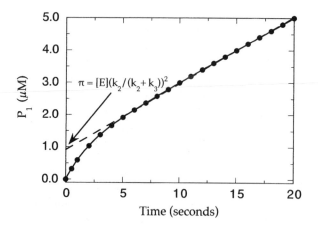

Figure 7.4 Examples of a progress curve for an enzyme demonstrating burst phase kinetics. The dashed line represents the linear fit of the data in the steady state phase of the reaction; the y intercept from this fitting give an estimate of π, as defined here and in the text. The solid curve represents the best fit of the entire time course to an equation for a two-step kinetic process.

active site titration in some cases represent a lower limit on the true active enzyme concentration. This complication can be avoided by use of substrates that form irreversible covalent adducts with the enzyme (i.e., substrates for which $k_3 = 0$). Application of these "suicide substrates" for active site titration of enzymes has been discussed by Schonbaum et al. (1961) and, more recently, by Silverman (1988).

It is also important to ensure that the reaction goes to completion—that is, the plateau must be reached when $[P] = [S]_0$, the starting concentration of substrate. In certain situations, the progress curve plateaus well before full substrate utilization. The enzyme, the substrate, or both, may be unstable under the conditions of the assay, leading to a premature termination of reaction (see Section 7.6 for a discussion of enzyme stability and inactivation). The presence of an enzyme inactivator, or slow binding inhibitor, can also cause the reaction to curve over or stop prior to full substrate utilization (see Chapter 10). In some cases, the product formed by the enzymatic reaction can itself bind to the enzyme in inhibitory fashion. When such *product inhibition* is significant, the buildup of product during the progress curve can lead to premature termination of the reaction. Finally, limitations on the detection method used in an assay may restrict the concentration range over which it is possible to measure product formation. Specific examples are described later in this chapter. In summary, before one proceeds to more detailed analysis using initial velocity measurements, it is important to establish that the full progress curve of the enzymatic reaction is well behaved, or at least that the cause of deviation from the expected behavior is understood.

As we have stated, in the early portion of the progress curve, substrate and product concentrations track linearly with time, and it is this portion of the progress curve that we shall use to determine the initial velocity. Of course, the duration of this linear phase must be determined empirically, and it can vary with different experimental conditions. It is thus critical to verify that the time interval over which the reaction velocity is to be determined displays good signal linearity with time under all the experimental conditions that will be used. Changes in enzyme or substrate concentration, temperature, pH, and other solution conditions can change the duration of the linear phase significantly. One cannot assume that because a reaction is linear for some time period under one set of conditions, the same time period can be used under a different set of conditions.

7.1.3 Continuous Versus End Point Assays

Once an appropriate linear time period has been established, the researcher has two options for obtaining a velocity measurement. First, the signal might be monitored at discrete intervals over the entire linear time period, or some convenient portion thereof. This strategy, referred to as a *continuous assay*, provides the safest means of accurately determining reaction velocity from the slope of a plot of signal versus time.

It is not always convenient to assay samples continuously, however, especially when one is using separation techniques, such as high performance liquid chromatography (HPLC) or electrophoresis. In such cases a second strategy, called *end point* or *discontinuous assay*, is often employed. Having established a linear time period for an assay, one measures the signal at a single specific time point within the linear time period (most preferably, a time point near the middle of the linear phase). The reaction velocity is then determined from the difference in signal at that time point and at the initiation of the reaction, divided by the time:

$$v = \frac{\Delta I}{\Delta t} = \frac{I_t - I_0}{t_{\text{reading}}} \tag{7.10}$$

where the intensity of the signal being measured at time t and time zero is given by I_t and I_0, respectively, and t_{reading} is the time interval between initiation of the reaction and measurement of the signal.

In many instances it is much easier to take a single reading than to make multiple measurements during a reaction. Inherent in the use of end point readings, however, is the danger of assuming that the signal will track linearly with time over the period chosen, under the conditions of the measurement. Changes in temperature, pH, substrate, and enzyme concentrations, as well as the presence of certain types of inhibitor (see Chapter 10) can dramatically change the linearity of the signal over a fixed time window. Hence, end point readings can be misleading. Whenever feasible, then, one should use continuous assays to monitor substrate loss or product formation. When this is impractical, end point readings can be used, but cautiously, with careful controls.

7.1.4   Initiating, Mixing, and Stopping Reactions

In a typical enzyme assay, all but one of the components of the reaction mixture are added to the reaction vessel, and the reaction is started at time zero by adding the missing component, which can either be the enzyme or the substrate. The choice of the initiating component will depend on the details of the assay format and the stability of the enzyme sample to the conditions of the assay. In either case, the other components should be mixed well and equilibrated in terms of pH, temperature, and ionic strength. The reaction should then be initiated by addition of a small volume of a concentrated stock solution of the missing component. A small volume of the initiating component is used to ensure that its addition does not significantly perturb the conditions (temperature, pH, etc.) of the overall reaction volume. Unless the reaction mixture and initiating solutions are well matched in terms of buffer content, pH, temperature, and other factors, the initiating solution should not be more than about 5–10% of the total volume of the reaction mixture.

Samples should be mixed rapidly after addition of the initiating solution, but vigorous shaking or vortex mixing is denaturing to enzymes and should be avoided. Mixing must, however, be complete; otherwise there will be artifactual deviations from linear initial velocities as mixing continues during the measurements. One way to rapidly achieve gentle, but complete, mixing is to add the initiating solution to the side of the reaction vessel as a "hanging drop" above the remainder of the reaction mixture, as illustrated in Figure 7.5. With small volumes (say, $< 50\,\mu$L), the surface tension will hold the drop in place above the reaction mixture. At time zero the reaction is initiated by gently inverting the closed vessel two or three times to mix the solutions. Figure 7.5 illustrates this technique for a reaction taking place in a microcentrifuge tube. It is also convenient to place the initiating solution in the tube cap, which then can be closed, permitting the solutions to be mixed by inversion as illustrated in Figure 7.5. For optical spectroscopic assays (see Section 7.2.1), the reaction can be initiated directly in the spectroscopic cuvette by the same technique, using a piece of Parafilm and one's thumb to seal the top of the cuvette during the inversions.

Regardless of how the reacting and initiating solutions are mixed, the mixing must be achieved in a short period of time relative to the time interval between measurements of the reaction's progress. With a little practice one can use the inversion method just described to achieve this mixing in 10 seconds or less. This is usually fast enough for assays in which measurements are to be made in intervals of 1 minute or longer time. A number of parameters, such as temperature and enzyme concentration (see Section 7.4), can be adjusted to

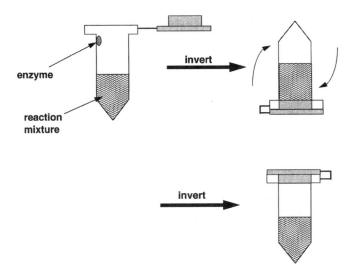

Figure 7.5 A common strategy for initiating an enzymatic reaction in a microcentrifuge tube.

ensure that the reaction velocity is slow enough to allow mixing of the solutions and making of measurements on a convenient time scale. In some rare cases, the enzymatic velocity is so rapid that it cannot be measured conveniently in this way. Then one must resort to specialized rapid mixing and detection methods, such as stopped-flow techniques (Roughton and Chance, 1963; Kyte, 1995); these methods are also used to measure pre–steady state enzyme kinetics, as described in Chapter 5 (Kyte, 1995).

For assays in which samples are removed from the reaction vessel at specific times for measurement, one can start the timer at the point of mixing and make measurements at known time intervals after the initiation point. In many spectroscopic assays, however, one measures changes in absorption or fluorescence with time. For most modern spectrometers, the detection is initiated by pressing a button on an instrument panel or depressing a key on a computer keyboard. Thus to start an assay one must mix the solutions, place the cuvette, or optical cell, in the holder of the spectrometer, and start the detection by pressing the appropriate button. The delay between mixing and actually starting a measurement can be as much as 20 seconds. Thus the time point recorded by the spectrometer as zero will not be the true zero point (i.e., mixing point) of the reaction. Again, with practice one can minimize this delay time, and in most cases the assay can be set up to render this error insignificant.

As we shall see, there should always be two control measurements: one in which all the reaction components except the enzyme are present, and a separate one in which everything but the substrate is present. (In these controls, buffer is added to make up for the volumes that would have been contributed by the enzyme or substrate solutions.) With these two control measurements one can calculate what the absorption or fluorscence should be for the reaction mixture at the true time zero. If the first spectrometer reading (i.e., the time point recorded as time zero by the spectrometer) is significantly different from this calculated value, it is necessary to correct the time points recorded by the spectrometer for the time delay between the start of mixing and the initiation of the detection device. A laboratory timer or stopwatch can be used to determine the time gap.

Many nonspectroscopic assays require measurement times that are long in comparison to the rate of the enzymatic reaction being monitored. Suppose, for example, that we wish to measure the amount of product formed every 5 minutes over the course of a 30-minute reaction and assay for product by an HPLC method. The HPLC measurement itself might take 20–30 minutes to complete. If the enzymatic reaction is continuing during the measurement time, the amount of product produced during specific time intervals cannot be determined accurately. In such cases it is necessary to quench or stop the reaction at a specific time, to prevent further enzymatic production of product or utilization of substrate.

Methods for stopping enzymatic reactions usually involve denaturation of the enzyme by some means, or rapid freezing of the reaction solution. Examples of quenching methods include immersion in a dry ice–ethanol slurry to rapidly freeze the solution, and denaturation of the enzyme by addition of strong acid or base, addition of electrophoretic sample buffer, or immersion in a boiling water bath. In addition to these methods, reagents can be added that interfere in a specific way with a particular enzyme. For example, the activity of many metalloenzymes can be quenched by adding an excess of a metal chelating agent, such as ethylenediaminetetraacetic acid (EDTA).

Three points must be considered in choosing a quenching method for an enzymatic reaction. First, the technique used to quench the reaction must not interfere with the subsequent detection of product or substrate. Second, it must be established experimentally that the quenching technique chosen does indeed completely stop the reaction. Finally, the volume change that occurs upon addition of the quenching reagent to the reaction mixture must be accounted for. Similarly, measurement of product or substrate concentration must be corrected to compensate for the dilution effects of quencher addition.

7.1.5   The Importance of Running Controls

Regardless of the detection method used to follow an enzymatic reaction, it is always critical to perform control measurements in which enzyme and substrate are separately left out of the reaction mixture. These control experiments permit the analyst to correct the experimental data for any time-dependent changes in signal that might occur independent of the action of the enzyme under study, and to correct for any static signal due to components in the reaction mixture. To illustrate these points, let us follow a hypothetical

Table 7.2   Volumes of stock solutions to prepare experimental and control samples for a hypothetical enzyme-catalyzed reaction

	Volumes Added (μL)		
Stock Solutions	Experimental	No Substrate Control	No Enzyme Control
10 × Substrate	100	0	100
10 × Buffer	100	100	100
Distilled water	790	890	800
Enzyme	10	10	0
Total volume	1.0 mL	1.0 mL	1.0 mL

enzymatic reaction by tracking light absorption decrease at some wavelength, as substrate is converted to product. Let us say that there is some low rate of spontaneous product formation in the absence of the enzyme, and that the enzyme itself imparts a small, but measurable absorption at the analytical wavelength. We might set up an experiment in which all the reaction components are placed in a cuvette, and the reaction is initiated by the addition of a small volume of enzyme stock solution. For this illustration, let us say that the reaction mixture is prepared by addition of the volumes of stock solutions listed in Table 7.2. The strategy for preparing the reaction mixtures in Table 7.2 is typical of what one might use in a real experimental situation.

Figure 7.6A illustrates the time courses we might see for the hypothetical solutions from Table 7.2. For our experimental run, the true absorption readings are displaced by about 0.1 unit, as a result of the absorption of the enzyme itself ("No substrate" trace in Figure 7.6A). To correct for this, we subtract this constant value from all our experimental data points. If we were to now determine the slope of our corrected experimental trace, however, we would be overestimating the velocity of our reaction because such a slope would reflect both the catalytic conversion of substrate to product and the spontaneous absorption change seen in our "No enzyme" control trace. To correct for this, we subtract these control data points from the experimental points at each measurement time to yield the difference plot in Figure 7.6B. Measuring the slope of this difference plot yields the true reaction velocity.

As illustrated in Figure 7.6A, the correction for the spontaneous absorption change may appear at first glance to be trivial. However, the velocity measured for the uncorrected data differs from the corrected velocity by more than 10% in this example. In some cases the background signal change is even more substantial. Hence, the types of control measurement discussed here are essential for obtaining meaningful velocity measurements for the catalyzed reaction under study.

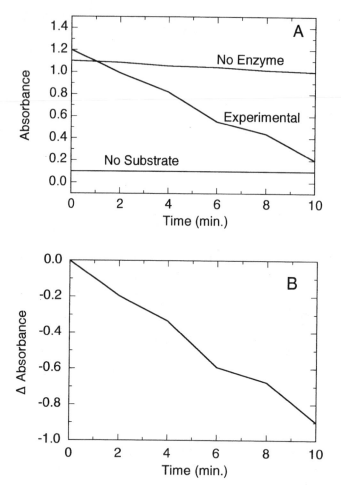

Figure 7.6 The importance of running blank controls. (A) Time courses of absorption for a hypothetical enzymatic reaction (experimental trace), along with two control samples: one with all the reaction mixture components except enzyme; and the other with all of the reaction mixture components except substrate. In this example, the enzyme absorbs minimally at the analytical wavelength, but enough to displace the time zero measurement by about 0.1 absorption unit. (B) The two control readings at each time point have been subtracted from the experimental measurement to yield the true reaction time course of the reaction.

7.2 DETECTION METHODS

A wide variety of physicochemical methods have been used to follow the course of enzymatic reactions. Some of the more common techniques are described here, but our discussion is far from comprehensive. Any signal that differentiates the substrates or products of the reaction from the other components of the reaction mixture can, in principle, form the basis of an enzyme assay; the only limit is the creativity of the investigator.

7.2.1 Assays Based on Optical Spectroscopy

Two very common means of following the course of an enzymatic reaction are absorption spectroscopy and fluorescence spectroscopy. Both methods are based on the changes in electronic configuration of molecules that result from their absorption of light energy of specfic wavelengths. Molecules can absorb electromagnetic radiation, such as light, causing transitions between various energy levels. Energy in the infrared region, for example, can cause transitions between vibrational levels of a molecule. Microwave energy can induce transitions among rotational energy levels, while radiofrequency energy, which forms the basis of NMR spectroscopy, can cause transitions among nuclear spin levels. The energy differences between electronic levels of a molecule are so large that light energy in the ultraviolet (UV) and visible regions of the electromagnetic spectrum is required to induce transitions among these states. If a molecule is irradiated with varying wavelengths of light (of similar intensity), only light of specific wavelengths will be strongly absorbed by the sample. At these wavelengths, the energy of the light matches the energy gap between two electronic states of the molecule, and the light is absorbed to induce such a transition. Figure 7.7 is a hypothetical absorption spectrum of a molecule in which light of wavelength λ_{max} induces an electron redistribution to bring the molecule from its ground π state to an excited π^* state. This process is illustrated as a potential energy diagram in Figure 7.8A.

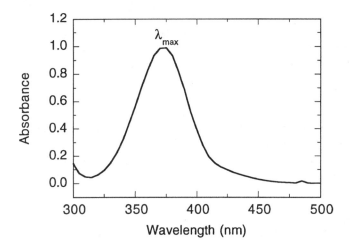

Figure 7.7 A typical absorption spectrum of a molecule with an absorption maximum at 375 nm.

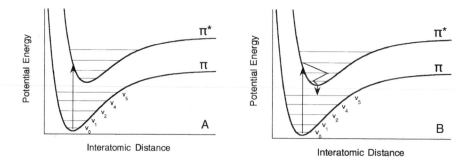

Figure 7.8 Energy level diagrams for (A) a light-induced transition from a π electronic ground state to an excited π^* state of a molecule, and (B) the relaxation of the excited state molecule back to the ground state by photon emission (fluorescence).

7.2.2 Absorption Measurements

The value of absorption spectroscopy as an analytical tool is that the absorption of a molecule at a particular wavelength can be related to the concentration of that molecule in solution, as described by Beer's law:

$$A = \varepsilon c l \tag{7.11}$$

where A is the absorbance of the sample at some wavelength, c is the concentration of sample in molarity units, l is the path length of sample the light beam traverses (in centimeters), and ε is an intrinsic constant of the molecule, known as the extinction coefficient, or molar absorptivity.

Since absorbance is a unitless quantity, ε must have units of reciprocal molarity times reciprocal path length (typically expressed as $M^{-1} \cdot cm^{-1}$ or $mM^{-1} \cdot cm^{-1}$).

Thus if we know the value of ε for a particular molecule, and the path length of the cuvette, we can calculate the concentration of that molecule in a solution by measuring the absorption of that solution. As we have seen in the examples of Figures 7.1 and 7.6, we can use the unique absorption features of a substrate or product of an enzymatic reaction to follow the course of such a reaction. Using Beer's law, we can convert our measured ΔA values at different time points to the concentration of the molecule being followed, and thus report the reaction velocity in terms of change in molecular concentration as a function of time.

For example, let us say that the substrate of the reaction illustrated in Figure 7.6 had an extinction coefficient of $2.5\ mM^{-1} \cdot cm^{-1}$ and that these measurements were made in a cuvette having a path length of 1 cm. We see from Figure 7.6B that after data corrections, the absorption of our reaction mixture changes by 0.89 over the course of 10 minutes, or 0.089/min. Our reaction velocity is thus given as follows:

$$v = -\frac{d[S]}{dt} = -\frac{\Delta A}{\varepsilon l \Delta t} = -\frac{0.89}{2.5 \times 1 \times 10} = -0.0356\,\text{mM/min} \quad (7.12)$$

Note that the units here, molarity per unit time (i.e., moles per liter per unit time) are commonly used in reporting enzyme velocities. Some workers instead report velocity in units of moles per unit time. The two units are easily interconverted by making note of the total volume of the reaction mixture for which the velocity is being measured.

7.2.3 Choosing an Analytical Wavelength

The wavelength used for following enzyme kinetics should be one that gives the greatest difference in absorption between the substrate and product molecules of the reaction. In many cases, this will correspond to the wavelength maximum of the substrate or product molecule. However, when there is significant spectral overlap between the absorption bands of the substrate and product, the most sensitive analytical wavelength may *not* be the same as the wavelength maximum. This concept is illustrated in Figure 7.9. In inspecting the spectra for the substrate and product of the hypothetical enzymatic reaction of Figure 7.9A, note that at the wavelength maximum for each molecule, the other molecule displays significant absorption. The wavelength at which the largest difference in signal is observed can be determined by calculating the difference spectrum between these two spectra, as illustrated in Figure 7.9B. Thus, in our hypothetical example, the wavelength maxima for the substrate and product are 374 and 385 nm, respectively, but the most sensitive wavelengths for following loss of substrate or formation of product would be 362 and 402 nm, respectively.

7.2.4 Optical Cells

Absorption measurements are most commonly performed with a standard spectrophotometer, and the samples are contained in specialized cells known as cuvettes. These specialized cells come in a range of path lengths and are constructed of various optical materials. Disposable plastic cuvettes that hold 1 or 3 mL samples are commercially available. Although these disposable units are very convenient and reduce the chances of sample-to-sample cross-contamination, their use should be restricted to the visible wavelength range (350–800 nm). For measurements at wavelengths less than 350 nm, high quality quartz cuvettes must be used, since both plastic and glass absorb too much light in the ultraviolet. Quartz cuvettes are available from numerous manufacturers in a variety of sizes and configurations. Regardless of the type of cuvette selected, its path length must be known to ensure correct application

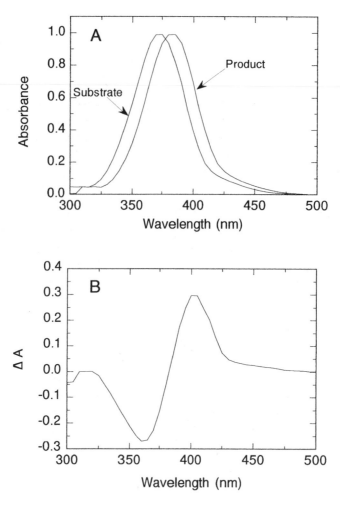

Figure 7.9 Example of the use of difference spectroscopy. (A) Absorption spectra of the substrate and product of some enzymatic reaction. Because of the high degree of spectral overlap between these two molecules, it would be difficult to quantify changes in one component of a mixture of the two species. (B) The mathematical difference spectrum (product minus substrate) for the two spectra in (A). The difference spectrum highlights the differences between the two spectra, making quantitation of changes more straightforward.

of Beer's law to the measurements. Usually, the manufacturer provides this information at the time of purchase. If, for any reason, the path length of a particular cuvette is not known with certainty, it can be determined experimentally by measuring the absorption of any stable chromophoric solution in a standard 1 cm path length cuvette and then measuring the absorption of the same sample in the cuvette of unknown path length. The ratio of the two absorption readings ($A_{unknown}/A_{1\,cm}$) yields the path length of the unknown cuvette in centimeters. A good standard solution for this purpose can be prepared as follows (Haupt, 1952). Weigh out 0.0400 g of potassium chromate and 3.2000 g of potassium hydroxide and dissolve both in 700 mL of distilled water. Transfer to a one-liter volumetric flask and bring the total volume to 1000 mL with additional distilled water. Mix the solution well. The resulting solution will have an absorption of 0.991 at 375 nm in a 1 cm cuvette.

Many workers now perform absorption measurements with 96-well microtiter plate readers. These devices have remarkable sensitivity (many can measure changes in absorption of as little as 0.001) and can greatly increase productivity by allowing up to 96 samples to be assayed simultaneously. Most plate readers are based on optical filters rather than monochromators, hence providing only a finite number of analytical wavelengths for measurements. It is usually not difficult, however, to purchase an optical filter for a particular wavelength and to install it in the plate reader. Because most commercially available microtiter plates are constructed of plastic, plate reader measurements are normally restricted to the visible region of the spectrum.

Recently, however, monochromator-based plate readers have come on the market that allow one to make measurements at any wavelength between 250 and 750 nm. The manufacturers of these instruments also provide special quartz-bottomed plates for performing measurements in the ultraviolet. As with nonstandard cuvettes, it is necessary to determine the path length through which one is making measurements in a microtiter plate well. The path length in this situation will depend on the total volume of reaction mixture in the well. For example, a 200 μL solution in a well of a 96-well plate has an approximate path length of 0.7 cm. The exact path length under given experimental conditions should be determined empirically, as described above.

Filter-based plate readers may show an apparent extinction coefficient of a molecule that differs from corresponding measurements made with a conventional spectrometer. Such discrepancies are due in part to the broader bandwidths of the optical filters used in plate readers. The absorption readings thus survey a wider range of wavelengths than the corresponding spectrometer measurements. Therefore, it is important to use the plate reader under the same conditions as the experimental measurements to determine a standard curve of absorption as a function of chromophore concentration.

Mixing of samples can be problematic in a 96-well plate, since the inversion method (Figure 7.5) cannot conveniently be applied. Instead, many commercial plate readers have automatic built-in shaking devices that allow for sample mixing. Another way to ensure good mixing is to alter the general method just described as follows. The components of the reaction mixture, except for the initiating reagent, are mixed together in sufficient volume for all the wells of the plate to be used. This solution is placed in a multichannel pipet reservoir. The initiating solution is added to the individual wells of the plate, and the remainder of the reaction solution is added to the wells by means of a multichannel pipet. The solutions are mixed by repeatedly pulling up and dispensing the reaction mixture with the pipettor. An entire row of 12 wells can be mixed in this way in less than 10 seconds.

7.2.5 Errors in Absorption Spectroscopy

A common error associated with absorption measurements is deviation from Beer's law. The form of Beer's law suggests that the absorption of a sample will increase linearly with the concentration of the molecule being analyzed, and indeed, this is the basis for the use of absorption spectroscopy as an analytical tool. Experimentally, however, one finds that this linear relationship holds only over a finite range of absorption values. As illustrated in Figure 7.10, absorption readings greater than 1.0 in general should not be trusted to accurately reflect the concentration of analyte in solution. Thus, experiments should be designed so that the maximum absorption to be measured is less than 1.0. With a few preliminary trials, it usually is possible to adjust conditions so that the measurements fall safely below this limit. Additionally, the amount of instrumental noise in a measurement is affected by the overall absorption of the sample. For this reason it is more difficult to measure a small absorption change for a sample of high absorption. Empirically it turns out that the best compromise between minimizing this noise and having a reasonable signal to

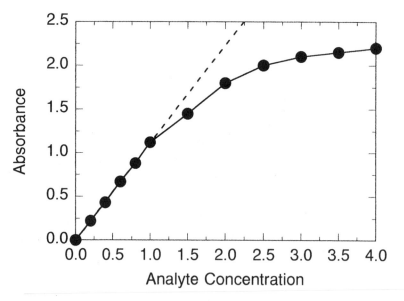

Figure 7.10 Deviation from Beer's law. Over a small concentration range, the absorption at some analytical wavelength tracks linearly with analyte concentration, as expected from Beer's law (Equation 7.11). When the analyte concentration increases to the point at which $A = 1.0$, however, significant deviations from this straight-line behavior begin to appear.

follow occurs when the sample absorption is in the vicinity of 0.5. This is usually a good target absorption for following small absorption changes.

The lamps used to generate the UV and visible light for absorption spectrometers must be given ample time to warm up. The light intensity from these sources varies considerably shortly after the lamps are turned on, but stabilizes after about 30–90 minutes. Since the amount of warm-up time needed to stabilize the lamp output will vary from instrument to instrument, and from lamp to lamp within the same instrument, it is best to determine the required warm-up time for one's own instrument. This is easily done by measuring the signal from a sample of low absorption (say, ~ 0.05–0.1) as a function of time after turning the lamp on, and noting how long it takes for the signal to reach a stable, constant reading.

Another source of error in absorption measurements is sample turbidity. Particulate matter in a solution will scatter light that is detected as increased absorption by the sample. If settling of such particles occurs during kinetic measurements, significant noise in the data may result, and in severe cases there will appear to be an additional kinetic component of the data. The best way to avoid these complications is to ensure that the sample is free of particles by filtering all the solutions through $0.2\,\mu m$ filters or by centrifugation (see Copeland, 1994, for further details.)

7.2.6 Fluorescence Measurements

Light of an appropriate wavelength can be absorbed by a molecule to cause an electronic transition from the ground state to some higher lying excited state, as we have discussed. Because of its highly energetic nature, the excited state is short-lived (excited state lifetimes are typically less than 50 ns), and the molecule must find a means of releasing this excess energy to return to the ground state electronic configuration. Most of the time this excess energy is released through the dissipation of heat to the surrounding medium. Some molecules, however, can return to the ground state by emitting the excess energy in the form of light. Fluorescence, the most common and easily detected of these emissive processes, involves singlet excited and ground electronic states. The energetic processes depicted in Figure 7.8 are characteristic of molecular fluorescence. First, light of an appropriate wavelength is absorbed by the molecule, exciting it to a higher lying electronic state (Figure 7.8A). The molecule then decays through the various high energy vibrational substates of the excited electronic state by heat dissipation, finally, relaxing from its lowest vibrational level to the ground electronic state with release of a photon (Figure 7.8B).

Because of the differences in equilibrium interatomic distances between the ground and excited states, and because of the loss of energy during the decay through the higher energy vibrational substates, the emitted photon is far less energetic than the corresponding light energy required to excite the molecule in the first place. For these reasons, the fluorescence maximum of a molecule is always at a longer wavelength (less energy) than the absorption maximum; this difference in wavelength between the absorption and fluorescence maxima of a molecule is referred to as the Stokes shift. For example, the amino acid tryptophan absorbs light maximally at about 280 nm and fluoresces strongly between 325 and 350 nm (Copeland, 1994). To take advantage of this behavior, fluorescence instruments are designed to excite a sample in a cuvette with light at the wavelength of maximal absorption and detect the emitted light at a different (longer) wavelength. To best detect the emitted light with minimal interference from the excitation light beam, most commercial fluorometers are designed to collect the emitted light at an angle of 90° from the excitation beam path. Thus, unlike cells for absorption spectroscopy, fluorescence cuvettes must have at least two optical quality widows at right angles to one another; all four sides of most fluorescence cuvettes have polished optical surfaces.

The strategies for following enzyme kinetics by fluorescence are similar to those just described for absorption spectroscopy. Many enzyme substrate–product pairs are naturally fluorescent and provide convenient signals with which to follow their loss or production in solution. If these molecules are not naturally fluorescent, it is often possible to covalently attach a fluorescent group without significantly perturbing the interactions with the enzyme under study. Fluorescence measurements offer two key advantages over absorption measurements for following enzyme kinetics. First, fluorescence instruments are very sensitive, permitting the detection of much lower concentration changes in substrate or product. Second, since many fluorophores have large Stokes shifts, the fluorescence signal is typically in an isolated region of the spectrum, where interferences from signals due to other reaction mixture components are minimal.

Fluorescence signals track linearly with the concentration of fluorophore in solution over a finite concentration range. In principle, fluorescence signals should vary with fluorophore concentration by a relationship similar to Beer's law, where the extinction coefficient is replaced by the molar quantum yield (ϕ). In practice, however, it is difficult to calculate sample concentrations by means of applying tabulated values of ϕ to experimental fluorescence measurement. This limitation is in part due to the nature of the instrumentation and the measurements (see Lackowicz, 1983, for more detail). Thus, to convert fluorescence intensity measurements into concentration units, it is necessary to prepare a standard curve of fluorescence signal as a function of fluorophore concentration, using a set of standard solutions for which the fluorophore concentration has been determined independently. The standard curve data points must be collected at the same time as the experimental measurements, however, since day-to-day variations in lamp intensity and other instrumental factors can greatly affect fluorescence measurements.

Sometimes the fluorophore is generated only as a result of the enzymatic reaction, and it is difficult to obtain a standard sample of this molecule for construction of a standard curve. In such cases it may not be possible to report velocity in true concentration units, and units of relative fluorescence per unit time must be used instead. It is still important to quantify this fluorescence relative to some standard fluorescent molecule, to permit comparisons of relative fluorescence measurements from one day to the next and from one laboratory to another. A good standard for this purpose is quinine sulfate. A dilute solution of quinine sulfate in an aqueous sulfuric acid solution can be excited at any wavelength between 240 and 400 nm to yield a strong fluorescence signal that maximizes at 453 nm (Fletcher, 1969). Russo (1969) suggests the following protocol for preparing a quinine sulfate solution as a standard for fluorescence spectroscopy:

- Weigh out 5 mg of quinine sulfate dihydrate and dissolve in 100 mL of 0.1 N H_2SO_4.
- Measure the absorption of the sample at 366 nm, and adjust the concentration with 0.1 N H_2SO_4 so that the solution has an absorption of 0.40 at this wavelength in a 1 cm cuvette.
- Dilute a sample of this solution 1/10 with 0.1 N H_2SO_4 and use the solution to record the fluorescence spectrum.

The relative fluorescence of a sample can then be reported as the fluorescence intensity of the sample at some wavelength, divided by the fluorescence intensity of the quinine sulfate standard at 453 nm, when the same fluorometer is used to excite both sample and standard, at the same wavelength. Of course, both sample and standard measurements must be made under the same set of experimental conditions (monochrometer slit width, lamp voltage, dwell time, etc.), and the second set should be made soon after the first.

7.2.7 Internal Fluorescence Quenching and Energy Transfer

If a molecule absorbs light at the same wavelength at which another molecule fluoresces, the fluorescence from the second molecule can be absorbed by the first molecule, leading to a diminution or *quenching* of the observed fluorescence intensity from the second molecule. (Note that this is only one of numerous means of quenching fluorescence; see Lackowicz, 1983, for a more comprehensive treatment of fluorecence quenching.) The first molecule may then decay back to its ground state by radiationless decay (e.g., heat dissipation), or it may itself fluoresce at some characteristic wavelength. We refer to the first process as quenching because the net effect is a loss of fluorescence intensity. The second situation is described as "resonance energy transfer" because here excitation at the absorption maximum of one molecule leads to fluorescence by the other molecule (Figure 7.11).

Both these processes depend on several factors, including the spatial proximity of the two molecules. This property has been exploited to develop fluorescence assays for proteolytic enzymes based on synthetic peptide substrates. The basic strategy is to incorporate a fluorescent group (the donor)

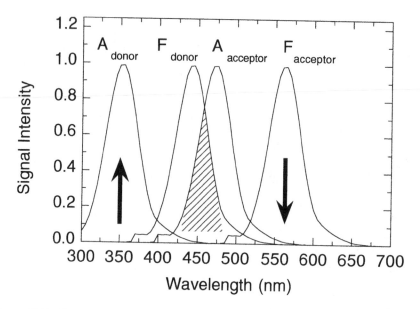

Figure 7.11 Resonance energy transfer. In an energy transfer experiment, the sample is excited with light at the wavelength of the donor absorption band (A_{donor}) to induce fluorescence of the donor molecule at the wavelength indicated by F_{donor}. The absorption band of the acceptor molecule ($A_{acceptor}$) occurs in a wavelength range where it overlaps with the fluorescence band of the donor; the area of overlap between these two features is shown by the hatched region. Because of this spectral overlap (and other factors), the light that would have been emitted as F_{donor} is reabsorbed by $A_{acceptor}$. This indirect excitation of the acceptor molecule can lead to fluorescence by the acceptor at the wavelengths corresponding to $F_{acceptor}$. Experimentally, one excites at the wavelength indicated by the up-pointing arrow, and the fluorescence signal is measured at the wavelength indicated by the down-pointing arrow.

into a synthetic peptide on either the N- or C-terminal side of the scissile peptide bond that is recognized by the target enzyme. A fluorescence quencher or energy acceptor molecule (both referred to hereafter as the acceptor molecule) is also incorporated into the peptide on the other side of the scissile bond. When the peptide is intact, the donor and acceptor molecules are covalently associated and remain apart at a relatively fixed distance, able to energetically interact. Once hydrolyzed by the enzyme, however, the two halves of the peptide will diffuse away from each other, thus eliminating the possibility of any interaction between the donor and acceptor. The observed effect of this hydrolysis will be an increase in the fluorescence from the donor molecule, and, in the case of energy transfer, a concomitant decrease in the fluorescence of the acceptor molecule with exication under the absorption maximum of the donor.

These approaches have been used to follow hydrolysis of peptide substrates for a large variety of proteases (e.g., see Matayoshi et al., 1990; Knight et al., 1992; Knight, 1995; and Packard et al., 1997). Table 7.3 summarizes some donor–acceptor pairs that are commonly used in synthetic peptide substrates

Table 7.3 Donor–acceptor pairs[a] for quenching by resonance energy transfer in peptide substrates of proteolytic enzymes

		Wavelengths (nm)	
Quencher	Fluorophore	Excitation	Emission
Dabcyl	Edans	336	490
Dansyl	Trp	336	350
DNP	Trp	328	350
DNP	MCA	328	393
DNP	Abz	328	420
Tyr(NO$_2$)	Abz	320	420

[a]Dabcyl, 4-(4-dimethylaminophenylazo)benzoic acid; Edans, 5-[(2-amino-ethyl)amino]naphthalene-1-sulfonic acid; Dansyl, (5-dimethylaminonaphtha-lene-1-sulfonyl); DNP, 2,4-dinitrophenyl; MCA, 7-methoxycoumarin-4 acetic acid; Abz, o-aminobenzyl; Tyr(NO$_2$), 3-nitrotyrosine.

for proteases. Another good source for information on donor–acceptor pairs is the Internet site of the Molecular Probes Company,* a company specializing in fluorescence tools for biochemical and biological research.

One example will suffice to illustrate the basic approach used in these assays. Knight et al. (1992) described the incorporation of the fluorescent molecule 7-methoxycoumarine-4-yl acetyl (MCA) at the N-terminus of a peptide designed to be a substrate for the matrix metalloprotease stromelysin; then, immediately after the scissile Gly-Leu peptide bond that is hydrolyzed by the enzyme, the quencher N-3-(2,4-dinitrophenyl)-L-2,3-diaminopropionyl (DPA) was incorporated as well. The complete peptide sequence is:

MCA-Pro-Leu-*Gly-Leu*-DPA-Ala-Arg-NH$_2$

MCA absorbs maximally at 328 nm and fluoresces maximally at 393 nm. The DPA group has a strong absorption band at 363 nm with a prominent shoulder at 410 nm. This shoulder overlaps with the fluorescence band of MCA and leads to significant fluorescence quenching; a 1 μM solution of MCA-Pro-Leu (the product of enzymatic hydrolysis) was found to be 130 times more fluorescent that a comparable solution of the MCA-Pro-Leu-Gly-Leu-DPA-Ala-Arg-NH$_2$ with excitation and emission at 328 and 393 nm, respectively (Knight et al., 1992). Enzymatic hydrolysis of this peptide results in separation of the MCA and DPA groups, hence a large increase in MCA fluorescence. This fluorescence increase could be followed over time as a measure of the reaction velocity, allowing the investigators to establish the values of k_{cat}/K_m of this substrate for several members of the matrix metalloprotease family. This

*www.probes.com/handbook

assay was used recently to determine the potency of potential inhibitors of stromelysin by measuring the effects of the inhibitors on the initial velocity of the enzyme reaction (Copeland et al., 1995).

Recently, fluorescence resonance energy transfer (FRET) has been applied to the study of enzymatic group transfer reactions, and to the study of protein–protein interactions in solution. Space does not permit a review of these applications. The interested reader is referred to the online handbook from the Molecular Probes Company (cited in an earlier footnote) for more information and literature examples of biochemical applications of FRET technology.

7.2.8 Errors in Fluorescence Measurements

Most of the caveats described for absorption spectroscopy hold for fluorescence measurements as well. Samples must be free of particulate matter, since light scattering is a severe problem in fluorescence. Many of the commonly used fluorophores emit light in the visible region but must be excited at wavelengths in the near ultraviolet, necessitating the use of quartz cuvettes for these measurements. Also, any fluorescence due to buffer components and so on must be measured and corrected for to ensure that meaningful data are obtained.

In addition to these more common considerations are several sources of error unique to fluorescence measurements. First, many fluorescent molecules are prone to photodecomposition after long exposure to light. Hence, fluorescent substrates and reagents should be stored in amber glass or plastic, and the containers should be wrapped in aluminum foil to minimize exposure to environmental light. Second, the quantum yield of fluorescence for any molecule is highly dependent on sample temperature. We shall see shortly that temperature affects enzyme kinetics directly, but this is distinct from the general influence of temperature on fluorescence intensity. In general the fluorescence signal increases with decreasing temperature, as competing nonradiative decay mechanisms for return to the ground state become less efficient. Hence, good temperature control of the sample must be maintained. Most commercial fluorometers provide temperature control by means of jacketed sample holders that attach to circulating water baths.

Finally, a major source of error in fluorescence measurements is light absorption by the sample at high concentrations. Individual molecules in a sample may be excited by the excitation light beam and caused to fluoresce. To be detected, these emitted photons must traverse the rest of the sample and escape the cuvette to impinge on the surface of the detection device (typically a photomultiplier tube or diode array). Any such photon will be lost from detection, however, if before escaping the sample it encounters another molecule that is capable of absorbing light at that wavelength. As the sample concentration increases, the likelihood of such encounters and instances of light reabsorption increases exponentially. This phenomenon, referred to as the *inner*

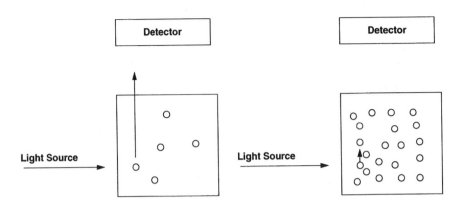

Figure 7.12 Schematic diagram illustrating the inner filter effect. When a dilute sample (left) of a fluorescent molecule is excited at an appropriate wavelength, a detector stationed at 90° relative to the excitation source will detect the emitted light that emerges from the sample container. If, however, the sample is very concentrated, emitted light from one molecule in a sample may encounter and be reabsorbed by another molecule before emerging from the sample compartment (right). These reabsorbed photons, of course, will not be detected. The likelihood of this self-absorption, or inner filter effect, increases with increasing sample concentration.

filter effect (Figure 7.12), can dramatically reduce the fluorescence signal observed from a sample. Consider Figure 7.13, which plots the apparent fluorescent product yield after a fixed amount of reaction time as a function of substrate concentration for the fluorogenic MCA/DPA peptide described in Section 7.2.7 in an assay for stromelysin activity. Instead of the rectangular hyperbolic fit expected from the Henri–Michaelis–Menten equation (Chapter 5), we observe an initial increase in fluorescence yield with increasing substrate, followed by a rapid diminution of signal as the substrate concentration is further increased. At first glance, this behavior might appear to be the result of substrate or product inhibition, as described in Chapter 5. In this case, however, the loss of fluorescence at higher substrate concentrations is an artifact of the inner filter effect. This can be verified by remeasuring the fluorescence of the higher substrate samples after a large dilution with buffer. If, for example, the sample were diluted 20-fold with buffer, the observed fluorescence would not be 20-fold less than that of the undiluted sample; rather, it would show much higher fluorescence than expected on the basis of the dilution factor.

The inner filter effect can be corrected for if the absorption of the sample is known at the excitation and emission wavelengths used in the fluorescence measurement. The true, or corrected fluorescence F_{corr} can be calculated from the observed fluorescence F_{obs} as follows (Lackowicz, 1983):

$$F_{corr} = F_{obs} \times 10^{(A_{ex} \times A_{em})/2} \tag{7.13}$$

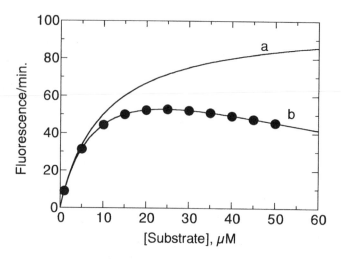

Figure 7.13 Errors in enzyme kinetic measurements due to fluorescence inner filter effects: the rate of fluorescence from a fluorescent peptide substrate of stromelysin is plotted as a function of substrate concentration. Instead of the rectangular hyperbolic behavior expected from the Henri–Michaelis–Menten equation (curve a), we see a diminution of the expected signal at high substrate concentrations (curve b). One might interpret this result as indicating substrate inhibition. In this case, however, the deviation is due to the inner filter effects that become significant at high substrate concentrations. The correct interpretation can be reached by measuring the fluorescence of the higher substrate sample at several dilutions, as discussed in the text.

where A_{ex} and A_{em} are the sample absorptions at the excitation and emission wavelengths, respectively. This correction works only over a limited sample absorption range. If the sample absorption is greater than about 0.1, the correction will not be adequate. Hence, a good rule of thumb is to begin with samples that have absorption values of about 0.05 at the excitation wavelength. The sample concentration can be adjusted from this starting point to optimize the signal-to-noise ratio, with care taken to not introduce a significant inner filter effect.

7.2.9 Radioisotopic Measurements

The basic strategy for the use of radioisotopes in enzyme kinetic measurements is to incorporate into the structure of the substrate a radioactive species that will be retained in the product molecule after catalysis. Using an appropriate technique for separating the substrate from the product (see Section 7.3 on separation methods), one can then measure the amount of radioactivity in the substrate and product fractions, and thus quantify substrate loss and product production. Most of the isotopes that are used commonly in enzyme kinetic measurements decay through emission of β particles (Table 7.4). The decay

Table 7.4 **Properties of radioisotopes that are commonly used in enzyme kinetic assays**

Isotope	Decay Process	Half-life
Carbon-14	$_6^{14}C \rightarrow _{-1}\beta^0 + _7^{14}N$	5700 years
Phosphorus-32	$_{15}^{32}P \rightarrow _{-1}\beta^0 + _{16}^{32}S$	14.3 days
Sulfur-35	$_{16}^{35}S \rightarrow _{-1}\beta^0 + _{17}^{35}Cl$	87.1 days
Tritium	$_1^3H \rightarrow _{-1}\beta^0 + _2^3He$	12.3 years

process follows first-order kinetics, and the loss (or disintegration) of the starting material is thus associated with a characteristic half-life for the parent isotope. The standard unit of radioactivity is the curie (Ci), which originally defined the rate at which 1 gram of radium-226 decays completely. Relating this to other isotopes, a more useful working definition of the curie is that quantity of any substance that decays at a rate of 2.22×10^{12} disintegrations per minute (dpm).

Solutions of p-terphenyl or stilbene, in xylene or toluene, will emit light when in contact with a radioactive solute. This light emission, known as *scintillation*, is most commonly measured with a scintillation counter, an instrument designed around a photomultiplier tube or other light detector. Radioactivity on flat surfaces, such as thin-layer chromatography (TLC) plates and gels can be measured by scintillation counting after the portion of the surface containing the sample has been scraped or cut out and immersed in scintillation fluid.

Another common means of detecting radioactivity on such surfaces entails placing the surface in contact with a sheet of photographic film. The radioactivity darkens the film, making a permanent record of the location of the radioactive species on the surface. This technique, called *autoradiography*, is one of the oldest methods known for detecting radioactivity. Today computer-interfaced phosphor imaging devices also are commonly used for locating and quantifying radioactivity on two-dimensional surfaces (dried gels, TLC plates, etc.).

Radioactivity in a sample is quantified by measuring the dpm's of a sample using one of the methods just described. Since, however, no detector is 100% efficient, any instrumental reading obtained experimentally will differ from the true dpm of the sample. The experimental units of radioactivity are referred to as counts per minute (cmp's: events detected or counted by the instrument per minute). For example, a 1 μCi sample would display 2.22×10^6 dpm. If the detector used to measure this sample had an efficiency of 50%, the experimental value obtained would be 1.11×10^6 cpm. To convert this experimental reading into true dpm's, it would be necessary to measure a standard sample of the isotope of interest, of known dpm's. This information would permit the calibration of the efficiency of the instrument and the ready conversion of the cpm values of samples into dpm units.

When radiolabeled substrates are used in enzyme kinetic studies, the labeled substrate is usually mixed with "cold" (i.e., unlabeled) substrate to achieve a particular total substrate concentration without having to use high quantities of radioactivity. It is important, however, to quantify the proportion of radiolabeled molecules in the substrate sample. Quantification is commonly expressed in terms of the specific radioactivity of the sample (*Note:* "Specific activity" in this case refers to the radioactivity of the sample and should not be confused with the specifc activity of an enzyme sample, which is defined later.) Specific radioactivity is given in units of radioactivity per mass or molarity of the sample. Common units of specific radioactivity include dpm/μmol and μCi/mg. With the specific radioactivity of a substrate sample defined, one can easily convert into velocity units radioactivity measurements taken during an enzymatic reaction.

A worked example will illustrate the foregoing concept. Suppose that we wished to study the conversion of dihydroorotate to orotic acid by the enzyme dihydroorotate dehydrogenase. Let us say that we have obtained a ^{14}C-labeled version of the substrate dihydroorotate and have mixed it with cold dihydroorotate to prepare a stock substrate solution with a specific radioactivity of 1000 cpm/nmol. The final concentration is 1 mM substrate in a reaction mixture with a total volume of 100 μL. Let us say that we initiate the reaction with enzyme and allow the reaction mixture to incubate at 37°C. Every 10 minutes we remove 10 μL of the reaction mixture and add it to 10 μL of 6 N HCl to denature the enzyme, thus stopping the reaction. The total 20 μL is then spotted onto a TLC plate and the product is separated from the substrate. Let us say that we scraped the product spot from the TLC plate and measured the radioactivity by scintillation counting (in this hypothetical assay, a control sample of 1 nmol of substrate in the reaction mixture buffer is spotted onto the TLC plate, the substrate spot is scraped off and counted, and the reading is 1000 cpm). Table 7.5 and Figure 7.14 show the results of this sequence of steps.

From Figure 7.14 we see that the slope of our plot of cpm versus time is 100 cpm/min. Since our substrate had a specific radioactivity (SRA) of 1000 cpm/nmol, this slope value can be directly converted into a velocity value of 0.1 nmol product/min. Since the volume of reaction mixture spotted onto the TLC plate per measurement was 10 μL (i.e., 1×10^{-5} L), the velocity in molarity units is obtained by dividing the velocity in nanomoles product per minute by 1×10^{-5} μL to yield a velocity of 10 μM/min. These calculations are summarized as follows.

$$\frac{\text{slope}}{\text{SRA}} = \frac{\text{cpm/min}}{\text{cpm/nmol}} = \frac{\text{product mass}}{\text{unit time}} = \frac{\text{nmol}}{\text{min}}$$

$$\frac{\text{nmol/min}}{\text{L}} = \frac{\text{product mass/unit time}}{\text{reaction volume}} = \frac{\text{molarity}}{\text{unit time}} = \mu\text{M/min}$$

Table 7.5 Results of a hypothetical reaction of dihydroorotate dehydrogenase with [¹⁴C]dihydroorotate substrate

Incubation Time (min)	Radioactivity (cpm)
10	980
20	2010
30	3050
40	3900
50	5103
60	5952

As the example illustrates, good bookkeeping is essential in these assays. The amount of total substrate used will be dictated by the purpose of the experiment and its K_m for the enzyme. The specific radioactivity, on the other hand, should be adjusted to ensure that the amount of radioactivity used is the minimum that will provide good signal-over-background readings. Guidelines for sample preparation using different radioisotopes can be found in the reviews by Oldham (1968, 1992). The other point illustrated in our example is that good postreaction separation of the labeled substrate and product molecules is critical to the use of radiolabels for following enzyme kinetics.

Radiolabeled substrates are commonly used in conjunction with chromatographic and electrophoretic separation methods. When the substrate and/or the product is a protein, as in some assays for kinases and proteases, bulk precipitation or capture on nitrocellulose membranes can be used to separate

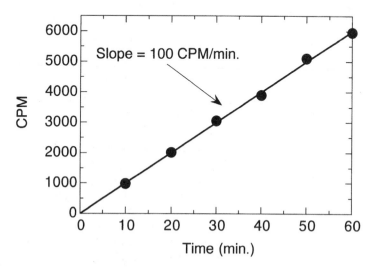

Figure 7.14 Radioassay for dihydroorotate dehydrogenase, measuring the incorporation of ¹⁴C into the product, orotic acid.

the macromolecule from the other solution components. Some of these methods are discussed separately in Section 7.3.1.

7.2.10   Errors in Radioactivity Measurements

Aside from errors associated with bookkeeping, the most commonly encountered cause of inaccurate radioactivity measurements is self-absorption. When the separation method used in conjunction with the assay involves a solid separation medium, such as paper or thin-layer plate chromatography, gel electrophoresis, or capture on activated charcoal, the solid material in the sample may absorb some of the emitted radiation, preventing the signals from reaching the detection device. This self-absorption is best corrected for by measuring all samples and standards at a constant density in terms of milligrams of material per milliliter. Segel (1976) suggests using an inert material such a gelatin to adjust the density of all samples for this purpose. Because scintillation counting measures light emission, the same interferences discussed for fluorescence measurements can occur. In particular, if the sample is highly colored, quenching of the signal due to the equivalent of an inner filter effect may be observed When possible, this should be correct for by adjusting the optical density of the samples and standards with a similarly colored inert material (Segel, 1976).

7.2.11   Other Detection Methods

Absorption, fluorescence, and radioactivity are by far the most common means of following enzyme kinetics, but a wide variety of other techniques have been utilized as well. Immunologic detection, for example, has been applied to follow proteolytic cleavage of a protein substrate by Western blotting, using antibodies raised against that protein substrate. Recently, antibodies have been developed that react exclusively with the phosphorylated forms of peptides and proteins; these reagents have been widely used to follow the enzymatic activity of the kinases and phosphatases using Western and dot blotting as well as ELISA-type assays. Reviews of immunologic detection methods can be found in Copeland (1994) and in Harlow and Lane (1988).

Polarographic methods have also been used extensively to follow enzyme reactions. Many oxidases utilize molecular oxygen during their turnover, and the accompanying depletion of dissolved O_2 from the solutions in which catalysis occurs can be monitored with an O_2-specific electrode. Very sensitive pH electrodes can be used to follow proton abstraction or release into solution during enzyme turnover. Enzymes that perform redox chemistry as part of their catalytic cycle can also be monitored by electrochemical means. Reviews of these methods can be found in the text by Eisenthal and Danson (1992).

The variety of detection methods that have been applied to enzyme activity measurements is too broad to be covered comprehensively in any one volume. Our discussion should provide the reader with a good overview of the more common techniques employed in this field. The references given can provide more in-depth accounts. Another very good source for new and interesting enzyme assay methods, the journal *Analytical Biochemistry* (Academic Press), has historically been a repository for papers dealing with the development and improvement of enzyme assays. Finally, the series *Methods in Enzymology* (Academic Press) comprises volumes dedicated to in-depth reviews of varying topics in enzymology. This series details assay methods for many of the enzymes one is likely to work with and very frequently will indicate at least an assay for a related enzyme that can serve as a starting point for development of an individual assay method.

7.3 SEPARATION METHODS IN ENZYME ASSAYS

For many enzyme assays the detection methods described thus far are applicable only after the substrate or product has been separated from the rest of the reaction mixture components, as is the case when the method of detection would not, by itself, discriminate between the analyte and other species. For example, if the optical properties of the substrate and products of an enzymatic reaction are similar, measuring the spectrum of the reaction mixture alone will not provide a useful means of monitoring changes in the concentration of the individual components. This section briefly describes some of the common separation techniques that are applied to enzyme kinetics. These techniques are usually combined with one of the detection methods already covered to develop a useful assay for the enzyme of interest.

7.3.1 Separation of Proteins from Low Molecular Weight Solutes

A number of assay strategies involve measuring the incorporation of a radioactive or optical label into a protein substrate, or release of a labeled peptide fragment from the protein. For these assays it is convenient to separate the protein from the bulk solution prior to detection. This can be accomplished in several ways (see also Chapter 4, Section 4.7). Proteins can be precipitated out of solution by addition of strong acids or organic denaturants, followed by centrifugation. A 10% solution of trichloroacetic acid (TCA) is commonly used for this purpose (see Copeland, 1994, for details). For dilute protein solutions ($< 5 \mu g$ total), the TCA is often supplemented with the detergent deoxycholate (DOC) to effect more efficient precipitation. Proteins can also be precipitated by high concentrations of ammonium sulfate; most proteins precipitate from solution when the ammonium sulfate concentration is at 80% of saturation. Organic solvents such as acetone, acetonitrile, methanol, or some combination of these solvents also are used to denature and precipitate proteins. For example, mixing $100 \mu L$ of an aqueous protein solution with $900 \mu L$ of a 1:1 acetone/acetonitrile mixture will precipitate most proteins with good efficiency. In addition to these general precipitation methods, specific proteins can be separated from solution with an immobilized antibody (e.g., an antibody linked directly to an agarose bead, or indirectly to a Protein A or Protein G bead) that has been raised against the target protein (Harlow and Lane, 1988).

Proteins also can be separated from low molecular weight solutes by selective binding to nitrocellulose membranes. Nitrocellulose and certain other membrane materials bind proteins strongly, while allowing the other components of the solution to pass. Hence, one can capture the protein molecules in a solution by fitration or centrifugation through a nitrocellulose membrane. For example, many kinase assays are based on the enzymatic incorporation of ^{32}P into a protein substrate. After incubation, the protein substrate is captured on a disk filter of nitrocellulose. After the filter has been washed to remove adventitiously bound ^{32}P, the radioactivity that is retained on the filter can be measured by scintillation counting. Of course the kinase, being a protein itself, is also captured on the nitrocellulose by this method. Since, however, the mass of enzyme in a typical assay is very small relative to that of the protein substrate, the background due to any radiolabel on the enzyme is insignificant and can be subtracted out by performing the appropriate control measurements.

A similar strategy can be used with nominal molecular weight cutoff filters. These filters, which come in a variety of formats, are constructed of a porous material whose pore size distribution permits passage only of molecules having a molecular weight below some critical value; larger molecular weight species, such as proteins, are retained by these filters. One word of caution is in order with regard to these filters: the molecular weight cutoffs quoted by the manufacturers represent the median value of a normal distribution of filtrate molecular weights; thus to avoid significant losses, it is best to use a filter that has a much lower molecular weight cutoff than the molecular weight of the protein being studied. Manufacturers' descriptions of the individual filters should be carefully read before the products are put to use.

Another method for separating proteins from low molecular weight molecules is size exclusion chromatography. Small disposable size exclusion columns, commonly referred to as *desalting columns*, are commercially available for removing low molecular weight solutes from protein solutions. The resins used in these columns are chosen to ensure that macromolecules, such as proteins, elute at the void volume of the column; low molecular weight solutes, such as salts, elute much later. Desalting columns typically are run by gravity, since the large molecular weight differences between the proteins and small solutes allow for separation without the need for high chromatographic resolution.

These and other methods for separating macromolecules from low molecular weight solutes have been described in greater detail in Copeland (1994) and references therein.

7.3.2 Chromatographic Separation Methods

The three most commonly used chromatographic separation methods in modern enzymology laboratories are paper chromatography, TLC, and HPLC. Before HPLC instrumentation became widely available, the paper and thin-layer modes of chromatography were very commonly used for the separation of low molecular weight substrates and products of enzymatic reactions. Today these methods have largely been replaced by HPLC. There is one exception, however: separations involving radiolabeled low molecular weight substrates and products. Since paper and TLC separating media are disposable, and the separation can be performed in a restricted area of the laboratory, these methods are still preferable for work involving radioisotopes.

The theory and practice of paper chromatography and TLC will be familiar to most readers from courses in general and organic chemistry. Basically, separation is accomplished through the differential interactions of molecules in the sample with ion exchange or silica-based resins that are coated onto paper sheets or plastic or glass plates. A capillary tube is used to spot samples onto the medium at a marked location near one end of the sheet, which is placed in a developing tank with some solvent system (typically a mixture of aqueous and organic solvents) in contact with the end of the sheet closest to the spotted samples (Figure 7.15, steps 1 and 2). The tank is sealed, and the solvent moves up the sheet through capillary action, bringing different solutes in the sample along at different rates depending on their degree of interaction with the stationary phase media components. After a fixed time the sheet is removed from the tank and dried. The locations of solutes that have migrated during the chromatography are observed by autoradiography, by illuminating the sheet with ultraviolet light, or by spraying the sheet with a chemical (e.g., ninhydrin) that will react with specific solutes to form a colored spot (Figure 7.15, step 3). The spot locations are then marked on the sheet, and the spots can be cut out or scraped off for counting in a scintillation counter. Alternatively, the radioactivity of the entire sheet can be quantified by two-dimensional radioactivity scanners, as described earlier.

In our discussion of radioactivity assays, we used the example of a TLC-based assay for following the conversion of [^{14}C]dihydroorotate to [^{14}C]-orotic acid by the enzyme dihydroorotate dehydrogenase. Figure 7.16 shows the separation of these molecules on TLC and their detection by autoradiography. This figure and the example given in Section 7.2.9 well illustrate the use of TLC-based assays. More complete descriptions of the uses of paper chromatography and TLC in enzyme assays can be found in the reviews by Oldham (1968, 1977).

HPLC has been used extensively to separate low molecular weight substrates and products, as well as the peptide-based substrates and products of proteolytic enzymes. The introduction of low compressibility resins, typically based on silica, has made it possible to run liquid chromatography at greatly elevated pressures. At these high pressures (as much as 5000 psi) resolution is greatly enhanced; thus much faster flow rates can be used, and the time required for a chromatographic run is shortened. With modern instrumentation, a typical HPLC separation can be performed in less than 30 minutes. The three most commonly used separation mechanisms used in enzyme assays are reversed phase, ion exchange, and size exclusion HPLC.

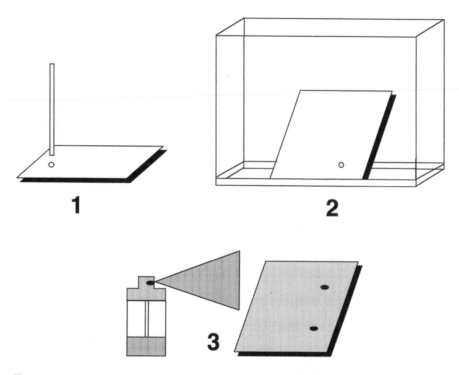

Figure 7.15 Schematic diagram of a TLC-based enzyme assay. In step 1 a sample of reaction mixture is spotted onto the TLC plate. Next the plate is dried and placed in a development tank (step 2) containing an appropriate mobile phase. After the chromatography, the plate is removed from the tank and dried again. Locations of substrate and/or product spots are then determined by, for example, spraying the plate with an appropriate visualizing stain (step 3), such as ninhydrin.

In reversed phase HPLC separation is based on the differential interactions of molecules with the hydrophobic surface of a stationary phase based on alkyl silane. Samples are typically applied to the column in a polar solvent to maximize hydrophobic interactions with the column stationary phase. The less polar a particular solute is, the more it is retained on the stationary phase. Retention is also influenced by the carbon content per unit volume of the stationary phase. Hence a C_{18} column will typically retain nonpolar molecules more than a C_8 column, and so on. The stationary phase must therefore be selected carefully, based on the nature of the molecules to be separated. Molecules that have adhered to the stationary phase are eluted from the column in solvents of lower polarity, which can effectively compete with the analyte molecules for the hydrophobic surface of the stationary phase. Typically methanol, acetonitrile, acetone, and mixtures of these organic solvents with water are used for elution. Isocratic and gradient elutions are both commonly used, depending on the details of the separation being attempted.

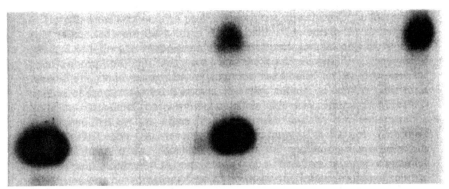

Figure 7.16 Autoradiograph of a TLC plate demonstrating separation of ^{14}C-labeled dihydro-orotate and orotic acid, the substrate and product of the enzyme dihydroorotate dehydrogenase: left lane contained, [^{14}C]dihydroorotate; right lane, [^{14}C]orotic acid; middle lane, a mixture of the two radiolabeled samples (demonstrating the ability to separate the two components in a reaction mixture).

A typical reversed phase separation might involve application of the sample to the column in 0.1% aqueous trifluoroacetic acid (TFA) and elution with a gradient from 100% of this solvent to 100% of a solvent composed of 70% acetonitrile, 0.085% TFA, and water. As the percentage of the organic solvent increases, the more tightly bound, hydrophobic molecules will begin to elute. As the various molecules in the sample elute from the column, they can be detected with an in-line absorption or fluorescence detector (other detection methods are used, but these two are the most common). The detector response to the elution of a molecule will produce on the strip chart a Gaussian–Lorentzian band of signal as a function of time. The length of time between application of the sample to the column and appearance of the signal maximum, referred to as the *retention time*, is characteristic of a particular molecule on a particular column under specified conditions (Figure 7.17).

To quantify substrate loss or product formation by HPLC, one typically measures the integrated area under a peak in the chromatograph and compares it to a calibration curve of the area under the peak as a function of mass for a standard sample of the analyte of interest. Let us again use the reaction of dihydroorotate dehydrogenase as an example. Both the substrate, dihyroorotate, and the product, orotic acid, can be purchased commercially in high purity. Ittarat et al. (1992) developed a reversed phase HPLC assay for following dihydroorotate dehydrogenase activity based on separation of dihydro-orotate and orotic acid on a C_{18} column using isocratic elution with a mixed mobile phase (water/buffer/methanol) and detection by absorption at 230 nm. When a pure sample of dihydroorotate (DHO) was injected onto this column and eluted as described earlier, the resulting chromatograph displayed a single peak that eluted 4.9 minutes after injection. A pure sample of orotic acid (OA), on the other hand, displayed a single peak that eluted after 7.8 minutes under

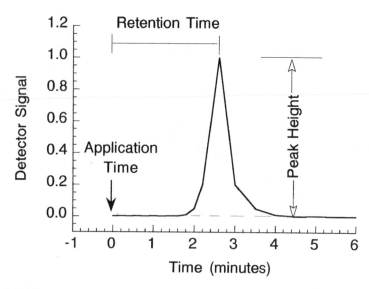

Figure 7.17 Typical signal from an HPLC chromatograph of a molecule. The sample is applied to the column at time zero and elutes, depending on the column and mobile phase, after a characteristic retention time. The concentration of the molecule in the sample can be quantified by the integrated area under the peak, or from the peak height above baseline, as defined in this figure.

the same conditions. Using these pure samples, these workers next measured the area under the peaks for injections of varying concentrations of DHO and OA and, from the resulting data constructed calibration curves for each of these analytes.

Note that the area under a peak will correlate directly with the mass of the analyte injected onto the column. Hence calibration curves are usually constructed with the y axis representing integrated peak area in some units of area [mm^2, absorption units (AU), etc.] and the x axis representing the injected mass of analyte in nanograms, micrograms, nanomoles, and so on. Since the volume of sample injected is known, it is easy enough to convert these mass units into standard concentration units. In this way, Ittarat et al. (1992) determined that the area under the peaks tracks linearly with concentration for both DHO and OA over a concentration range of 0–200 μM. With these results in hand, it was possible to then measure the concentrations of substrate (DHO) and product (OA) in samples of a reaction mixture containing dihydroorotate dehydrogenase and a known starting concentration of substrate, as a function of time after initiating the reaction. From a plot of DHO or OA concentration as a function of reaction time, the initial velocity of the reaction could thus be determined.

With modern HPLC instrumentation, integration of peak area is performed by built-in computer programs for data analysis. If a computer-interfaced instrument is lacking, two commonly used alternative methods are available to quantify peaks from strip-chart recordings. The first is to measure the peak height rather than integrated area as a measure of analyte mass. This is done by drawing with a straightedge a line that connects the baseline on either side of the peak of interest. Next one draws a straight line, perpendicular to the x axis of the recording, from the peak maximum to the line drawn between the baseline points. The length of this perpendicular line can be measured with a ruler and records the peak height (Figure 7.17). This procedure is repeated with each standard sample to construct the calibration curve.

The second method involves estimating the integrated area of the peak by again drawing a line between the baseline points. The two sides of the peak and the drawn baseline define an approximately triangular area, which is carefully cut from the strip-chart paper with scissors. The excised piece of paper is weighted on an analytical balance, and its mass is taken as a reasonable estimate of the relative peak area.

Obviously, the two manual methods just described are prone to greater error than the modern computational methods. Nevertheless, these traditional methods served researchers long before the introduction of laboratory computers and can still be used successfully when a computer is not readily available.

While reversed phase is probably the chromatographic mode most commonly employed in enzyme assays, ion exchange and size exclusion HPLC are also widely used. In ion exchange chromatography the analyte binds to a charged stationary phase through electrostatic interactions. These interactions can be disrupted by increasing the ionic strength (i.e., salt concentration) of the mobile phase; the stronger the electrostatic interactions between the analyte and the stationary phase, the greater the salt concentration of the mobile phase required to elute the analyte. Hence, multiple analytes can be separated and quantified by their differential elution from an ion exchange column.

The most common strategy for elution is to load the sample onto the column in a low ionic strength aqueous buffer and elute with a gradient from low to high salt concentration (typically NaCl or KCl) in the same buffer system. In size exclusion chromatography (also known as gel filtration), analyte molecules are separated on the basis of their molecular weights. This form of chromatography is not commonly used in conjunction with enzyme assays, except for the analysis of proteolytic enzymes when the substrate and products are peptides or proteins. For most enzymes that catalyze the reactions of small molecules the molecular weight differences between substrates and products tend to be too small to be measured by this method.

Size exclusion stationary phases are available in a wide variety of molecular weight fractionation ranges. In choosing a column for size exclusion, the ideal is to select a column for which the molecular weights of the largest and smallest analytes (i.e., substrate and product) span much of the fractionation range of the stationary phase. At the same time, the higher molecular weight analyte must lie well within the fractionation range and must not be eluted in the void volume of the column. By following these guidelines, one will obtain good separation between the analytes on the column and be able to quantify all of the analyte peaks. For example, a column with a fractionation range of 8000–500 would be ideal to study the hydrolysis by a protease of a 5000 Da peptide into two fragments of 2000 and 3000 Da, since all three analytes would be well resolved and within the fractionation range of the column. On the other hand, a column with a fractionation range of 5000–500 would not be a good choice, since the substrate molecular weight is near the limit of the fractionation range; thus the substrate peak would most likely elute with the void volume of the column, potentially making quantitation difficult. Size exclusion column packing is available in a wide variety of fractionation ranges from a number of vendors (e.g., BioRad, Pharmacia). Detailed information to guide the user in choosing an appropriate column packing and in handling and using the material correctly is provided by the manufacturers.

The analysis of peaks from ion exchange and size exclusion columns is identical to that described for reversed phase HPLC. More detailed descriptions of the theory and practice of these HPLC methods can be found in a number of texts devoted to this subject (Hancock, 1984; Oliver, 1989).

7.3.3 Electrophoretic Methods in Enzyme Assays

Electrophoresis is most often used today for the separation of macromolecules in hydrated gels of acrylamide or agarose. The most common electrophoretic technique used in enzyme assays is sodium dodecyl sulfate/polyacrylamide gel electrophoresis (SDS-PAGE), which serves to separate proteins and peptides on the basis of their molecular weights. In SDS-PAGE, samples of proteins or peptides are coated with the anionic detergent SDS to give them similar anionic charge densities. When such samples are applied to a gel, and an electric field is applied across the gel, the negatively charged proteins will migrate toward the positively charged electrode. Under these conditions, the migration of molecules toward the positive pole will be retarded by the polymer matrix of the gel, and the degree of retardation will depend on the molecular weight of the species undergoing electrophoresis. Hence, large molecular weight species will be most retarded, showing minimal migration over a fixed period of time, while smaller molecular weight species will be less retarded by the gel matrix and will migrate further during the same time period. This is the basis of resolving protein and peptide bands by SDS-PAGE. Examples of the use of SDS-PAGE can be found for enzymatic assays of proteolytic enzymes, kinases, DNA-cleaving nucleases, and similar materials.

The purpose of the electrophoresis in a protease assay is to separate the protein or peptide substrate of the enzymatic reaction from the products. The fractionation range of SDS-PAGE varies with the percentage of acrylamide in the gel matrix (see Copeland, 1994, for details). In general, acrylamide percentages between 5 and 20% are used to fractionate globular proteins of molecular weights between 10,000 and 100,000 Da. Higher percentage acrylamide gels are used for separation of lower molecular weight peptides (typically 20–25% gels). In a typical experiment, the substrate protein or peptide is incubated with the protease in a small reaction vial, such as a microcentrifuge tube. After a given reaction time, a volume of the reaction mixture is removed and mixed with an equal volume of $2 \times$ SDS-containing sample buffer to denature the proteins and coat them with anionic detergent (Copeland, 1994). This buffer contains SDS to unfold and coat the proteins, a disulfide bond reducing agent (typically mercaptoethanol), glycerol to give density to the solution, and a low molecular weight, inert dye to track the progress of the electrophoresis in the gel (typically bromophenol blue). The sample mixture is then incubated at boiling water temperature for 1–5 minutes and loaded onto a gel of an appropriate percentage acrylamide to effect separation. Current is applied to the gel from a power source, and the electrophoresis is allowed to continue for some fixed period of time until the bromophenol blue dye front reaches the bottom of the gel. (For a 10% gel, a typical electrophoretic run would be performed at 120 V constant voltage for 1.5–3 h, depending on the size of the gel).

After electrophoresis, protein or peptide bands are visualized with a peptide-specific stain, such as Coomassie Brilliant Blue or silver staining (Hames and Rickwood, 1990; Copeland, 1994). A control lane containing the substrate protein or peptide alone is always run, loaded at the same concentration as the starting concentration of substrate in the enzymatic reaction. When possible, a second control lane should be run containing samples of the expected product(s) of the enzymatic reaction. A third control lane, containing commercial molecular weight markers (a collection of proteins of known molecular weights) is commonly run on the same gel also. The amounts of substrate remaining and product formed for a particular reaction can be quantified by densitometry from the stained bands on the gel. A large number of commercial densitometers are available for this purpose (from BioRad, Pharmacia, Molecular Devices, and other manufacturers).

Figure 7.18 illustrates a hypothetical protease assay using SDS-PAGE. In this example, the protease cleaves a protein substrate of 20 kDa to two unequal fragments (12 and 8 kDa). As the reaction time increases, the amount of substrate remaining diminishes, and the amount of product formed increases. Upon scanning the gel with a densitometer, the relative amounts of both substrate and products can be quantified by ascertaining the degree of staining of these bands. As illustrated by Figure 7.18, it is fairly easy to perform this type of relative quantitation. To convert the densitometry units into concentration units of substrate or product is, however, less straightforward. For substrate loss, one can run a similar gel with varying loads of the substrate (at known concentrations) and establish a calibration curve of staining density as a function of substrate concentration. One can do the same for the product of the enzymatic reaction when a genuine sample of that product is available. For synthetic peptides, this is easily accomplished. A standard sample for protein products can sometimes be obtained by producing the product protein recombinantly in a bacterial host. This is not always a convenient option,

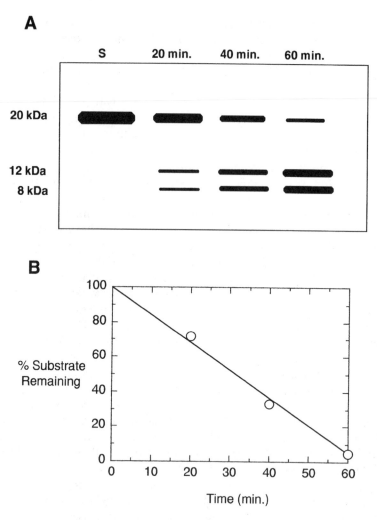

Figure 7.18 Schematic diagram of a protease assay based on SDS-PAGE separation of the protein substrate (20 kDa) and products (12 and 8 kDa) of the enzyme. (A) Typical SDS-PAGE result of such an experiment: the loss of substrate could then be quantified by dye staining or other visualization methods, combined with such techniques as densitometry or radioactivity counting. (B) Time course of substrate depletion based on staining of the substrate band in the gel and quantitation by densitometry.

however, and in such cases one's report may be limited to relative concentrations based on the intensity of staining.

The foregoing assay would work well for a purified protease sample, where the only major protein bands on the gel would be from substrate and product. When samples are crude enzymes—for example, early in the purification of a target enzyme — contaminating protein bands may obscure the analysis of the substrate and product bands on the gel. A common strategy in these cases is to perform Western blotting analysis using an antibody that recognizes specfically the substrate or product of the enzymatic reaction under study. Detailed protocols for Western blotting have been described (Harlow and Lane, 1988; Copeland, 1994; see also technical bulletins from manufacterers of electrophoretic equipment such as BioRad, Pharmacia, and Novex).

Briefly, in Western blotting an SDS-PAGE gel is run under normal electrophoretic conditions. Afterward, the gel is soaked in a buffer designed to optimize electrophoretic migration of proteins out of the gel matrix. The gel is then placed next to a sheet of nitrocellulose (or other protein binding surface), and protein bands are transferred electrophoretically from the gel to the nitrocellulose. After transfer, the remaining protein binding sites on the nitrocellulose are blocked by means of a large quantity of some nonspecific protein (typically, nonfat dried milk, gelatin, or bovine serum albumin). After blocking, the nitrocellulose is immersed in a solution of an antibody that specifically recognizes the protein or peptide of interest (i.e., in our case, the substrate or product of the enzymatic reaction). This antibody, referred to as the *primary antibody*, is obtained by immunizing an animal (typically a mouse or a rabbit) with a purified sample of the protein or peptide of interest (see Harlow and Lane, 1988, for details).

After treatment with the primary antibody, and further blocking with nonspecific protein, the nitrocellulose is treated with a *secondary antibody* that recognizes primary antibodies from a specific animal species. For example, if the primary antibody is obtained by immunizing rabbits, the secondary antibody will be an anti-rabbit antibody. The secondary antibody carries a label that provides a simple and sensitive method of detecting the presence of the antibody. Secondary antibodies bearing a variety of labels can be purchased. A popular strategy is to use a secondary antibody that has been covalently labeled with biotin, a ligand that binds tightly and specifically to streptavidin, which is commercially available as a conjugate with enzymes such as horseradish peroxidase or alkaline phosphatase. The biotinylated secondary antibody adheres to the nitrocellulose at the binding sites of the primary antibody. The location of the secondary antibody on the nitrocellulose is then detected by treating the nitrocellulose with a solution containing a streptavidin-conjugated enzyme. After the streptavidin–enzyme conjugate has been bound to the blot, the blot is treated with a solution containing chromophoric substrates for the enzyme linked to the streptavidin. The products of the enzymatic reaction form a highly colored precipitate on the nitrocellulose blot wherever the enzyme–streptavidin conjugate is present. In this roundabout fashion, the presence of a protein band of interest can be specifically detected from a gel that is congested with contaminating proteins.

SDS-PAGE is also used in enzyme assays to follow the incorporation of phosphate into a particular protein or peptide that results from the action of a specific kinase. There are two common strategies for following kinase activity by gel electrophoresis. In the first, the reaction mixture includes a ^{32}P- or ^{33}P-labeled phosphate source (e.g., ATP as a cosubstrate of the kinase) that incorporates the radiolabel into the products of the enzymatic reaction. After the reaction has been stopped, the reaction mixture is fractionated by SDS-PAGE. The resulting gel is dried, and the ^{32}P- or ^{33}P-containing bands are located on the gel by autoradiography or by digital radioimaging of the dried gel. The second strategy uses commercially available antibodies that specifically recognize proteins or peptides that have phosphate modifications at specific types of amino acid residues. Antibodies can be purchased that recognize phosphotyrosine or phosphoserine/phosphothreonine, for example. These antibodies can be used as the primary antibody for Western blot analysis as described earlier. Since the antibodies recognize only the phosphate-containing proteins or peptides, they provide a very specific measure of kinase activity.

Aside from their use in quantitative kinetic assays, electrophoretic methods also have served in enzymology to identify protein bands associated with specific enzymatic activities after fractionation on gels. This technique, which relies on specific staining of enzyme bands in the gel, based on the enzymatic conversion of substrates to products, can be a very powerful tool for the initial identification of a new enzyme or for locating an enzyme during purification attempts. For these methods to work, one must have a staining method that is specific to the enzymatic activity of interest, and the enzyme in the gel must be in its native (i.e., active) conformation.

Since SDS-PAGE is normally denaturing to proteins, measures must be taken to ensure that the enzyme will be active in the gel after electrophoresis: either the electrophoretic method must be altered so that it is not denaturing, or a way must be found to renature the unfolded enzyme in situ after electrophoresis.

Native gel electrophoresis is commonly used for these applications. In this method, SDS and disulfide-reducing agents are excluded from the sample and the running buffers, and the protein samples are not subjected to denaturing heat before application to the gel. Under these conditions most proteins will retain their native conformation within the gel matrix after electrophoresis. The migration rate during electrophoresis, however, is no longer dependent solely on the molecular weight of the proteins under native conditions. In the absence of SDS, the proteins will not have uniform charge densities; hence, their migration in the electric field will depend on a combination of their molecular weights, total charge, and general shape. It is thus not appropriate to compare the electrophoretic mobility of proteins under the denaturing and native gel forms of electrophoresis.

Sometimes enzymes can be electrophoresed under denaturing conditions and subsequently refolded or renatured within the gel matrix. In these cases the gel is usually run under nonreducing conditions (i.e., without mercaptoethanol or other disulfide-reducing agents in the sample buffer), since proper re-formation of disulfide bonds is often difficult inside the gel. A number of methods for renaturing various enzymes after electrophoresis have been reported, and these were reviewed by Mozhaev et al. (1987). The following protocol, provided by Novex, Inc., has been found to work well for many enzymes in the author's laboratory. Since, however, not all enzymes will be successfully renatured after the harsh treatments of electrophoretic separation, the appropriateness of any such method must be determined empirically for each enzyme.

GENERAL PROTOCOL FOR RENATURATION OF ENZYMES AFTER SDS-PAGE

1. After electrophoresis, soak gel for 30 minutes at room temperature, with gentle agitation, in 100 mL of 2.5% (v:v) Triton X-100 in distilled water.
2. Decant the solution and replace with 100 mL of an aqueous buffer containing 1.21 g/L Tris Base, 6.30 g/L Tris HCl, 11.7 g/L NaCl, 0.74 g/L $CaCl_2$, and 0.02% (w:v) Brig 35 detergent. Equilibrate the gel in this solution for 30 minutes at room temperature, with gentle agitation. Replace the solution with another 100 mL of the same buffer and incubate at 37°C for 4–16 hours.

The electrophoresis text by Hames and Rickwood (1990) provides an extensive list of enzymes ($\sim$200) that can be detected by activity staining after native gel electrophoresis and gives references to detailed protocols for each of the listed enzymes. Figure 7.19 illustrates activity staining after native gel electrophoresis for human dihydroorotate dehydrogenase (DHODase), the enzyme that uses the redox cofactor ubiquinone to catalyze the conversion of dihydroorotate to orotic acid. As is true of many other dehydrogenases, it is possible to couple the activity of this enzyme to the formation of an intensely colored formazan product by reduction of the reagents nitroblue tetrazolium (NBT) or methyl thiazolyl tetrazolium (MTT); the formazan product precipitates on the gel at the sites of enzymatic activity. The left-hand panel of Figure 7.19 shows a native gel of a detergent extract of human liver mitochondrial membranes stained with Coomassie Brilliant Blue. As one would expect, there are a large number of proteins present in this sample, displaying a congested pattern of protein bands on the gel. The right-hand panel of Figure 7.19 displays another native gel of the same sample that was soaked after electrophoresis in a solution of 100 μM dihydroorotate, 100 μM ubiquinone, and 1 mM NBT (in a 50 mM Tris buffer, pH 7.5). There is a single dark band due to the NBT staining of the enzymatically active protein in the sample. Thus, it is seen that the enzymatic activity in a complex sample can be associated with a specific protein or set of proteins. The active band(s) can be excised from the gel for further analysis, such as N-terminal sequencing, or to serve as part of a purification protocol for a particular enzyme; alternatively they can be used for the production of antibodies against the enzyme of interest.

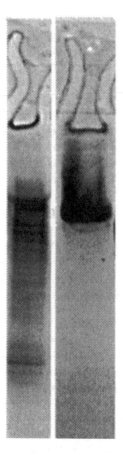

Figure 7.19 Example of activity staining of enzyme after gel electrophoresis. Left lane: Native gel of a detergent extraction of human liver mitochondrial membranes stained with Coomassie Brilliant Blue; note the large number of proteins of varied electrophoretic mobility in the sample. Right lane: Native gel of the same sample (run under conditions identical to those used in the left lane) stained with nitroblue tetrazolium (NBT) in the presence of the substrates of the enzyme dihydroorotate dehydrogenase; the single protein band that is stained intensely represents the active dihydroorotate dehydrogenase.

In the case of proteolytic enzymes, an alternative to activity staining is a technique known as gel zymography. In this method the acrylamide resolving gel is cast in the presence of a high concentration of a protein-based substrate of the enzyme of interest (casein, gelatin, collagen, etc.). The polymerized gel is thus impregnated with the protein throughout. Samples containing the proteolytic enzyme are then electrophoresed on the gel. If denaturing conditions are used, the enzymes are renatured by the protocol described earlier, and the gel is then stained with Coomassie Brilliant Blue. Because there is a high concentration of protein (i.e., substrate) throughout the gel, the entire field will

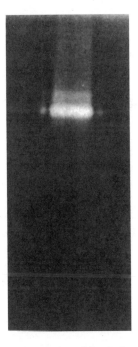

Figure 7.20 Gelatin zymography of a whole cell lysate from Sf9 insect cells that had been infected with a baculovirus construct containing the gene for human 92 kDa gelatinase (MMP9). The location of the active enzyme is easily observed from the loss of Coomassie staining of the gelatin substrate in the gel. (Figure kindly provided by Henry George, DuPont Merck Research Laboratories.)

be stained bright blue. Where there has been significant proteolysis of the protein substrate, however, the intensity of blue staining will be greatly diminished. Hence, the location of proteolytic enzymes in the gel can be determined by the *reverse staining* (i.e., the absence of Coomassie staining), as illustrated in Figure 7.20 for the metalloprotease gelatinase (MMP9).

A related, less direct method of protease detection has also been reported. In this "sandwich gel" technique, an agar solution is saturated with the protein substrate and allowed to solidify in a petri dish or another convenient container. A standard acrylamide gel is used to electrophorese the protease-containing sample. After electrophoresis (and renaturation in the case of denaturing gels) the substrate-containing agar is overlaid with the protease-containing acrylamide gel, and the materials are left in contact with each other for 30–90 minutes at 37°C. The sites of proteolytic activity can then be determined by treating the agar with an ammonium sulfate solution, trichloro-acetic acid, or some other protein-precipitating agent. After this treatment, the bulk of the agar will turn opaque as a result of protein precipitation. The proteolysis sites, however, will appear as clear zones against the opaque field of the agar. Methods for the detection of enzymatic activity after gel electrophoresis have been reviewed in the text by Hames and Rickwood (1990) and by Gabriel and Gersten (1992).

7.4   FACTORS AFFECTING THE VELOCITY OF ENZYMATIC REACTIONS

The velocity of an enzymatic reaction can display remarkable sensitivity to a number of solution conditions (e.g., temperature, pH, ionic strength, specific cation and anion concentration). Failure to control these parameters can lead to significant errors and lack of reproducibility in velocity measurements. Hence it is important to keep these parameters constant from one measurement to the next. In some cases, the changes in velocity that are observed with controlled changes in some of these conditions can yield valuable information on aspects of the enzyme mechanism. In this section we discuss five of these parameters: enzyme concentration, temperature, pH, viscosity, and solvent isotope makeup. Each of these can affect enzyme velocities in well-understood ways, and each can be controlled by the investigator to yield important information.

7.4.1   Enzyme Concentration

In Chapter 5, in our discussion of the Henri–Michaelis–Menten equation, we saw how the concentration of substrate can affect the velocity of an enzymatic reaction. At the end of Chapter 5 we recast this equation, replacing the term V_{max} with the product of k_{cat} and [E], the total concentration of enzyme in the sample (Equation 5.22). From this equation we see that the velocity of an enzyme-catalyzed reaction should be linearly proportional to the concentration of enzyme present at constant substrate concentration.

Over a finite range, a plot of velocity as a function of [E] should yield a straight line, as illustrated in Figure 7.21, curve a. The range over which this linear relationship will hold depends on our ability to measure the true initial velocity of the reaction at varying enzyme concentrations. Recall from Chapter 5 that initial velocity measurements are valid only in the range of substrate depletion between 0 and 10% of the total initial substrate concentration. As we add more and more enzyme, the velocity can increase to the point at which significant amounts of the total substrate concentration are being depleted during the time window of our assay. When substrate depletion becomes significant, further increases in enzyme concentration will no longer demonstrate as steep a change in reaction velocity as a function of [E]. As a result, we may observe a plot of velocity as a function of $[E_t]$ that is linear at low [E] but then curves over and may even show saturation effects at higher values of [E], as in curve b of Figure 7.21.

In general, as stated in Chapter 5, one should work at enzyme concentrations very much lower than the substrate concentration. This range will vary

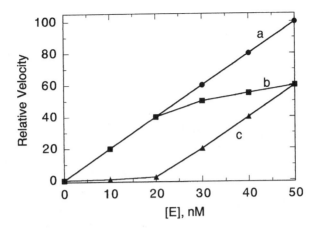

Figure 7.21   The relative velocity of an enzymatic reaction, under controlled conditions, as a function of total enzyme concentration [E]. The straight-line relationship of curve a is the expected behavior. Curve b illustrates the type of behavior observed when substrate depletion becomes significant at the higher enzyme concentrations. Curve c illustrates the behavior that would be observed for an enzyme sample that contained a reversible inhibitor. See text for further details.

from system to system; but in a typical assay substrate is present in micromolar to millimolar concentrations, and enzyme is present in picomolar to nanomolar concentrations. Within this range of [E] ≪ [S], initial velocity measurements must be made over a number of enzyme concentrations to determine the range of [E] over which substrate depletion is not significant.

Substrate depletion is not the only cause of a downward-curving velocity–[E] plot like that represented by cuvre b of Figure 7.21. The same type of behavior also results from saturation of the detection system at the higher velocity values seen at high [E]. We have discussed some of these problems in this chapter. For example, suppose that we measured the velocity of an enzymatic reaction as an end point absorption reading, following product formation. As we increase [E], the velocity increases, and thus the amount of product formed over the fixed time window of our end point assay increases. If the concentration of product increases until the sample absorption is beyond the Beer's law limit (see the discussion of optical methods of detection in Section 7.2.4), we observe an apparent saturation of velocity at high value of [E]. As with substrate depletion, detector saturation effects lead to down-curving velocity–[E] plots, not as a result of any intrinsic property of our enzyme system, but rather because of a failure to measure the true initial velocity of the reaction under conditions of high [E].

Plots of velocity as a function of enzyme concentration also can display upward curvature, as illustrated by curve c of Figure 7.21. Potential causes of this type of behavior can be inadequate temperature equilibration, as discussed shortly, and the presence of an inhibitor or enzyme activator in the reaction mixture. If, for example, a small amount of an irreversible inhibitor (see Chapter 10) is present in one of the components of the reaction mixture, additions of low concentrations of enzyme will result in complete inhibition of the enzyme, and no activity will be observed. The enzymatic activity will be realized in such a system only after enzyme has been added to a concentration that exceeds that of the irreversible inhibitor. Hence, at low values of [E] one observes zero or minimal velocity, while above some critical concentration, the velocity–[E] curve is steeper. Another potential cause of upward curvature is the presence in the enzyme stock solution of an enzyme activator or cofactor that is missing in the remainder of the reaction mixture components. Suppose that the enzyme under study requires a dissociable cofactor for full activity (as we saw in Chapter 3, many enzymes fall into this category). The concentration of free enzyme $[E_f]$ and free cofactor $[C_f]$ will be in equilibrium with that of the active enzyme–cofactor complex [EC], and the concentration of [EC] present under any set of solution conditions will be defined by the equilibrium constant K_c:

$$[EC] = \frac{[E_f][C_f]}{K_c} \tag{7.14}$$

In the enzyme stock solution, the concentrations of enzyme and cofactor will be in some specific proportion. When we dilute a sample of this stock solution into our reaction mixture, the total amount of enzyme added will be the sum of free enzyme and enzyme–cofactor complex; that is, $[E] = [E_f] + [EC]$. Hence, the concentration of cofactor added to the reaction mixture from the enzyme stock solution will be proportional to the amount of total enzyme added: that is, $[C] = \alpha[E]$. It can be shown (Tipton, 1992) that the amount of active EC complex in the final reaction mixture will depend on the total enzyme added and the enzyme–cofactor equilibrium constant as follows:

$$[EC] = \frac{[E]^2}{[E] + (K_c/\alpha)} \tag{7.15}$$

We can see from Equation 7.15 that the amount of activated enzyme (i.e., [EC]) will not track linearly with the amount of total enzyme added at low values of [E], and thus an upward-curving plot, as in curve c of Figure 7.21, will result.

If one is aware of a cofactor requirement for the enzyme under study, these effects can often be avoided by supplementing the reaction mixture with an excess of the required cofactor. For example, the enzyme prostaglandin synthase is a heme-requiring oxidoreductase that binds the heme cofactor in a noncovalent, dissociable fashion. The apoenzyme (without heme) is inactive, but it can be reconstituted with excess heme to form the active holoenzyme. The activity of the enzyme can be followed by diluting a stock solution of the holoenzyme into a reaction mixture containing a redox active dye and measuring the changes in dye absorption following initiation of the reaction with arachidonic acid, the substrate of the enzyme. To observe full enzymatic activity, it is necessary to supplement the reaction mixture with heme so that the final heme concentration is in excess of the total enzyme concentration. As long as this precaution is taken, well-behaved plots of linear velocity versus [E] are observed for prostaglandin synthase over a fairly broad range of enzyme concentrations (Copeland et al., 1994).

In summary, when the true initial velocity of the reaction is measured, the velocity of an enzyme-catalyzed reaction will increase linearly with enzyme concentration. Deviations from this linear behavior can be seen when the analyst's ability to measure the true initial velocity is compromised by instrumental or solution limitations. Deviations from linearity are observed also when certain inhibitors or enzyme activators are present in the reaction mixture. A more comprehensive discussion of cases of deviation from the expected linear response can be found in the text by Dixon and Webb (1979).

7.4.2 pH Effects

The pH of an enzyme solution can affect the overall catalytic activity in a number of ways. Like all proteins, enzymes have a native tertiary structure that is sensitive to pH, and in general denaturation of enzymes occurs at extremely low and high pH values. There are a number of physical methods for following protein denaturation. Loss of secondary structure can be followed by circular dichroic spectroscopy, and changes in tertiary structure can often be observed by absorption and fluorescence spectroscopy (Copeland, 1994). Many proteins aggregate or precipitate upon pH-induced denaturation, and this behavior can be observed by light scattering methods and sometimes by visual inspection. The pH range over which the native state of an enzyme will be stable varies from one such protein to the next. While most enzymes are most stable near physiological pH (~ 7.4), some display maximal activity at much lower or higher pH values. The appropriate range for a specific enzyme must be determined empirically.

Typically, one finds that protein conformation can be maintained over a relatively broad pH range, say 4–5 pH units. Within this range, however, the velocity of the enzymatic reaction varies with pH. Figure 7.22 shows a typical profile of the velocity of an enzymatic reaction as a function of pH, within the pH range over which protein denaturation is not a major factor. What is most obvious from this figure is the narrow range of pH values over which enzyme catalytic efficiency is typically maximized. For most general assays of enzyme activity then, one will wish to maintain the solution pH at the optimum for catalysis. To keep within this range, the reaction mixture must be buffered by a component with a pK_a at or near the desired solution pH value.

A buffer is a species whose presence in solution resists changes in the pH of that solution due to additions of acid or base. For enzymatic studies, a number

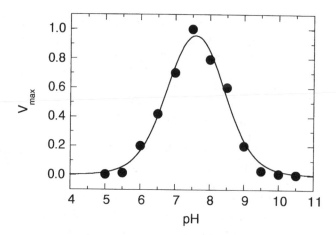

Figure 7.22   The effects of pH on the velocity of a typical enzymatic reaction.

of useful buffers are available commercially; some of these are listed in Table 7.6. The buffering capacity of these and other buffers declines as one moves away from the pK_a value of the substance. In general these buffers provide good buffering capacity from one pH unit below to one pH unit above their pK_a values. Thus, for example, HEPES buffer can be used to stabilize the pH of a solution between pH values of 6.55 and 8.55 but would not be an appropriate buffer below pH 6.5 or above pH 8.6.

The buffers listed in Table 7.6 span a broad range of pK_a values, providing a selection of single-component buffers for maintaining specific solution pH

Table 7.6   Some buffers that are useful in enzyme studies

Common Name	Molecular Weight	pK_a at 25°C[a]	$\Delta pK_a/\Delta°C$
MES	195.2	6.15	−0.011
PIPES	324.3	6.80	−0.0085
Imidazole	68.1	7.00	−0.020
MOPS	231.2	7.20	−0.013
TES	251.2	7.50	−0.020
HEPES	260.3	7.55	−0.014
HEPPS	252.3	8.00	−0.015
Tricine	179.2	8.15	−0.021
Tris	121.1	8.30	−0.031
CHES	207.3	9.50	−0.029
CAPS	221.3	10.40	−0.032

[a]Values listed for the buffers at an infinite dilution.

values. Note, however, that the pK_a values listed in Table 7.6 are for the buffers at 25°C and at infinite dilution. The temperature, buffer concentration, and overall ionic strength can perturb these pK_a values, hence altering the pH of the final solution. In most enzymatic studies, the buffers will be present at final concentrations of 0.05–0.1 M and solution ionic strength is typically between 0.1 and 0.2 M (near physiological conditions). Typically one will have a high concentration stock solution of the pH-adjusted buffer in the laboratory that will be diluted to prepare the final reaction mixture. It is important to measure the final solution pH to determine the extent of pH change that accompanies dilution. These effects are usually relatively small, and minor adjustments can be made if necessary.

Another potential problem is the change in pK_a due to changes in solution temperature. In some cases the pH of a buffered solution can change dramatically between temperatures of 4 and 37°C. Table 7.6 lists the change in pK_a per change in degree Celsius of the tabulated buffers. In principle, one could calculate the change in solution pH that will accompany a temperature change, but this is a tedious task and undertaking it often is impractical. Instead, if the assays are to be run at elevated temperatures (e.g., 37°C), the pH meter should be calibrated at the assay temperature and all pH measurements performed at that temperature as well. This will ensure that the pH values measured reflect accurately the true pH values under the assay conditions. In some cases one may wish to measure enzyme activity over a range of temperatures while maintaining the pH at a fixed value (see later). For such studies it is best to use a buffer with a low $\Delta pK_a/\Delta °C$ value, to keep the change in pH over the temperature range of interest minimal. From Table 7.6, PIPES ($pK_a = 6.8$; $\Delta pK_a/\Delta °C = -0.0085$) and MOPS ($pK_a = 7.20$; $\Delta pK_a/\Delta °C = -0.013$) would be good choices for this application.

The pH dependence of the activity of an enzyme is of practical importance in optimizing assay conditions, but the dependency is largely phenomenological. On the other hand, useful mechanistic information regarding the role of acid–base groups involved in enzyme turnover can be gleaned from properly performed pH studies. By measuring the velocity as a function of substrate concentration at varying pH, one can simultaneously determine the effects of pH on the k_{cat}, K_m, and k_{cat}/K_m values for an enzyme-catalyzed reaction. If titration of ionizable groups on the substrate molecule does not occur over the pH range being studied, these pH profiles will make possible some general conclusions about the roles of acid–base groups within the enzyme molecule. In general the pH dependence of K_m reflects the involvement of acid–base groups that are essential to initial substrate binding event(s) that precede catalysis. Effects of pH on k_{cat} mainly reflect acid–base group involvement in the catalytic steps of substrate to product conversion; that is, these ionization steps occur in the enzyme–substrate complex.

Finally, a plot of k_{cat}/K_m as a function of pH is said to reflect the essential ionizing groups of the free enzyme that play a role in both substrate binding and catalytic processing (Palmer, 1985). As an example, let us consider the pH profile of the serine protease chymotrypsin. As described in Chapter 6, the active site of α-chymotrypsin contains a catalytic triad (Figure 7.23A) composed of Asp 102, His 57, and Ser 195. Both acylation of Ser 195 to form the intermediate state and hydrolysis of the peptidic substrate depend on hydrogen-bonding and proton transfer steps among the residues within this active site triad.

From idealized pH profiles for the k_{cat}, K_m, and k_{cat}/K_m for α-chymotrypsin (Figure 7.23B–D), we see that both K_m and k_{cat} display pH profiles that are well fit by the Henderson–Hasselbalch equation introduced in Chapter 2 (Equation 2.12), but with different profiles. In the case of the profile for K_m versus pH, substrate binding affinity decreases (i.e., K_m increases) with increasing pH with an apparant pK_a value of 9.0. This pK_a has been shown to reflect ionization of an N-terminal isoleucine residue, which must be protonated for the enzyme to adopt a conformation capable of binding substrate. The value of k_{cat} for this enzyme increases with increasing pH and displays an apparent pK_a of 6.8. This pK_a value has been alternatively ascribed to Asp 102 and His 57 of the active site triad. It is now thought that this pK_a is more correctly associated with the catalytic triad as a whole, rather than with an individual amino acid residue. In Figure 7.23D, the idealized pH profile of k_{cat}/K_m for α-chymotrypsin, we do not observe the expected "S-shaped" curve associated with the Henderson–Hasselbalch equation; instead, there is a bell-shaped curve. This plot represents the cumulative effects of two titratable groups that influence the catalytic efficiency of the enzyme in opposite ways (i.e., one group facilitates catalysis in its conjugate base form, while the other facilitates catalysis in its Brønsted–Lowry acid form). A pH profile such as that seen in Figure 7.23D can be fit by the following equation:

$$y = \frac{y_{max}}{\left(\dfrac{10^{-pH}}{10^{-pK_{a1}}}\right) + \left(\dfrac{10^{-pK_{a2}}}{10^{-pH}}\right) + 1} \tag{7.16}$$

where y is the experimental measure that is plotted on the y axis (in this case k_{cat}/K_m), y_{max} is the observed maximum value of that experimental measure, and pK_{a1} and pK_{a2} refer to the pK_a values for the two relevant acid–base groups being titrated. A fit of the curve in Figure 7.23D to Equation 7.16 yields values of pK_{a1} and pK_{a2} of 6.8 and 9.0, respectively. Thus both the pK_a values that were found to influence k_{cat} and K_m, respectively, are reflected in the pH profile of k_{cat}/K_m for this enzyme.

The type of data presented in Figure 7.23 is often used to predict the identities of key amino acid residues participating in acid–base chemistry during catalysis. Some caution must be exercised, however, in making such predictive statements. As we have seen for chymotrypsin, in some cases the pK_a value that is measured cannot be correctly ascribed to a particular amino acid, but rather reflects a specific set of residue interactions within an enzyme

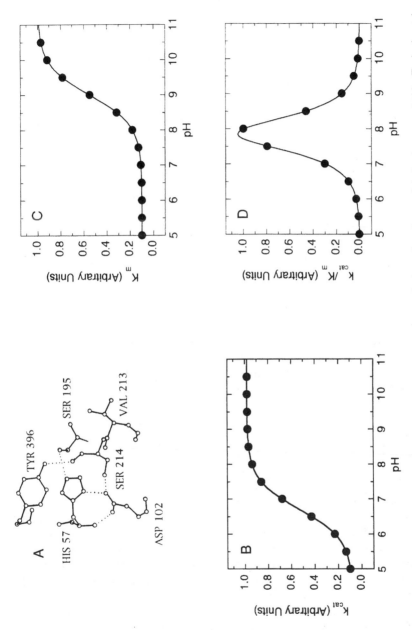

Figure 7.23   (A) Cartoon of the active site structure of α-chymotrypsin, based on the crystal structure reported by Tsukada and Blow (1985), showing the active site triad of amino acids. (B) Idealized pH profile of k_{cat} for α-chymotrypsin, showing an apparent pK_a of 6.8. (C) Idealized pH profile of K_m for α-chymotrypsin, showing an apparent pK_a of 9.0. (D) Idealized pH profile of k_{cat}/K_m for α-chymotrypsin, showing a bell-shaped curve that can be fit to Equation 7.16, with pK_a values of 6.8 and 9.0.

Table 7.7   Examples of amino acid residues with perturbed side chain pK_a values

	pK_a in		
Side Chain	Free Amino Acid	Enzyme	Enzyme
Glu	3.9	6.5	Lysozyme
His	6.8	3.4	Papain
His	6.8	5.2	Ribonuclease
Cys	8.3	4.0	Papain
Lys	10.8	5.9	Acetoacetate Dehydrogenase

molecule that create in situ a unique acid–base center. Also, the hydrophobic interior of enzyme molecules can greatly perturb the pK_a values of amino acid side chains relative to their typical pK_a values in aqueous solution. Some examples of such perturbations of amino acid side chain pK_a are listed in Table 7.7. These examples should make it clear that one cannot rely simply on a comparison between the measured pK_a of an enzymatic kinetic parameter and the pK_a values of amino acid side chains in solution. Thus, for example, a k_{cat}–pH profile for a particular enzyme that displays an apparent pK_a value of 6.8 may reflect the ionization of an essential histidine residue, but it may equally well represent the ionization of a perturbed glutamic acid residue, or yet some other residue within the specialized environment of the protein interior.

The effects of pH on the kinetic parameters k_{cat} and K_m also have been analyzed by plotting the value of the kinetic constant on a logarithmic scale as a function of pH (Dixon and Webb, 1979). For a single titration event, such plots appear as the superposition of two linear functions, one with a slope of zero and the other with a unit slope value. Similarly, a kinetic parameter that undergoes two titration events over a specific pH range will yield a plot that appears as the superposition of three straight lines: one with a positive unit slope value, one with a slope of zero, and one with a negative unit slope value. Figure 7.24 is an example of such a plot.

An advantage of these plots is that the pK_a value can be determined graphically without resorting to nonlinear curve-fitting routines; rather, the pK_a is defined by the point of intersection of the two straight lines drawn through the data points in the regions of minimal curvature of the plot. A second advantage is that the number of acid–base groups participating in the ionization event can be estimated: the slope of the line in the transition region of the plot reflects the number of ionizable groups that are titrated over this pH range. Thus, a slope of 1 indicates that a single group is being titrated, a slope of 2 indicates the involvement of two ionizable species, and so on. An expanded discussion of these plots, and examples of their application to specific enzymes, can be found in the text by Dixon and Webb (1979).

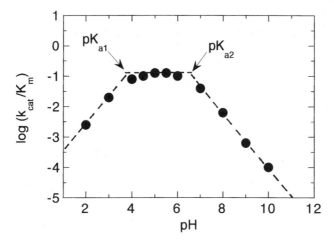

Figure 7.24 Plot of log(k_{cat}/K_m) as a function of pH for a typical enzymatic reaction. The two pK_a values are determined from the intersection of the straight lines drawn throughout the data in different regions of the pH range. See text for further details.

In designing experiments to measure the effects of pH on the steady state kinetics of an enzymatic reaction, it is critical for the researcher to ensure that the changes in solution pH are not made in a way that causes simultaneous changes in other solution conditions, thus confounding the analysis of the experiments. For example, a change in the species used to buffer the solution could, in principle, effect a change in the kinetics by itself. Since these studies are typically conducted over a broad range of pH values, no one buffer will have sufficient buffering capacity over the entire range of study. Hence, more than one buffering species is needed in these experiments.

One way to check that buffer-specific effects are not influencing the pH profile is to use buffers with overlapping pH ranges and perform duplicate measurements in the overlap regions. For example, to cover the pH range from 5.5 to 8.5 one might choose to use MES buffer (useful range ~5.15–7.15) at the lower pH values and HEPES buffer (useful range ~6.55–8.55) for the higher pH values. In the range of overlapping buffering capacity, between 6.55 and 7.15, one should make measurements with both buffers independently. If there are no significant differences between the measured values for the two buffer systems at the same pH values, it is fairly safe to assume that no major buffer-specific effects are occurring.

A better way to perform these measurements is to use a mixed buffer system that will have good buffering capacity throughout the entire pH range of study. Gomori (1992) has provided recipes for preparing buffer systems composed of two or more buffering species that are appropriate for enzyme studies and span several different ranges of pH values. For example, Gomori recommends the use of a mixed Tris-maleate buffer for work in the pH range between 5.2 and

Table 7.8   Volume of 0.2 M NaOH to be added to 50 mL of 0.2 M Tris-maleate stock to produce a 0.05 M Tris-maleate buffer at the indicated pH after dilution to 200 mL with distilled water

x (mL)	pH	x (mL)	pH
7.0	5.2	48.0	7.0
10.8	5.4	51.0	7.2
15.5	5.6	54.0	7.4
20.5	5.8	58.0	7.6
26.0	6.0	63.5	7.8
31.5	6.2	69.0	8.0
37.0	6.4	75.0	8.2
42.5	6.6	81.0	8.4
45.0	6.8	86.5	8.6

Source: Data from Gomori (1992).

8.6. This buffer system requires two stock solutions. The first solution is composed of 24.2 g of tris(hyroxymethyl)aminomethane and 23.2 g of maleic acid dissolved in 1 L of distilled water (0.2 M Tris-maleate). The second solution is 0.2 M NaOH in distilled water. To prepare a 0.05 M buffer at a given pH, 50 mL of the Tris-maleate stock is mixed with x mL of the 0.2 M NaOH stock, according to Table 7.8, and diluted with distilled water to a final volume of 200 mL. Recipes for other mixed buffer systems for use in lower and higher pH ranges can be found in the compilation by Gomori (1992) and references therein.

7.4.3   Temperature Effects

It is often stated that the rate of a chemical reaction generally doubles with every 10°C increase in reaction temperature. Most chemical catalysts display such an increase in activity with increasing temperature, and enzymes are no exception. Enzymes, however, are also proteins, and like all proteins they undergo thermal denaturation at elevated temperatures. Hence the enhancement of catalytic efficiency with increasing temperature is compromised by the competing effects of general protein denaturation at high temperatures. For this reason, the activity of a typical enzyme will increase with temperature over a finite temperature range, and then diminish significantly above some critical temperature that is characteristic of the denaturation of the protein (Figure 7.25). As with the pH profile, the information gleamed from plots such as Figure 7.25 is of practical value in designing enzyme assays. One will wish to measure the activity at a temperature that supports high enzymatic activity, does not lead to significant protein denaturation, and is experimentally convenient. Balancing these factors, one finds that the majority of enzyme

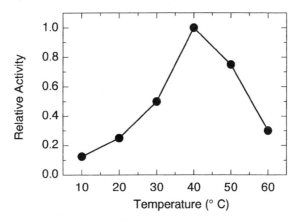

Figure 7.25 Typical profile of the relative activity of an enzyme as a function of temperature.

assays reported in the literature are conducted at either 25 or 37°C (i.e., physiological temperature).

If one restricts attention to the temperature range over which protein denaturation is not significant, an analysis of the changes in enzyme activity that accompany changes in temperature can be mechanistically informative. Recall from Chapter 2 that the rate of a reaction can be related to the activation energy for attaining the reaction transition state E_a by the Arrhenius equation (Equation 2.7). This relationship holds true for enzyme catalysis as well, as long as protein denaturation is not a complicating factor in the temperature range being studied. We can relate the kinetic constant k_{cat} to the activation energy as follows:

$$k_{cat} = A \exp\left(-\frac{E_a}{RT}\right) \qquad (7.17)$$

where R is the ideal gas constant (1.98×10^{-3} kcal/mol·degree), T is the temperature in degrees Kelvin, and for our purpose, we can treat the pre-exponential term A as a constant of proportionality (see Chapter 2 for a more explicit definition of A). Taking the $\log_{10}$ of both sides of Equation 7.17, we obtain:

$$\log(k_{cat}) = -E_a \frac{1}{2.3RT} + \log(A) \qquad (7.18)$$

Thus if we plot the $\log(k_{cat})$ of an enzymatic reaction as a function of $1/2.3RT$, Equation 7.18 predicts that we will obtain a straight-line relationship with a slope equal to $-E_a$, the activation energy in units of kcal/mol. Note that the relationships described by Equations 7.17 and 7.18 hold for V_{max} at constant

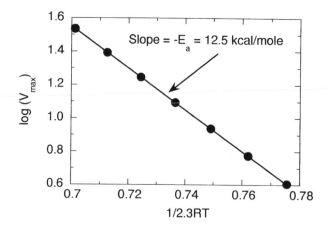

Figure 7.26 Arrhenius plot of $\log(V_{max})$ of an enzymatic reaction as a function of $1/(2.3RT)$. The slope of the line in such a plot gives an estimate of the negative value of E_a, the activation energy of the reaction.

enzyme concentration as well as for k_{cat}. When an accurate estimate of the enzyme concentration in a stock solution is lacking, one can hold the amount of that stock solution used in the assays constant and determine the activation energy of the reaction by using the measured values of V_{max} in place of k_{cat} in Equation 7.18. Figure 7.26 is such an Arrhenius plot for a hypothetical enzyme with an activation energy of 12.5 kcal/mol, measured over a temperature range of 10–40°C.

Since the temperature of the reaction mixture can have such a dramatic effect on the kinetic parameters of an enzyme-catalyzed reaction, it is critical to carefully control temperature during measurements of initial velocity. For reactions initiated by addition of substrate, the reaction mixture and substrate solution should be equilibrated to the same temperature before mixing. For a reaction initiated by addition of enzyme, it is sometimes desirable to maximize protein stability by maintaining the enzyme stock solution at 4°C (ice temperature) prior to the assay. In such situations it is best to use a stock of enzyme so concentrated that only a small volume needs to be added to the overall reaction mixture, which should already be equilibrated at the assay temperature. In a typical assay of this type, one might add 10–50 μL of enzyme stock (at 4°C) to a 1.0 mL reaction mixture that is equilibrated at 25 or 37°C. This small volume will not significantly perturb the temperature of the overall reaction mixture, and the enzyme will come to the assay temperature during mixing. If too large a volume of enzyme stock is used for these assays, full temperature equilibrium may not be achieved during the mixing time. This will be reflected as a lag phase in the initial velocity measurements, as illustrated in Figure 7.27. Lag phases of this type can significantly compromise the accuracy of end point type assays, where the occurrence of a lag phase might

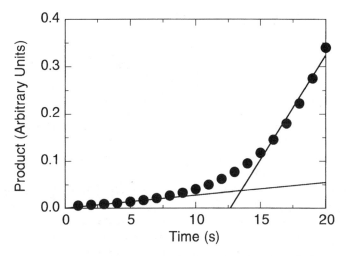

Figure 7.27 Example of a reaction progress showing a long lag phase before reaching the true steady state rate of reaction. Such a lag phase can be caused by several factors, including insufficient temperature equilibrium of the enzyme and reaction mixture solutions. See text for further details.

be missed. Control measurement at several time points should be performed in these cases, to ensure that such effects of insufficient temperature equilibration are not affecting the measurements.

7.4.4 Viscosity Effects

When a diffusional event, such as initial collisional encounter of E and S to form the ES complex or dissociation of product from the EP complex is rate limiting, solution microviscosity can affect the overall rate of reaction. The microviscosity of a solution refers to the resistance to motion that is experienced by a molecule in the solution. This is in contrast to the macroviscosity measured by conventional viscometers, which is a bulk property of the solution. Because increases in microviscosity increase the resistance to molecular motions in solution, the frequency of diffusional events is slowed down. Polymeric viscogenes, such as polyethylene glycol, influence the macroviscosity only, while monomeric viscogenes, such as sucrose and glycerol, affect both the macro- and microviscosty. Hence, the simplest way to increase the microviscosity of an enzyme solution is by addition of monomeric viscogenes. The viscosities of solutions of different sucrose or glycerol composition have been tabulated and can be found in references such as *The CRC Handbook of Physics and Chemistry*. The actual viscosity of final solutions containing these viscogenes should be determined empirically with a standard laboratory viscometer (such as a Cannon–Fenskie or Ostwald viscometer), available from any commercial laboratory supply company.

If an enzymatic reaction is diffusion limited, the value of k_{cat}/K_m will depend on the solution viscosity as follows:

$$\left(\frac{k_{cat}}{K_m}\right) = \left(\frac{k_{cat}}{K_m}\right)^0 \left(\frac{\eta^0}{\eta}\right) \qquad (7.19)$$

where η is the viscosity and the superscript 0 refers to the values of k_{cat}/K_m and η in the absence of added viscogene. From Equation 7.19 we see that a plot of $(k_{cat}/K_m)^0/(k_{cat}/K_m)$ as a function of (η/η^0) should yield a straight line with slope and y-intercept values of 1.00 each. Behavior of this type is an indication that the enzyme under study is completely rate-limited by a diffusional process. A good control to run for such an experiment is to measure k_{cat}/K_m over the same range of η using a polymeric viscogene, such as polyethylene glycol. Since only the macroviscosity is affected, no change in k_{cat}/K_m should be observed.

Figure 7.28 illustrates an example of this approach for the study of the enzyme triosephosphate isomerase (Blacklow et al., 1988). This enzyme is known to run at near kinetic perfection, with $(k_{cat}/K_m)^0 > 10^8 \text{ M}^{-1} \text{ s}^{-1}$. Thus scientists have speculated that the reaction velocity is rate-limited by the diffusion-controlled formation of the ES complex. To test this, Blacklow et al. measured the effects of viscosity changes on the reaction of triosephosphate isomerase brought about by glycerol (affecting both macro- and microviscosity) and polyethylene glycol (affecting macroviscosity only). As seen in Figure 7.28, they obtained the expected results: a linear dependence of k_{cat}/K_m on microviscosity, but no effect from changing macroviscosity only. As pointed out by the authors, these data do not prove that the rate-limiting step in

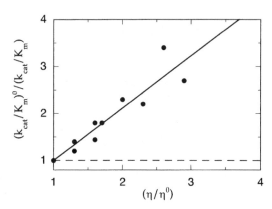

Figure 7.28 Effects of microviscosity on k_{cat}/K_m for an enzymatic reaction that is rate-limited by a diffusional process. Solid circles represent data points for the enzyme triosephosphate isomerase when the viscosity is adjusted with glycerol, thus affecting both the macro- and microviscosity of the solution. Dashed line shows the effects of changing solution viscosity with polyethylene glycol, where only the macroviscosity is affected. [Data redrawn from Blacklow et al. (1988).]

catalysis for triosephosphate isomerase is ES complex formation. Rather, the data indicate only that whatever step is rate limiting for this enzyme, it is likely to be a diffusion-controlled step.

7.4.5   Isotope Effects in Enzyme Kinetics

When an enzyme-catalyzed reaction is rate-limited by a group transfer step, a slowing down of the reaction rate will be observed if the group being transferred is isotopically enriched with a heavy isotope. Such kinetic isotope effects can be used to identify the atoms of a substrate molecule that are undergoing transfer during catalysis by an enzyme. To perform such an analysis, the investigator must synthesize a version of the substrate that is isotopically labeled at a specific atom. Since protons are perhaps the most commonly transferred atoms in enzymatic reactions, we shall focus our attention on the use of heavy isotopes of hydrogen in such studies.

Why is it that a heavier isotope leads to a diminution of the reaction rate for proton transfer reactions? To answer this question, let us consider a reaction in which a hydrogen is transfered from a carbon atom to some general base:

$$C\!-\!H \quad :B \rightleftharpoons [C^{\delta^-} \cdots H \cdots B^{\delta^+}]^{\ddagger} \rightarrow C^- \quad HB$$

As we saw in Chapter 2, the electronic state of the reactant can be represented as a potential energy well that has built upon it a manifold of vibrational substates (see Figure 2.9). Among these vibrational substrates will be potential energies associated with the stretching of the C—H bond. The transition state of the reaction is reached by elongating this C—H bond prior to bond rupture. Thus, in going from the reactant state to the transition state, the potential energy of the C—H stretching vibration is converted to transitional energy that contributes to the overall energy of activation for the reaction.

Now the potential energy minimum (i.e., the very bottom of the well) of the reactant state is characteristic of the electronic configuration of the reactant molecule when all the atoms in the molecule are at their equilibrium distances (i.e., when the vibrations of the bonds are "frozen out"). If we were to replace the proton on the carbon with a deuteron, the electronic configuration of the molecule would not be changed, and thus the bottom of the potential well for the reactant state would be unchanged. The vibrational substates involving the C—H stretching mode would, however, be affected by the isotopic change. The potential energy of a vibrational mode is directly proportional to the frequency v, at which the bond vibrates. In the case of a vibration that stretches a bond between two atoms, as in our C—H bond, the vibrational frequency can be expressed in terms of the force constant for that vibration (a measure similar to the tension or resistance to compression of a macroscopic spring) and the masses of the two atoms of the bond by:

$$v = \frac{k}{\sqrt{m_r}} \tag{7.20}$$

where k is the force constant, and m_r is the reduced mass of the diatomic system involved in the vibration. The reduced mass is related to the masses of the two atoms in the system (m_1 and m_2) as follows:

$$\frac{1}{m_r} = \frac{1}{m_1} + \frac{1}{m_2}$$ (7.21)

The activation energy associated with the transition between the reactant state and the transition state here is most correctly measured as the energetic distance between the vibrational ground state of the reactant potential well and the transition state. The energy difference between the vibrational ground state and the potential well minimum is referred to as the *zero-point energy* and is given the symbol Δe. The value of Δe is directly proportional to ν, which in turn is inversely proportional to the masses of the atoms involved in the vibration, according to Equation 7.21.

The frequency of a C—H bond stretching vibration can be measured by infrared or Raman spectroscopy, and has a typical value of about $2900\ \text{cm}^{-1}$. If we replace the proton with a deuteron (C—D, or C—^{2}H), this vibrational frequency shifts to roughly $2200\ \text{cm}^{-1}$. These frequencies correspond to Δe values of 4.16 and 3.01 kcal/mol for the C—H and C—D bonds, respectively. Therefore, the zero-point energy for a C—D bond will be 1.15 kcal/mol lower than that for a C—H bond (Figure 7.29). If all the vibrational potential energy of the reactant ground state is converted to transitional energy in achieving the transition state, this difference in zero-point energy corresponds to a 1.15 kcal/mol increase in overall activation energy for the C—D bond over that for the C—H bond. As we saw in Chapter 2, an increase in activation energy corresponds to a decrease in reaction rate, and thus the lowering of zero-point

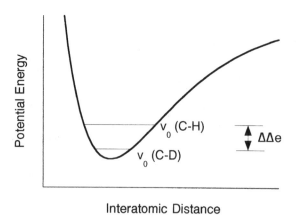

Figure 7.29 Potential energy diagram for an electronic state of a molecule illustrating the difference in zero point enegry, $\Delta\Delta e$, for C—H and C—D bonds.

energy for a heavier isotope explains the reduction in reaction rate observed in kinetic experiments.

The effects of deuterium isotope substitution on the rate of reactions is typically expressed as the ratio V_{max}^H/V_{max}^D or k_{cat}^H/k_{cat}^D. Based on the difference in zero-point energy for a C—H bond and a C—D bond, we would expect the difference in activation energy for these two group transfers to be 1.15 kcal/mol if all the vibrational potential energy is converted to transitional energy. From Equation 2.7 we would thus expect the kinetic isotope effect here to be:

$$\frac{k_{cat}^{C-H}}{k_{cat}^{C-D}} = \exp\left(-\frac{\Delta\Delta G^\ddagger}{RT}\right) = 7 \qquad (7.22)$$

Note that the isotope effect will be realized in the measured kinetics only if the hydrogen transfer step is rate limiting (or partially rate limiting) in the overall reaction. Also, the magnitude of the isotope effect will vary from enzyme to enzyme, depending on the degree to which the transition state converts the vibrational potential energy of the ground reactant state to transitional energy. In proton transfer reactions one also finds that the magnitude of the kinetic isotope effect is influenced by the pK_a of the general base group that participates in the transfer step. As a rule, the largest kinetic isotope effects occur when the pK_a of the general base is well matched to that of the carbon acid of the proton donor; the magnitude of the kinetic isotope effects diminishes as the difference in these pK_a values increases.

Kinetic isotope effects can be very useful in identifying the specific atoms that participate in rate-limiting group transfer steps during catalysis. A common strategy is to synthesize substrate molecules in which a specific atom is isotopically labeled and then compare the rate of reaction for this substrate with that for the unlabeled molecule. When a group that participates in a rate-limiting transfer step is labeled, a kinetic isotope effect is observed. This information not only can be used to determine what groups are involved in particular transfer reactions, but can also help to identify the rate-determining steps in the catalytic mechanism of an enzyme. Comprehensive treatments of the use of kinetic isotope effects in elucidation of enzyme mechanisms can be found in the reviews by Cleland and coworkers (Cleland et al., 1977) and by Northrop (1975), and in the recent compilation of selected articles from *Methods in Enzymology* edited by Purich (1996).

Isotopic substitution of the solvent water hydrogens can affect the kinetics of enzyme reactions if the solvent itself serves as a proton donor during catalysis, or if the proton donor groups on the enzyme or substrate can rapidly exchange with the solvent; these effects are referred to as *solvent isotope effects*. In simple enzyme systems, the solvent isotope effects can be used to determine the number of protons that are transferred during the rate-determining step of catalysis. This is done by measuring the velocity of the reaction as a function of the atom fraction of deuterium (n), or the percentage of D_2O in a mixed

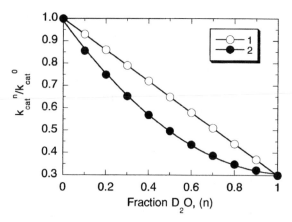

Figure 7.30 Proton inventory plot for reactions involving transfer of one (open circles) and two (solid circles) protons during the rate-limiting step in catalysis. The y axis is the ratio of k_{cat} in some mixture of D_2O and H_2O (k_{cat^n}) and the k_{cat} value in 100% H_2O (k_{cat^0}). Data for the one-proton reaction fit by a linear function; data for the two-proton reaction fit by a quadratic function. Reactions involving three-proton transfers in the rate-limiting step would be fit by a cubic function. Reactions involving more than three protons usually are fit by an exponential function.

H_2O/D_2O solvent system. A plot of V_{max} or k_{cat} as a function of n, or % D_2O, will show a diminution in these kinetic parameters as the amount of D_2O in the solvent system increases. If a single proton transfer event is responsible for the solvent isotope effect, the data in such a plot will be well fit by a linear function. If two protons are transferred, the data will be best fit by a quadratic equation. For three protons, a cubic equation will be required to fit the data, and so on (Figure 7.30).

Generally, the involvement of more than three protons yields a plot that is best fit by an exponential function, which would suggest an "infinite number" of proton transfers during the rate-limiting step. This *proton inventory* method does not provide any insight into the structures or locations of the proton transfering groups, but does allow one to quantify the number of groups participating in the rate-limiting step.

Some caution must be exercised in interpreting these data. The interpretations based on curve fitting assume that a single step, the rate-limiting step, is responsible for the entire observed solvent isotope effect. In most simple enzyme systems this generally holds true. In more complex, multi–transition state systems, however, the assumption may not be valid. The validity of the data also depends on having all other solution conditions held constant as the percentage of D_2O in the solvent system is varied. One must remember, for example, that there is a difference between the true pD value of a D_2O solution and the value measured with a conventional pH meter (for a pure D_2O solution, the true pD = pH meter reading + 0.41 at 25°C); these effects must be accounted for in the preparation of solutions for the measurement of proton

inventories. The reader interested in applying these techniques wou
advised to refer to more comprehensive treatments of the subject (Schow
Schowen, 1982; Venkatasubban and Schowen, 1984).

7.5   REPORTING ENZYME ACTIVITY DATA

As we have seen in the preceding section, many solution conditions can affect
the overall activity of an enzyme catalyzed reaction. Thus, for investigators in
different laboratories to reproduce one another's results it is critical that the
data be reported in meaningful units, and be accompanied by sufficient
information on the details of the assay used. In reporting activity measurements, one should always specify the buffer system used in the reaction mixture,
the pH and temperature at which the assay was recorded, the time interval over
which initial velocity measurements were made, and the detection method used.
Initial velocities and V_{max} values should always be reported in units of molarity
(of substrate or product) change per unit time, while K_m and k_{cat} values should
be reported in molarity units and reciprocal time (min^{-1}, or s^{-1}), respectively.
Turnover numbers are typically reported in terms of molarity change per unit
time per molarity of enzyme, moles of substrate lost or product produced per
unit time per mole of enzyme, or, equivalently, molecules of substrate lost or
product produced per unit time per molecule of enzyme.

Many times it will be necessary to measure the enzymatic activity of samples
that contain proteins other than the enzyme of interest. During the initial
purification of an enzyme, for example, it is often helpful to follow the activity
of the enzyme at various stages of the purification process, where multiple
contaminating proteins will be present in the sample also. To standardize
the reporting of activities in such cases, the International Union of Biochemistry has adapted the *international unit* (IU) as the standard measure
of enzyme activity: one international unit is the amount of enzyme (or
crude enzyme sample) required to catalyze the transformation of one microm-
ole of substrate per minute or, where more than one bond of each substrate
molecule is attacked, one microequivalent of the group concerned, under a
specific set of defined solution conditions. The definition allows the individual
researcher to specify the solution conditions, but the IUB recommended that
units be reported for measurements made at 25°C. The specific solution
conditions have no intrinsic significance, but they must to be reported to
ensure reproducibility.

In crude enzyme samples the total mass of protein can be determined by a
number of analytical methods (see Copeland, 1994 for details), but it is often
difficult to measure specifically the mass or concentration of the enzyme of
interest in such samples. To quantify the amount of enzyme present, researchers often report the *specific activity* of the sample: that is, the number of
international units of enzymatic activity per milligram of total protein under a
specific set of solution conditions. Most typically, specific activity values are

...ns of saturating substrate (i.e., where $v \sim V_{max}$) and ...ons (i.e., pH, temperature, etc.). As the purification of ...d more and more of the total protein mass of the ...e enzyme of interest, the specific activity of the sample ...e.

7.6 ENZYME STABILITY

One of the most common practical problems facing the experimental biochemist is the loss of enzymatic activity in a sample due to enzyme instability. Enzymes, like most proteins, are prone to denaturation under many laboratory conditions, and specific steps must be taken to stabilize these macromolecules as much as possible. Recommendations for the general handling of proteins for maximum stability have been described in detail in several texts devoted to proteins (see, e.g., Copeland, 1994). The general recommendations for the storage and handling of enzymes that follow can help to maintain the catalytic activities of these proteins.

7.6.1 Stabilizing Enzymes During Storage

Like all proteins, enzymes in their native states are optimally stabilized by specific solution conditions of pH, ionic strength, anion/cation composition, and so on. No generalities can be stated with respect to these conditions, and the best conditions for each enzyme individually must be determined empirically. Note, however, that the solution conditions that are optimal for protein stability may not necessarily the same as those for optimal enzymatic activity. When this caveat applies, enzyme stocks should be stored under the conditions that maximally promote stability, while the enzyme assays should be conducted under the conditions of optimal activity.

For long-term storage, enzymes should be kept at cryogenic temperatures in a $-70°C$ freezer or under liquid nitrogen. Conventional freezers operate at a nominal temperature of $-20°C$, but most of these cycle through higher temperatures to keep them "frost free." This can lead to unintentional freeze–thaw cycling of the enzyme sample, which can be extremely denaturing. If enzymes are stored in such a freezer, protein stability can be greatly enhanced by adding an equal volume of glycerol to the sample and mixing it well. This 50% glycerol solution will maintain the enzyme sample in the liquid phase at $-20°C$, and thus will prevent repeated freezing and thawing. In fact, many enzymes display optimal stability when stored at $-20°C$ as 50% glycerol solutions.

Before the samples are frozen, they should be sterile-filtered through a $0.22 \mu m$ filter composed of a low protein-binding material, and then placed in sterilized cryogenic tubes to avoid bacterial contamination. To avoid protein denaturation during the freeze–thaw process, it is critical that the samples be frozen quickly and thawed quickly. Rapid freezing is best accomplished by immersing the sample container in a slurry of dry ice and ethanol. Rapid thawing is best done by placing the sample in a 37°C water bath until most, but not all, of the sample is in the liquid state. When there is just a small bit of frozen material remaining, the sample should be removed from the bath and allowed to continue thawing on ice (i.e., 4°C).

Repeated freeze–thaw cycles are extremely denaturing to proteins and must be avoided. Thus, a frozen enzyme sample should be thawed once and used promptly. Sample remaining at the end of the experiment should not be refrozen. An enzyme that can be maintained in stable conditions for several days at 4°C, however, may be used in an experiment run soon after the first. If a particular enzyme is not stable under these conditions, any sample remaining at the end of an experiment must be discarded. To avoid wasting enzyme sample material, samples should be stored in small volume, high concentration aliquots. This way the volume of sample that is needed for each day's experiments can be thawed, while the bulk of the sample aliquots remain frozen. Once thawed, the enzyme should be kept at ice temperature (4°C) for as long as possible before equilibration to the assay temperature. Again, if the enzyme is stored at high concentration, only a small volume of the enzyme stock will be needed for dilution into the final reaction mixture.

For example, a typical enzyme assay might require a final concentration of enzyme in the reaction mixture of 10 nM. Suppose that an enzyme is in long-term storage at $-70°C$ as a 100 μM stock in 50 μL aliquots. On the day that assays are to be performed with the enzyme, a single aliquot might be thawed and diluted 1:100 with an appropriate buffer to make a 5 mL working stock of 1.0 μM enzyme. This stock would be stored on ice for the day (or potentially longer). The final reaction mixture would be prepared as a 1:100 dilution of this working stock to yield the desired final enzyme concentration of 10 nM.

Certain additives will enhance the stability of many enzymes for long-term storage at cryogenic temperatures and sometimes also for short-term storage in solution. Glycerol, sucrose, and cyclodextrans are often added to stabilize enzyme samples; the exact concentrations of these excipients that best stabilize a particular enzyme must be determined empirically. Some enzymes are greatly stabilized by the presence of cofactors, substrates, and even inhibitors that bind to their active sites. Again, the best storage conditions must be established for each enzyme individually.

Another common problem in handling enzyme solutions is the loss of enzymatic activity due to protein adsorption onto the surfaces of containers and pipet tips. Proteins bind avidly to glass, quartz, and polystyrene surfaces. Hence, containers made of these materials should not be used for enzyme samples. Containers and transfer devices constructed of low protein binding materials, such as polypropylene or polyethylene, should be used whenever possible; a wide variety of containers and pipet tips made of these materials are available commercially.

Even with low protein binding materials, one will still experience losses of protein due to adsorption. To minimize these effects, it is often possible to add a carrier protein to enzyme samples, as long as it has been established that the carrier protein does not interfere with the enzyme assay in any way. A carrier protein is an inert protein that is added to the enzyme solution at much higher concentrations than that of the enzyme. In this way potential protein binding surfaces will be saturated with the carrier protein, hence are not available for adsorption of the enzyme of interest. It is a very common practice among enzymologists to add carrier proteins to the enzyme stock solutions and to the final reaction mixtures. Bovine serum albumin (BSA), gelatin, and casein are commonly used proteins for this purpose. Our laboratory has found that gelatin, at a concentration of 1 mg/mL, is a particularly good carrier protein for many enzymes. The lack of aromatic amino acids in the gelatin makes this a useful carrier protein for enzyme studies utilizing ultraviolet absorption or fluorescence spectroscopy. Gelatin, casein, and BSA are available commercially in highly purified forms from a number of suppliers.

Some workers have found polyethylene glycol, molecular weight 8000 Da (PEG-8000), to be a useful alternative to carrier proteins for minimizing enzyme adsorption to container surfaces (Andrew M. Stern, personal communication). Addition of PEG-8000 to 0.1% has been used in this regard for a number of enzymes. If PEG-8000 is to be used for this application, a high grade (i.e., molecular biology grade or the equivalent) should be used, since lower grades of PEGs may contain impurities that can have deleterious effects on enzyme activity. Our own experience with the use of PEG-8000 suggests that this additive works well to stabilize some, but not all, enzyme activities. Hence, again, the reader is left to explore the utility of this approach on a case-by-case basis.

7.6.2   Enzyme Inactivation During Activity Assays

Certain enzymes that are stable under optimized conditions of long-term storage (as just described) will inactivate during the course of an activity assay. This behavior is characterized by progress curves that plateau early, before significant substrate loss has occurred (see Section 7.1.2 for other causes of this behavior). There are two common reasons for this type of enzyme inactivation. First, the active conformation of the enzyme may not be stable under the specific conditions (i.e., temperature, pH, ionic strength, and dilution of enzyme concentration) used in the assay. For example, if the active form of the enzyme is a dimer, dilution to low concentration at the initiation of an activity assay may cause simultaneous dissociation of the dimeric enzyme to monomers. If the time course of dimer dissociation is slow, hence similar to that of the enzymatic assay, a diminution of activity may be seen over the time course of the activity measurements. Sometimes minor adjustments in final enzyme concentration can help to ameliorate this situation. Likewise, minor adjustments in other solution conditions can help to extend the lifetime of the active enzyme species during activity assays. For multisubstrate enzymes (see Chapter 11), the stability of the enzyme can sometimes also be greatly augmented by preforming a binary enzyme–substrate complex and initiating the reaction by addition of a second substrate.

The second cause of activity loss during assay is spontaneous enzyme inactivation that results directly from catalytic turnover. For some enzymes, the chemistry associated with turnover can lead to inactivation of the enzyme by covalent adduct formation, or by destruction of a key active site amino acid residue or cofactor. For example, some oxidoreductases form highly damaging free radical species as a by-product of their catalytic activity. When this occurs, the radicals that build up during turnover can attack the enzyme active site, rendering it inactive. In these cases, the radical-based inactivation can sometimes be minimized by the addition of free radical scavengers, such as phenol, to the reaction mixture. Addition of a small amount of a peroxidase enzyme, such as catalase, can also sometimes help to stabilize the enzyme of interest from radical-based inactivation. Of course, it is critical to determine that addition of such species does not affect the measurement of enzyme activity in other ways.

Regardless of the cause, enzyme inactivation during activity assays can be diagnosed by two simple tests. The first test is to allow the progress curve to go to its premature plateau and then add a small volume of additional enzyme stock that would double the final enzyme concentration in the reaction mixture (i.e., addition of a mass of enzyme equal to the initial enzyme mass in the reaction mixture). If enzyme inactiviation during the assay is the cause of the premature plateau, a second phase of reaction should be realized after the addition of the second volume of fresh enzyme.

The second test, known as Selwyn's test (Selwyn, 1965), consists of measuring the reaction progress curve at several different concentrations of enzyme. The test makes use of the fact that regardless of its complexity for individual enzymatic reactions, the integrated rate equation has the general form:

$$[E]t = f([P]) \tag{7.23}$$

when all other conditions are held constant. Hence the concentration of product, $[P]$ is some constant function of the multiplicative product of enzyme concentration and assay time. The term $[E]$ in Equation 7.23 refers to the concentration of active enzyme molecules in solution. If the enzyme is stable over the course of the assay, a plot of $[P]$ as a function of $[E]t$ should give superimposable curves at all concentrations of enzyme (Figure 7.31A). If, however, the enzyme is undergoing unimolecular inactivation during the course of the activity assay, the concentration of active enzyme will itself show a first-order time dependence. Thus, the dependence of $[P]$ on $[E]$ will have

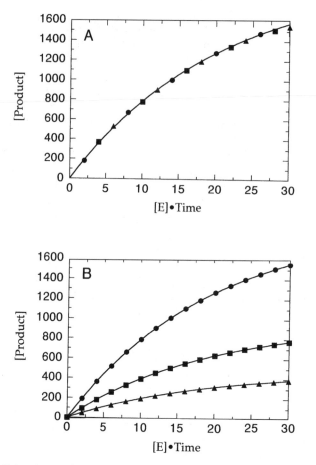

Figure 7.31 Selwyn's test for enzyme inactivation during an assay. (A) Data at several enzyme concentrations for an enzyme that is stable during the assay time course. Note that the data for different enzyme concentrations (represented by different symbols on the graph) are well fit by a single curve. (B) Corresponding data for an enzyme that undergoes inactivation during the course of the activity assay; the data for different enzyme concentrations cannot be fit by a single curve.

the more complex form of Equation 7.24:

$$[P] = \frac{k[E]}{\lambda}(1 - e^{-\lambda t}) \tag{7.24}$$

where k is a constant of proportionality and λ is the first-order decay constant for enzyme inactivation. Now plots of $[P]$ as a function $[E]t$ will vary with changing enzyme concentration (Figure 7.31B). The lack of superposition of the data plots, as seen in Figure 7.31B, is a clear indication that enzyme inactivation has occurred during the assay time period.

7.7 SUMMARY

In this chapter we have presented an overview of some of the common methodologies for obtaining initial velocity measurements of enzymatic reactions. The most common detection methods and techniques for separating substrate and product molecules after reaction were discussed. We saw that changes in reaction conditions, such as pH and temperature, can have dramatic effects on enzymatic reaction rate. We saw further that controlled changes in these conditions can be used to obtain mechanistic information about the enzyme of interest. Finally, some advice was provided for the proper storage and handling of enzymes to optimally maintain their catalytic activity in the laboratory.

REFERENCES AND FURTHER READING

Bender, M. L., Kezdy, F. J., and Wedler, F. C. (1967) *J. Chem. Educ.* **44**, 84.

Blacklow, S. C., Raines, R. T., Lim, W. A., Zamore, P. D., and Knowles, J. R. (1988) *Biochemistry*, **27**, 1158.

Cleland, W. W. (1979) *Anal. Biochem.* **99**, 142.

Cleland, W. W., O'Leary, M. H., and Northrop, D. B. (1977) *Isotope Effects on Enzyme-Catalyzed Reactions*, University Park Press, Baltimore.

Copeland, R. A. (1994) *Methods for Protein Analysis: A Practical Guide to Laboratory Protocols*, Chapman & Hall, New York.

Copeland, R. A., Williams, J. M., Giannaras, J., Nurnberg, S., Covington, M., Pinto, D., Pick, S., and Trzaskos, J. M. (1994) *Proc. Natl. Acad. Sci. USA* **91**, 11202.

Copeland, R. A., Lombardo, D., Giannaras, J., and Decicco, C. P. (1995) *Bioorg. Med. Chem. Lett.* **5**, 1947.

Cornish-Bowden, A. (1972) *Biochem. J.* **130**, 637.

Dixon, M., and Webb, E. C. (1979) *Enzymes*, 3rd ed., Academic Press, New York.

Duggleby, R. G. (1985) *Biochem. J.* **228**, 55.

Duggleby, R. G. (1994) *Biochim. Biophys. Acta*, **1205**, 268.

Duggleby, R. G., and Morrison, J. F. (1977) *Biochim. Biophys. Acta*, **481**, 297.

Easterby, J. S. (1973) *Biochim. Biophys. Acta*, **293**, 552.

Eisenthal, R., and Danson, M. J., Eds. (1992) *Enzyme Assays, A Practical Approach*, IRL Press, Oxford.

Fersht, A. (1985) *Enzyme Structure and Function*, 2nd ed., Freeman, New York.

Fletcher, A. N. (1969) *Photochem. Photobiol.* **9**, 439.

Gabriel, O., and Gersten, D. M. (1992) In *Enzyme Assays, A Practical Approach*, R. Eisenthal and M. J. Danson, Eds., IRL Press, Oxford, pp. 217–253.

Gomori, G. (1992) In *CRC Practical Handbook of Biochemistry and Molecular Biology*, G. D. Fasman, Ed., CRC Press, Boca Raton, FL, pp. 553–560.

Hames, B. D., and Rickwood, D. (1990) *Gel Electrophoresis of Proteins: A Practical Approach*, 2nd ed., IRL Press, Oxford.

Hancock, W. S. (1984) *Handbook of HPLC Separation of Amino Acids, Peptides, and Proteins*, CRC Press, Boca Raton, FL.

Harlow, E., and Lane, D. (1988) *Antibodies: A Laboratory Manual*, Cold Spring Harbor Laboratory, Cold Spring Harbor, NY.

Haupt, G. W. (1952) *J. Res. Natl. Bur. Stand.* **48**, 414.

Ittarat, I., Webster, H. K., and Yuthavong, Y. (1992) *J. Chromatography*, **582**, 57.

Kellershohn, N., and Laurent, M. (1985) *Biochem. J.* **231**, 65.

Knight, C. G. (1995) *Methods Enzymol.* **248**, 18–34.

Knight, C. G., Willenbrock, F., and Murphy, G. (1992) *FEBS Lett.* **296**, 263.

Kyte, J. (1995) *Mechanisms in Protein Chemistry*, Garland, New York.

Lackowicz, J. R. (1983) *Principles of Fluorescence Spectroscopy*, Plenum Press, New York.

Matayashi, E. D., Wang, G. T., Krafft, G. A., and Erickson, J. (1990) *Science*, **247**, 954.

Mozhaev, V. V., Berezin, I. V., and Martinek, K. (1987) *Methods Enzymol.* **135**, 586.

Northrop, D. B. (1975) *Biochemistry*, **14**, 2644.

Oldham, K. G. (1968) *Radiochemical Methods of Enzyme Analysis*, Amersham International, Amersham, U.K.

Oldham, K. G. (1977) In *Radiotracer Techniques and Applications*, Vol. 2, E. A. Evans and M. Muramatsu, Eds., Dekker, New York, pp. 823–891.

Oldham, K. G. (1992) In *Enzyme Assays, A Practical Approach*, R. Eisenthal and M. J. Danson, Eds., IRL Press, Oxford, pp. 93–122.

Oliver, R. W. (1989) *HPLC of Macromolecules: A Practical Approach*, IRL Press, Oxford.

Packard, B. Z., Toptygin, D. D., Komoriya, A., and Brand, L. (1997) *Methods Enzymol.* **278**, 15–28.

Palmer, T. (1985) Understanding Enzymes, Wiley, New York.

Purich, D. L., Ed. (1996) *Contemporary Enzyme Kinetics and Mechanisms*, 2nd ed., Academic Press, San Diego, CA.

Roughton, F. J. W., and Chance, B. (1963) In *Techniques of Organic Chemistry*, Vol. VIII, Part II, *Investigation of Rates and Mechanisms of Reactions*, S. L. Friess, E. S. Lewis, and A. Weissberger, Eds., Wiley, New York, pp. 703–792.

Rudolph, F. B., Baugher, B. W., and Beissmer, R. S. (1979) *Methods Enzymol.* **63**, 22.

Russo, S. F. (1969) *J. Chem. Educ.* **46**, 374.

Schonbaum, G. R., Zerner, B., and Bender, M. L. (1961) *J. Biol. Chem.* **236**, 2930.

Schowen, K. B., and Schowen, R. L. (1982) *Methods Enzymol.* **87**, 551.

Segel, I. H. (1976) *Biochemical Calculations*, 2nd ed., Wiley, New York.

Selwyn, M. J. (1965) *Biochim. Biophys. Acta*, **105**, 193.

Silverman, R. B. (1988) *Mechanism-Based Enzyme Inactivation: Chemistry and Enzymology*, Vols. I and II, CRC Press, Boca Raton, FL.

Storer, A. C., and Cornish-Bowden, A. (1974) *Biochem. J.* **141**, 205.

Tipton, K. F. (1992) In *Enzyme Assays, A Practical Approach*, R. Eisenthal and M. J. Danson, Eds., IRL Press, Oxford, pp. 1–58.

Tsukada, H., and Blow, D. M. (1985) *J. Mol. Biol.* **184**, 703.

Venkatasubban, K. S., and Schowen, R. L. (1984) *CRC Crit. Rev. Biochem.* **17**, 1.

Waley, S. G. (1982) *Biochem. J.* **205**, 631.

8

REVERSIBLE INHIBITORS

The activity of an enzyme can be blocked in a number of ways. For example, inhibitory molecules can bind to sites on the enzyme that interfere with proper turnover. We encountered the concept of substrate and product inhibition in Chapters 5, 6, and 7. For product inhibition, the product molecule bears some structural resemblance to the substrate and can thus bind to the active site of the enzyme. Product binding blocks the binding of further substrate molecules. This form of inhibition, in which substrate and inhibitor compete for a common enzyme species, is known as *competitive inhibition*. Perhaps less intuitively obvious are processes known as *noncompetitive* and *uncompetitive inhibition*, which define inhibitors that bind to distinct enzyme species and still block turnover. In this chapter, we discuss these varied modes of inhibiting enzymes and examine kinetic methods for distinguishing among them.

There are several motivations for studying enzyme inhibition. At the basic research level, inhibitors can be useful tools for distinguishing among different potential mechanisms of enzyme turnover, particularly in the case of multisubstrate enzymes (see Chapter 11). By studying the relative binding affinity of competitive inhibitors of varying structure, one can glean information about the active site structure of an enzyme in the absence of a high resolution three-dimensional structure from x-ray crystallography or NMR spectroscopy. Inhibitors occur throughout nature, and they provide important control mechanisms in biology. Associated with many of the proteolytic enzymes involved in tissue remodeling, for example, are protein-based inhibitors of catalytic action that are found in the same tissue sources as the enzymes themselves. By balancing the relative concentrations of the proteases and their inhibitors, an organism can achieve the correct level of homeostasis. Enzyme inhibitors have a number of commercial applications as well. For example, enzyme inhibitors form the basis of a number of agricultural products, such as insecticides and weed killers of certain types. Inhibitors are extensively used to control parasites and other pest organisms by selectively inhibiting an enzyme of the pest, while sparing the enzymes of the host organism. Many of the drugs that are prescribed by physicians to combat diseases function by inhibiting specific enzymes associated with the disease process (see Table 1.1 for some examples). Thus, enzyme inhibition is a major research focus throughout the pharmaceutical industry.

Inhibitors can act by irreversibly binding to an enzyme and rendering it inactive. This typically occurs through the formation of a covalent bond between some group on the enzyme molecule and the inhibitor. We shall discuss this type of inhibition in Chapter 10. Also, some inhibitors can bind so tightly to the enzyme that they are for all practical purposes permanently bound (i.e., their dissociation rates are very slow). These inhibitors, which form a special class known as *tight binding inhibitors*, are treated separately, in Chapter 9. In their most commonly encountered form, however, inhibitors are molecules that bind reversibly to enzymes with rapid association and dissociation rates. Molecules that behave in this way, known as classical reversible inhibitors, serve as the focus of our attention in this chapter.

Much of the basic and applied use of reversible inhibitors relies on their ability to bind specifically and with reasonably high affinity to a target enzyme. The relative potency of a reversible inhibitor is measured by its binding capacity for the target enzyme, and this is typically quantified by measuring the dissociation constant for the enzyme–inhibitor complex:

$$[E] + [I] \underset{K_d}{\rightleftharpoons} [EI]$$

$$K_d = \frac{[E][I]}{[EI]}$$

The concept of the dissociation constant as a measure of protein–ligand interactions was introduced in Chapter 4. In the particular case of enzyme–inhibitor interactions, the dissociation constant is often referred to also as the inhibitor constant and is given the special symbol K_i. The K_i value of a reversible enzyme inhibitor can be determined experimentally in a number of ways. Experimental methods for measuring equilibrium binding between proteins and ligands, discussed in Chapter 4, include equilibrium dialysis, and chromatographic and spectroscopic methods. New instrumentation based on surface plasmon resonance technology (e.g., the BIAcore system from Pharmacia Biosensor) also allows one to measure binding interactions between ligands and macromolecules in real time (Chaiken et al., 1991; Karlsson, 1994). While this method has been mainly applied to determining the binding affinities for antigen–antibody and receptor–ligand interactions, the same technology holds great promise for the study of enzyme–ligand interactions as well. For example, this method has already been used to study the interactions between protein-based protease inhibitors and their enzyme targets (see, e.g., Ma et al., 1994). Although these and many other physicochemical methods have been applied to the determination of K_i values for enzyme inhibitors, the most common and straightforward means of assessing inhibitor binding consists of determining its effect on the catalytic activity of the enzyme. By measuring the diminution of initial velocity with increasing concentration of the inhibitor, one can find the relative concentrations of free enzyme and enzyme–inhibitor complex at any particular inhibitor concentration, and thus calculate the relevant equilibrium constant. For the remainder of this chapter, we shall focus on the determination of K_i values through initial velocity measurements of these types.

8.1 EQUILIBRIUM TREATMENT OF REVERSIBLE INHIBITION

To understand the molecular basis of reversible inhibition, it is useful to reflect upon the equilibria between the enzyme, its substrate, and the inhibitor that can occur in solution. Figure 8.1 provides a generalized scheme for the potential interactions between these molecules. In this scheme, K_S is the equilibrium constant for dissociation of the ES complex to the free enzyme and the free substrate, K_i is the dissociation constant for the EI complex, and k_p is the forward rate constant for product formation from the ES or ESI complexes. The factor α reflects the effect of inhibitor on the affinity of the enzyme for its substrate, and likewise the effect of the substrate on the affinity of the enzyme for the inhibitor. The factor β reflects the modification of the rate of product formation by the enzyme that is caused by the inhibitor. An inhibitor that

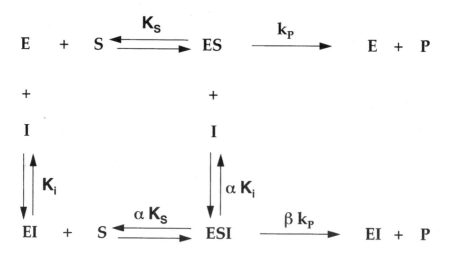

Figure 8.1 Equilibrium scheme for enzyme turnover in the presence and absence of an inhibitor.

completely blocks enzyme activity will have β equal to zero. An inhibitor that only partially blocks product formation will be characterized by a value of β between 0 and 1. An enzyme activator, on the other hand, will provide a value of β greater than 1.

The question is often asked: Why is the constant α the same for modification of K_S and K_i? The answer is that this constant must be the same for both on thermodynamic grounds. To illustrate, let us consider the following set of coupled reactions:

$$E + S \underset{K_S}{\rightleftharpoons} ES \qquad \Delta G = RT\ln(K_S) \tag{8.1}$$

$$ES + I \underset{\alpha K_i}{\rightleftharpoons} ESI \qquad \Delta G = RT\ln(\alpha K_i) \tag{8.2}$$

The net reaction of these two is:

$$E + S + I \rightleftharpoons ESI \qquad \Delta G = RT\ln(\alpha K_i K_S) \tag{8.3}$$

Now consider two other coupled reactions:

$$E + I \rightleftharpoons EI \qquad \Delta G = RT\ln(K_i) \tag{8.4}$$

$$EI + S \underset{aK_S}{\rightleftharpoons} ESI \qquad \Delta G = RT\ln(aK_S) \tag{8.5}$$

The net reaction here is:

$$E + S + I \rightleftharpoons ESI \qquad \Delta G = RT\ln(aK_S K_i) \tag{8.6}$$

Both sets of coupled reactions yield the same overall net reaction. Since, as we reviewed in Chapter 2, ΔG is a path-independent function, it follows that Equations 8.3 and 8.6 have the same value of ΔG. Therefore:

$$RT\ln(\alpha K_i K_S) = RT\ln(aK_S K_i) \tag{8.7}$$

$$\therefore \quad \alpha(K_i K_S) = a(K_i K_S) \tag{8.8}$$

$$\therefore \quad \alpha = a \tag{8.9}$$

Thus, the value of α is indeed the same for the modification of K_S by inhibitor and the modification of K_i by substrate.

The values of α and β provide information on the degree of modification that one ligand (i.e., substrate or inhibitor) has on the binding of the other ligand, and they define different modes of inhibitor interaction with the enzyme.

8.2 MODES OF REVERSIBLE INHIBITION

8.2.1 Competitive Inhibition

Competitive inhibition refers to the case of the inhibitor binding exclusively to the free enzyme and not at all to the ES binary complex. Thus, referring to the scheme in Figure 8.1, complete competitive inhibition is characterized by values of $\alpha = \infty$ and $\beta = 0$. In competitive inhibition the two ligands (inhibitor and substrate) compete for the same enzyme form and generally bind in a mutually exclusive fashion; that is, the free enzyme binds either a molecule of inhibitor or a molecule of substrate, but not both simultaneously. Most often competitive inhibitors function by binding at the enzyme active site, hence competing directly with the substrate for a common site on the free enzyme, as depicted in the cartoon of Figure 8.2A. In these cases the inhibitor usually shares some structural commonality with the substrate or transition state of the reaction, thus allowing the inhibitor to make similar favorable interactions with groups in the enzyme active site. This is not, however, the only way that a competitive inhibitor can block substrate binding to the free enzyme. It is also possible (although perhaps less likely) for the inhibitor to bind at a distinct site that is distal to the substrate binding site, and to induce some type of conformation change in the enzyme that modifies the active site so that substrate can no longer bind. The observation of competitive inhibition therefore cannot be viewed as prima facie evidence for commonality of binding sites for the inhibitor and substrate. The best that one can say from kinetic measurements alone is that the two ligands compete for the same form of the enzyme — the free enzyme.

When the concentration of inhibitor is such that less than 100% of the enzyme molecules are bound to inhibitor, one will observe residual activity due to the population of free enzyme. The molecules of free enzyme in this population will turn over at the same rate as in the absence of inhibitor, displaying the same maximal velocity. The competition between the inhibitor and substrate for free enzyme, however, will have the effect of increasing the concentration of substrate required to reach half-maximal velocity. Hence the presence of a competitive inhibitor in the enzyme sample has the kinetic effect of raising the apparent K_m of the enzyme for its substrate without affecting the value of V_{max}; this kinetic behavior is diagnositic of competitive inhibition. Because of the competition between inhibitor and substrate, a hallmark of competitive inhibition is that it can be overcome at high substrate concentrations; that is, the apparent K_i of the inhibitor increases with increasing substrate concentration.

8.2.2 Noncompetitive Inhibition

"Noncompetitive inhibition" refers to the case in which an inhibitor displays binding affinity for both the free enzyme and the enzyme–substrate binary

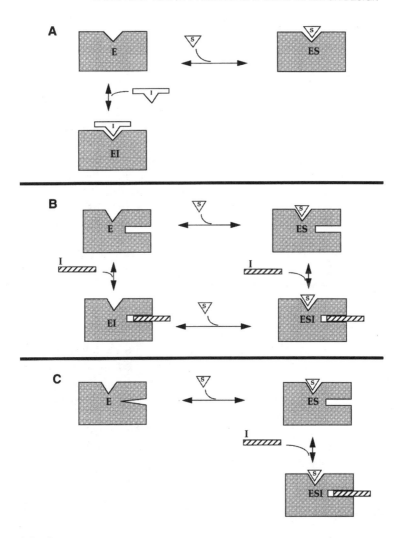

Figure 8.2 Cartoon representations of the three major forms of inhibitor interactions with enzymes: (A) competitive inhibition, (B) noncompetitive inhibition, and (C) uncompetitive inhibition.

complex. Hence, complete noncompetitive inhibition is characterized by a finite value of α and $\beta = 0$. This form of inhibition is the most general case that one can envision from the scheme in Figure 8.1; in fact, competitive and uncompetitive (see below) inhibition can be viewed as special, restricted cases of noncompetitive inhibition in which the value of α is infinity or zero, respectively. Noncompetitive inhibitors do not compete with substrate for binding to the free enzyme; hence they bind to the enzyme at a site distinct from the active site. Because of this, noncompetitive inhibition cannot be overcome by increasing substrate concentration. Thus, the apparent effect of a noncompetitive inhibitor is to decrease the value of V_{max} without affecting the apparent K_m for the substrate. Figure 8.2B illustrates the interactions between a noncompetitive inhibitor and its enzyme target.

The enzymological literature is somewhat ambiguous in its designations of noncompetitive inhibition. Some authors reserve the term "noncompetitive inhibition" exclusively for the situation in which the inhibitor displays equal affinity for both the free enzyme and the ES complex (i.e., $\alpha = 1$). When the inhibitor displays finite but unequal affinity for the two enzyme forms, these authors use the term "mixed inhibitors" (i.e., α is finite but not equal to 1). Indeed, the first edition of this book used this more restrictive terminology. In teaching this material to students, however, I have found that "mixed inhibition" is confusing and often leads to misunderstandings about the nature of the enzyme–inhibitor interactions. Hence, we shall use *noncompetitive inhibition* in the broader context from here out and avoid the term "mixed inhibition." The reader should, however, make note of these differences in terminology to avoid confusion when reading the literature.

8.2.3 Uncompetitive Inhibitors

Uncompetitive inhibitors bind exclusively to the ES complex, rather than to the free enzyme form. The apparent effect of an uncompetitive inhibitor is to decrease V_{max} and to actually decrease K_m (i.e., increase the affinity of the enzyme for its substrate). Therefore, complete uncompetitive inhibitors are characterized by $\alpha \ll 1$ and $\beta = 0$ (Figure 8.2C).

Note that a truly uncompetitive inhibitor would have *no* affinity for the free enzyme; hence the value of K_i would be infinite. The inhibitor would, however, have a measurable affinity for the ES complex, so that αK_i would be finite. Obviously this situation is not well described by the equilibria in Figure 8.1. For this reason many authors choose to distinguish between the dissociation constants for [E] and [ES] by giving them separate symbols, such as K_{iE} and K_{iES}, K_i and K_I, and K_{is} and K_{ii} (the subscripts in this latter nomenclature refer to the effects on the slope and intercept values of double reciprocal plots, respectively). Only rarely, however, does the inhibitor have no affinity whatsoever for the free enzyme. Rather, for uncompetitive inhibitors it is usually the case that $K_{iE} \gg K_{iES}$. Thus we can still apply the scheme in Figure 8.1 with the condition that $\alpha \ll 1$.

8.2.4 Partial Inhibitors

Until now we have assumed that inhibitor binding to an enzyme molecule completely blocks subsequent product formation by that molecule. Referring to the scheme in Figure 8.1, this is equivalent to saying that $\beta = 0$ in these cases. In some situations, however, the enzyme can still turn over with the inhibitor bound, albeit at a far reduced rate compared to the uninhibited enzyme. Such situations, which manifest *partial inhibition*, are characterized by

$0 < \beta < 1$. The distinguishing feature of a partial inhibitor is that the activity of the enzyme cannot be driven to zero even at very high concentrations of the inhibitor. When this is observed, experimental artifacts must be ruled out before concluding that the inhibitor is acting as a partial inhibitor. Often, for example, the failure of an inhibitor to completely block enzyme activity at high concentrations is due to limited solubility of the compound. Suppose that the solubility limit of the inhibitor is 10 μM, and at this concentration only 80% inhibition of the enzymatic velocity is observed. Addition of compound at concentrations higher that 10 μM would continue to manifest 80% inhibition, as the inhibitor concentration in solution (i.e., that which is soluble) never exceeds the solubility limit of 10 μM. Hence such experimental data must be examined carefully to determine the true reason for an observed partial inhibition. True partial inhibition is relatively rare, however, and we shall not discuss it further. A more complete description of partial inhibitors has been presented elsewhere (Segel, 1975).

8.3 GRAPHIC DETERMINATION OF INHIBITOR TYPE

8.3.1 Competitive Inhibitors

A number of graphic methods have been described for determining the mode of inhibition of a particular molecule. Of these, the double reciprocal, or Lineweaver–Burk, plot is the most straightforward means of diagnosing inhibitor modality. Recall from Chapter 5 that a double reciprocal plot graphs the value of reciprocal velocity as a function of reciprocal substrate concentration to yield, in most cases, a straight line. As we shall see, overlaying the double-reciprocal lines for an enzyme reaction carried out at several fixed inhibitor concentrations will yield a pattern of lines that is characteristic of a particular inhibitor type. The double-reciprocal plot was introduced in the days prior to the widespread use of computer-based curve-fitting methods, as a means of easily estimating the kinetic values K_m and V_{max} from the linear fits of the data in these plots. As we have described in Chapter 5, however, systematic weighting errors are associated with the data manipulations that must be performed in constructing such plots.

To avoid weighting errors and still use these reciprocal plots qualitatively to diagnose inhibitor modality, we make the following recommendation. To diagnose inhibitor type, measure the initial velocity as a function of substrate concentration at several fixed concentrations of the inhibitor of interest. To select fixed inhibitor concentrations for this type of experiment, first measure the effect of a broad range of inhibitor concentrations with [S] fixed at its K_m value (i.e., measure the Langmuir isotherm for inhibition (see Section 8.4) at $[S] = K_m$). From these results, choose inhibitor concentrations that yield between 30 and 75% inhibition under these conditions. This procedure will ensure that significant inhibitor effects are realized while maintaining sufficient signal from the assay readout to obtain accurate data.

With the fixed inhibitor concentrations chosen, plot the data in terms of velocity as a function of substrate concentration for each inhibitor concentration, and fit these data to the Henri–Michaelis–Menten equation (Equation 5.24). Determine the values of K_m^{app} (i.e., the *apparent* value of K_m at different inhibitor concentrations) and V_{max}^{app} directly from the nonlinear least-squares best fits of the untransformed data. Finally, plug these values of K_m^{app} and V_{max}^{app} into the reciprocal equation (Equation 5.34) to obtain a linear function, and plot this linear function for each inhibitor concentration on the same double-reciprocal plot. In this way the double-reciprocal plots can be used to determine inhibitor modality from the pattern of lines that result from varying inhibitor concentrations, but without introducing systematic errors that could compromise the interpretations.

Let's walk through an example to illustrate the method, and to determine the expected pattern for a competitive inhibitor. Let us say that we measure the initial velocity of our enzymatic reaction as a function of substrate concentration at 0, 10, and 25 μM concentrations of an inhibitor, and obtain the results shown in Table 8.1.

If we were to plot these data, and fit them to Equation 5.24, we would obtain a graph such as that illustrated in Figure 8.3A. From the fits of the data we would obtain the following apparent values of the kinetic constants:

$$[I] = 0 \; \mu M \qquad V_{max} = 100, \qquad K_m = 10.00 \; \mu M$$

$$[I] = 10 \; \mu M, \qquad V_{max}^{app} = 100, \qquad K_m^{app} = 30.00 \; \mu M$$

$$[I] = 25 \; \mu M, \qquad V_{max}^{app} = 100, \qquad K_m^{app} = 60.00 \; \mu M$$

Table 8.1   Hypothetical velocity as a function of substrate concentration at three fixed concentrations of a competitive inhibitor

	Velocity (arbitrary units)		
[S] (μM)	[I] = 0	[I] = 10 μM	[I] = 25 μM
1	9.09	3.23	1.69
2	16.67	6.25	3.23
4	28.57	11.77	6.25
6	37.50	16.67	9.09
8	44.44	21.05	11.77
10	50.00	25.00	14.29
20	66.67	40.00	25.00
30	75.00	50.00	33.33
40	80.00	57.14	40.00
50	83.33	62.50	45.46

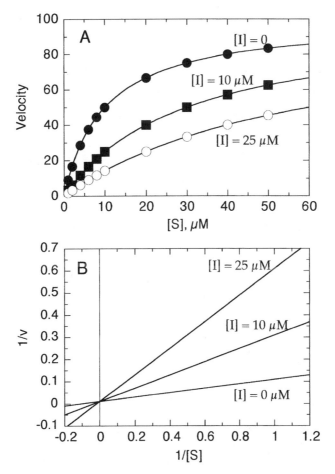

Figure 8.3 Untransformed (A) and double-reciprocal (B) plots for the effects of a competitive inhibitor on the velocity of an enzyme catalyzed reaction. The lines drawn in (B) are obtained by applying Equation 5.24 to the data in (A) and using the apparent values of the kinetic constants in conjunction with Equation 5.34. See text for further details.

If we plug these values of V_{max}^{app} and K_m^{app} into Equation 5.34 and plot the resulting linear functions, we obtain a graph like Figure 8.3B.

The pattern of straight lines with intersecting y intercepts seen in Figure 8.3B is the characteristic signature of a competitive inhibitor. The lines intersect at their y intercepts because a competitive inhibitor does not affect the apparent value of V_{max}, which, as we saw in Chapter 5, is defined by the y intercept in a double-reciprocal plot. The slopes of the lines, which are given by K_m^{app}/V_{max}^{app}, vary among the lines because of the effect imposed on K_m by the inhibitor. The degree of perturbation of K_m will vary with the inhibitor concentration and will depend also on the value of K_i for the particular inhibitor. The influence of these factors on the initial velocity is given by:

$$v = \frac{V_{max}[S]}{[S] + K_m\left(1 + \dfrac{[I]}{K_i}\right)} \tag{8.10}$$

or, taking the reciprocal of this equation, we obtain:

$$\frac{1}{v} = \frac{1}{V_{max}} + \left(\frac{1}{[S]}\frac{K_m}{V_{max}}\right)\left(1 + \frac{[I]}{K_i}\right) \tag{8.11}$$

Now, comparing Equation 8.11 to Equation 5.34, we see that the slopes of the double-reciprocal lines at inhibitor concentrations of 0 and i differ by the factor $(1 + [I]/K_i)$. Thus, the ratio of these slope values is:

$$\frac{slope_i}{slope_0} = 1 + \frac{[I]}{K_i} \tag{8.12}$$

or, rearranging:

$$K_i = \frac{[I]}{\left(\dfrac{slope_i}{slope_0}\right) - 1} \tag{8.13}$$

Thus, in principle, one could measure the velocity as a function of substrate concentration in the absence of inhibitor and at a single, fixed values of [I], and use Equation 8.13 to determine the K_i of the inhibitor from the double-reciprocal plots. This method can be potentially misleading, however, because it relies on a single inhibitor concentration for the determination of K_i.

A more common approach to determining the K_i value of a competitive inhibitor is to replot the kinetic data obtained in plots such as Figure 8.3A as the apparent K_m value as a function of inhibitor concentration. The x intercept of such a "secondary plot" is equal to the negative value of the K_i, as illustrated in Figure 8.4, using the data from Table 8.1.

In a third method for determining the K_i value of a competitive inhibitor suggested by Dixon (1953), one measures the initial velocity of the reaction as a function of inhibitor concentration at two or more fixed concentrations of substrate. The data are then plotted as $1/v$ as a function of [I] for each substrate concentration, and the value of $-K_i$ is determined from the x-axis value at which the lines intersect, as illustrated in Figure 8.5. The Dixon plot ($1/v$ as a function of [I]) is useful in determining the K_i values for other inhibitor types as well, as we shall see later in this chapter.

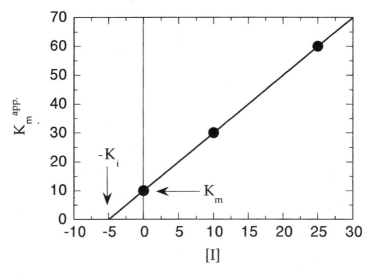

Figure 8.4   Secondary plot of K_m^{app} as a function of inhibitor concentration [I] for a competitive inhibitor. The value of the inhibitor constant K_i can be determined from the negative value of the x intercept of this type of plot.

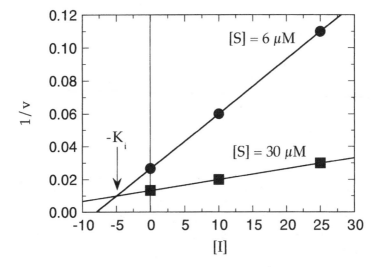

Figure 8.5   Dixon plot ($1/v$ as a function of [I]) for a competitive inhibitor at two different substrate concentrations. The K_i value for this type of inhibitor is determined from the negative of the x-axis value at the point of intersection of the two lines.

8.3.2 Noncompetitive Inhibitors

We have seen that a noncompetitive inhibitor has affinity for both the free enzyme and the ES complex; hence the dissociation constants from each of these enzyme forms must be considered in the kinetic analysis of these inhibitors. The most general velocity equation for an enzymatic reaction in the presence of an inhibitor is:

$$v = \frac{V_{max}[S]}{[S]\left(1 + \dfrac{[I]}{\alpha K_i}\right) + K_m\left(1 + \dfrac{[I]}{K_i}\right)} \qquad (8.14)$$

and this is the appropriate equation for evaluating noncompetitive inhibitors. Comparing Equations 8.14 and 8.10 reveals that the two are equivalent when α is infinite. Under these conditions the term $[S](1 + [I]/\alpha K_i)$ reduces to $[S]$, and Equation 8.14 hence reduces to Equation 8.10. Thus, as stated above, competitive inhibition can be viewed as a special case of the more general case of noncompetitive inhibition.

In the unusual situation that αK_i is exactly equal to K_i (i.e., α is exactly 1), we can replace the term αK_i by K_i and thus reduce Equation 8.14 to the following simpler form:

$$v = \frac{V_{max}[S]}{([S] + K_m)\left(1 + \dfrac{[I]}{K_i}\right)} \qquad (8.15)$$

Equation 8.15 is sometimes quoted in the literature as the appropriate equation for evaluating noncompetitive inhibition. As stated earlier, however, this reflects the more restricted use of the term "noncompetitive."

The reciprocal form of Equation 8.14 (after some canceling of terms) has the form:

$$\frac{1}{v} = \left(1 + \frac{[I]}{K_i}\right)\left(\frac{K_m}{V_{max}}\frac{1}{[S]}\right) + \frac{1 + \dfrac{[I]}{\alpha K_i}}{V_{max}} \qquad (8.16)$$

As described by Equation 8.16, both the slope and the y intercept of the double-reciprocal plot will be affected by the presence of a noncompetitive inhibitor. The pattern of lines seen when the plots for varying inhibitor concentrations are overlaid will depend on the value of α. When α exceeds 1, the lines will intersect at a value of $1/[S]$ less than zero and a value of $1/v$ of greater than zero (Figure 8.6A). If, on the other hand, $\alpha < 1$, the lines will intersect below the x and y axes, at negative values of $1/[S]$ and $1/v$ (Figure 8.6B). If $\alpha = 1$, the lines converge at $1/[S]$ less than zero on the x axis (i.e., at $1/[v] = 0$)

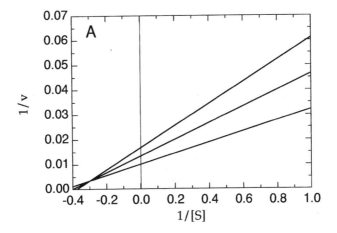

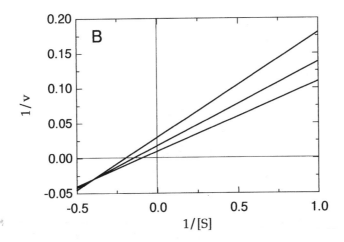

Figure 8.6 Patterns of lines in the double-reciprocal plots for noncompetitive inhibitors for (A) $\alpha > 1$ and (B) $\alpha < 1$.

To obtain the values of K_i and αK_i, two secondary plots must be constructed. The first of these is a Dixon plot of $1/V_{max}$ (i.e., at saturating substrate concentration) as a function of [I], from which the value of $-\alpha K_i$ can be determined as the x intercept (Figure 8.7A). In the second plot, the slope of the double-reciprocal lines (from the Lineweaver–Burk plot) are plotted as a function of [I]. For this plot, the x intercept will be equal to $-K_i$ (Figure 8.7B). Combining the information from these two secondary plots allows determination of both inhibitor dissociation constants from a single set of experimental data.

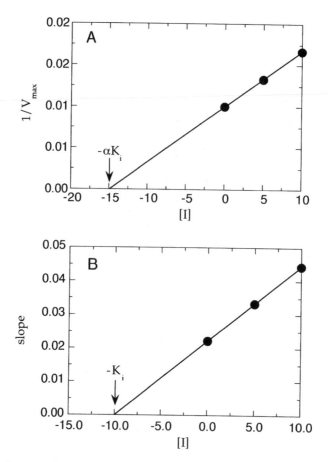

Figure 8.7 Secondary plots for the determination of the inhibitor constants for a noncompetitive inhibitor. (A) $1/V_{max}$ is plotted as a function of [I], and the value of $-\alpha K_i$ is determined from the x intercept of the line. (B) The value of $-K_i$ is determined from the x intercept of a plot of the slope of the lines from the double-reciprocal (Lineweaver–Burk) plot as a function of [I].

8.3.3 Uncompetitive Inhibitors

Both V_{max} and K_m are affected by the presence of an uncompetitive inhibitor. The form of the velocity equation therefore contains the dissociation constant αK_i in both the numerator and denominator:

$$v = \frac{\dfrac{V_{max}[S]}{1 + [I]/\alpha K_i}}{\dfrac{K_m}{1 + [I]/\alpha K_i} + [S]} \tag{8.17}$$

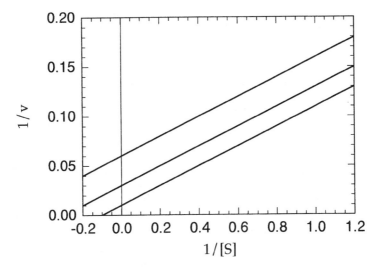

Figure 8.8   Pattern of lines in the double-reciprocal plot of an uncompetitive inhibitor.

If the numerator and denominator of Equation 8.17 are multiplied by $(1 + [I]/\alpha K_i)$, we can obtain the simpler form:

$$v = \frac{V_{max}[S]}{[S](1 + [I]/\alpha K_i) + K_m} \tag{8.18}$$

The reader will observe that Equation 8.18 is another special case of the more general equation given by Equation 8.14.

With a little algebra, it can be shown that the reciprocal form of Equation 8.17 is given by:

$$\frac{1}{v} = \frac{K_m}{V_{max}} \frac{1}{[S]} + \frac{1}{\dfrac{V_{max}}{1 + [I]/\alpha K_i}} \tag{8.19}$$

We see from equation 8.19 that the slope of the double-reciprocal plot is independent of inhibitor concentration and that the y intercept increases steadily with increasing inhibitor. Thus, the overlaid double-reciprocal plot for an uncompetitive inhibitor at varying concentrations appears as a series of parallel lines that intersect the y axis at different values, as illustrated in Figure 8.8.

For an uncompetitive inhibitor, the x intercept of a Dixon plot will be equal to $-\alpha K_i(1 + K_m/[S])$. At first glance this relationship may not look particularly convenient. If, however, one is working at saturating conditions, where $[S] \gg K_m$, the value of $K_m/[S]$ becomes very small and can be assumed to be zero. Under these conditions, the x intercept of the Dixon plot will be equal to

$-\alpha K_i$. Thus, *under conditions of saturating substrate*, one can determine the value of αK_i directly from the x intercept of a Dixon plot, as described earlier for the case of noncompetitive inhibition.

8.3.4 Global Fitting of Untransformed Data

The best method for determining inhibitor modality and the values of the inhibitor constant(s) is to fit directly and globally all the plots of velocity versus [S] at several fixed inhibitor concentrations to the untransformed equations for competitive (Equation 8.10), noncompetitive (Equation 8.14), and uncompetitive inhibition (Equation 8.18). From analysis of the statistical parameters for goodness of fit (typically χ^2), one can determine which model of inhibitor modality best describes the experimental data as a complete set and simultaneously determine the value of the inhibitor constant(s). This type of global fitting analysis has only recently become widely available. The commercial programs GraphFit and SigmaPlot, for example, allow this type of global fitting [i.e., fitting multiple curves that conform to the functional form $y = f(x, z)$, where x is substrate concentration and z is inhibitor concentration]. Cleland (1979) also published the source code for FORTRAN programs that allow this type of global data fitting. The reader is strongly encouraged to make use of these programs if possible.

8.4 DOSE–RESPONSE CURVES OF ENZYME INHIBITION

In many biological assays one can measure a specific signal as a function of the concentration of some exogenous substance. A plot of the signal obtained as a function of the concentration of exogenous substance is referred to as a *dose–response plot*, and the function that describes the change in signal with changing concentration of substance is known as a *dose–response curve* (Figure 8.9). These plots have the form of a Langmuir isotherm, as introduced in Chapter 4. We have already seen that such plots can be conveniently used to follow protein–ligand binding equilibria. The same plots are used to follow saturable events in a number of other biological contexts, such as effects of substances on cell growth and proliferation. Dose–response plots also can be used to follow the effects of an inhibitor on the initial velocity of an enzymatic reaction at a fixed concentration of substrate. The concentration of inhibitor required to achieve a half-maximal degree of inhibition is referred to as the IC_{50} value (for inhibitor concentration giving 50% inhibition), and the equation describing the effect of inhibitor concentration on reaction velocity is related to the Langmuir isotherm equation as follows:

$$\frac{v_i}{v_0} = \frac{1}{1 + \dfrac{[I]}{IC_{50}}} \tag{8.20}$$

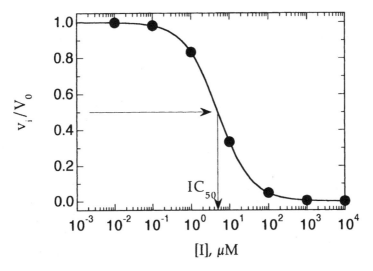

Figure 8.9 Dose–response plot of enzyme fractional activity as a function of inhibitor concentration. Note that the inhibitor concentration is plotted on a log scale. The value of the IC_{50} for the inhibitor can be determined graphically as illustrated.

where v_i is the initial velocity in the presence of inhibitor at concentration $[I]$ and v_0 is the initial velocity in the absence of inhibitor.

The observant reader will note two differences between the form of Equation 8.20 and that of the standard Langmuir isotherm equation (Equation 4.23). First, we have replaced the dissociation constant K_d (or in the case of enzyme inhibition, K_i) with the phenomenological term IC_{50}. This is because the concentration of inhibitor that displays half-maximal inhibition may be displaced from the true K_i by the influence of substrate concentration, as we shall describe shortly. The second difference between Equations 4.23 and 8.20 is that we have inverted the ratio of $[I]$ and IC_{50}. This is because the standard Langmuir isotherm equation tracks the fraction of ligand-bound receptor molecules. The term v_i/v_0 in Equation 8.20 is referred to as the *fractional activity* remaining at a given inhibitor concentration. This term reflects the fraction of free enzyme, rather than the fraction of inhibitor-bound enzyme. Considering mass conservation, the fraction of inhibitor-bound enzyme is related to the fractional activity as $1 - (v_i/v_0)$. Hence, we could recast Equation 8.20 in the more traditional form of the Langmuir isotherm as follows:

$$\text{fraction bound} = \left(1 - \frac{v_i}{v_0}\right) = \frac{1}{1 + \dfrac{IC_{50}}{[I]}} \tag{8.21}$$

Dose–response plots are very widely used for comparing the relative inhibitor potencies of multiple compounds for the same enzyme, under well-controlled conditions. The method is popular because it permits analysts to determine the IC_{50} by making measurements over a broad range of inhibitor concentrations at a single, fixed substrate concentration. A range of inhibitor concentrations spaning several orders of magnitude can be conveniently studied by means of the twofold serial dilution scheme described in Chapter 5 (Section 5.6.1), with inhibitor being varied in place of substrate here. This strategy is very convenient when many compounds of unknown and varying inhibitory potency are to be screened.

In the pharmaceutical industry, for example, one may wish to screen several thousand compounds as potential inhibitors to find those that have some potency against a particular target enzyme. These compounds are likely to span a wide range of IC_{50} values. Thus, one would set up a standard screening protocol in which the initial velocity of an enzymatic reaction is measured over five or more logs of inhibitor concentrations. In this way the IC_{50} values of many of the compounds could be determined without any prior knowledge of the range of concentrations required to effect potent inhibition of the enzyme.

The IC_{50} value is a practical readout of the relative effects on enzyme activity of different substances under a specific set of solution conditions. In many instances, it is the net effect of the inhibitor on enzyme activity, rather than its true dissociation constant for the enzyme, that is the ultimate criterion by which the effectiveness of a compound is judged. In some situations, a K_i value cannot be rigorously determined because of a lack of knowledge or control over the assay conditions; many times, in these cases, the only measure of relative inhibitor potency is an IC_{50} value. For example, consider the task of determining the relative effectiveness of a series of inhibitors for a target enzyme in a cellular assay. Often, in these cases, the inhibitor is added to the cell medium and the effects of inhibition are measured indirectly by a readout of biological activity that is dependent on the activity of the target enzyme. In a cellular situation like this, one often does not know either the substrate concentration in the cell or the relative amounts of enzyme and substrate (recall that in vitro we set up our steady state conditions so that $[S] \gg [E]$, but this is not necessarily the case in the cell). Also, in these situations, one does not truly know the effective concentration of inhibitor *within* the cell that is causing the degree of inhibition being measured. This is because the cell membrane may block the transport of the bulk of added inhibitor into the cell. Moreover, cellular metabolism may diminish the effective concentration of inhibitor that reaches the target enzyme. Because of these uncontrollable factors in the cellular environment, often it is necessary to report the effectiveness of an inhibitor as an IC_{50} value.

Despite their convenience and popularity, IC_{50} value measurements can be misleading if used inappropriately. The IC_{50} value of a particular inhibitor can change with changing solution conditions, so it is very important to report the details of the assay conditions along with the IC_{50} value. For example, in the case of competitive inhibition, the IC_{50} value observed for an inhibitor will depend on the concentration of substrate present in the assay, relative to the K_m of that substrate. This is illustrated in Figure 8.10 for a competitive

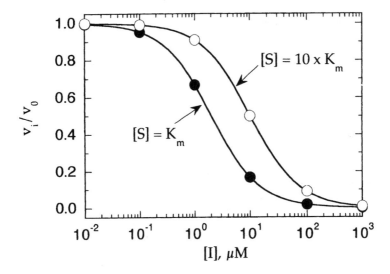

Figure 8.10 Effect of substrate concentration on the IC_{50} value of a competitive inhibitor.

inhibitor under conditions of $[S] = K_m$ and $[S] = 10 \times K_m$. Thus, in comparing a series of competitive inhibitors, it is important to ensure that the IC_{50} values are measured at the same substrate concentration. For the same reasons, it is not rigorously correct to compare the relative potencies of inhibitors of different modalities by use of IC_{50} values. The IC_{50} values of a noncompetitive and a competitive inhibitor will vary with substrate concentration, but in different ways. Hence, the relative effectiveness observed in vitro under a particular set of solution conditions may not be the same relative effectiveness observed in vivo, where the conditions are quite different. Whenever possible, therefore, the K_i values should be used to compare the inhibitory potency of different compounds.

It is possible to take advantage of the convenience of IC_{50} measurements and still report inhibitor potency in terms of true K_i values when the mode of inhibition for a series of compounds is known, as well as the values of $[S]$ and K_m. The relationship between the K_i, $[S]$, K_m, and IC_{50} values can be derived from the velocity equations already presented. The derivations have been described in detail by Cheng and Prusoff (1973) for competitive, noncompetitive, and uncompetitive inhibitors. The reader is referred to the original paper for the derivations. Here we shall simply present the final forms of the relationships

For competitive inhibitors:

$$K_i = \frac{IC_{50}}{1 + \dfrac{[S]}{K_m}} \tag{8.22}$$

For noncompetitive inhibitors:

$$IC_{50} = \frac{[S] + K_m}{\dfrac{K_m}{K_i} + \dfrac{[S]}{\alpha K_i}}$$

$$\text{if } \alpha = 1 \quad K_i = IC_{50} \tag{8.23}$$

For uncompetitive inhibitors:

$$\alpha K_i = \frac{IC_{50}}{1 + \dfrac{K_m}{[S]}} \qquad (\text{if } [S] \gg K_m, \text{ then } \alpha K_i \sim IC_{50}) \tag{8,24}$$

Equations 8.22–8.24, known as the Cheng and Prusoff relationships, can be conveniently used to convert IC_{50} values to K_i values. To ensure that the correct relationship can be applied, however, it is critical to know the mode of inhibition of the compounds being tested. It might thus seem that there is no great advantage to the use of the Cheng and Prusoff relationships if the mode of inhibition for each compound must be determined by Lineweaver–Burk analysis anyway. In many cases, however, one will wish to measure the relative inhibitory potency of a series of structurally related compounds. If these compounds represent small structural perturbations from a common parent molecule, it is often safe to assume that all the derivative molecules share the same mode of inhibition as the parent. In such situations, one could determine the mode of inhibition for the parent molecule only and then apply the appropriate Cheng and Prusoff relationship to the rest of the molecular series.

There is, of course, the possibility of an inadvertent change in the mode of inhibition as a result of the structural perturbations. This is usually not a great danger if the perturbations are minor, and one can spot-check by performing Lineweaver–Burk analysis on a subgroup of compounds representing a wide range of perturbations within the series. This is a common strategy used in development of structure–activity relationships for the determination of the key structural components in the inhibitory mechanism shared by a series of related compounds, as described next, in Section 8.6. Many scientists, however, consider the K_i values derived by application of the Cheng and Prusoff relationships to be less accurate than those obtained by the more traditional methods described earlier. There is lower confidence in the former results partly because the effects of the inhibitor are examined at only a single, fixed substrate concentration. Nevertheless, because of their convenience, the Cheng and Prusoff relationships are commonly used for high throughput inhibitor screening.

At the beginning of this chapter we mentioned that some inhibitors do not block completely the ability of the enzyme to turnover when bound to the inhibitor. These partial inhibitors will not display the same dose–response curves as full inhibitors because, for these compounds, one can never drive the reaction velocity to zero, even at very high inhibitor concentrations. Rather, the dose–response curve for a partial inhibitor will be best fit by a more generalized form of Equation 8.20, given by:

$$y = \frac{y_{max} - y_{min}}{1 + \left(\dfrac{[I]}{IC_{50}}\right)} + y_{min} \tag{8.25}$$

where y is the fractional activity of the enzyme in the presence of inhibitor at concentration $[I]$, y_{max} is the maximum value of y that is observed at zero inhibitor concentration (for fractional activity, this is 1.0), and y_{min} is the minimum value of y that can be obtained at high inhibitor concentrations.

Unlike the case of full inhibitors, the dose–response curve for a partial inhibitor will reach a minimum, nonzero value of v_i/v_0 at high inhibitor concentrations. In Figure 8.11A, for example, the value of β for our inhibitor is 0.05, so that even at very high inhibitor concentrations, the enzyme still displays 5% of its uninhibited velocity. When behavior of this type is observed, one must be very careful to ensure that the lack of complete inhibition is not an experimental artifact. For example, in densitometry measurements one often observes some finite background density that is difficult to completely subtract out and can give the appearance of partial inhibition when, in fact, full inhibition is taking place.

A more diagnostic signature of partial inhibition can be obtained by arranging the data as a Dixon plot. While all the full inhibitors discussed thus far yielded linear fits in Dixon plots, partial inhibitors typically display hyperbolic fits of the data in these plots (Figure 8.11B). In these cases one can extract the values of α, K_i, and β for the inhibitor, depending on the mode of partial inhibition that is taking place. These analyses are, however, beyond the scope of the present text. The reader who encounters this relatively unusual form of enzyme inhibition is referred to the text by Segel (1975) for a more comprehensive discussion of the data analysis.

8.5 MUTUALLY EXCLUSIVE BINDING OF TWO INHIBITORS

If two structurally distinct inhibitors, I and J, are found to both act on the same enzyme, it is possible for them to bind simultaneously to form an EIJ complex (or an ESIJ complex if both inhibitors are capable of binding to the ES complex). Alternatively, the two inhibitors may bind in a mutually exclusive fashion (i.e., competitive with each other) so that only an EI or an EJ complex can form. There are several tests by which it can be determined if the two inhibitors compete for binding to the enzyme.

The most direct way to measure exclusivity of inhibitor binding is by use of a radiolabeled or fluorescently labeled version of one of the inhibitors. If such labels are used to follow direct binding of the inhibitor to the enzyme, the

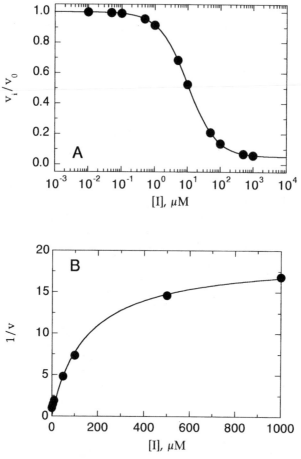

Figure 8.11 Dose–response (A) and Dixon (B) plots for a partial inhibitor. The value of v_i/v_0 in (A) reaches a nonzero plateau at high inhibitor concentrations. The hyperbolic nature of the Dixon plot in (B) is characteristic of partial inhibition.

ability of the second inhibitor to interfere with this binding can be directly measured as described in Chapter 4.

A number of kinetic measures have also been described to test the exclusivity of inhibitor interactions with a target enzyme (see Martines-Irujo et al., 1998, for a recent review). All these methods involve measuring the initial velocity of the enzyme at different combinations of the two inhibitors. The effects of two inhibitors on the velocity of an enzymatic reaction can be generally described by the following reciprocal relationship:

$$\frac{1}{v_{ij}} = \frac{1}{v_0}\left(1 + \frac{[I]}{K_i} + \frac{[J]}{K_j} + \frac{[I][J]}{\alpha K_i K_j}\right) \tag{8.26}$$

where v_{ij} is the initial velocity in the presence of both inhibitors, K_i and K_j are the dissociation constants for inhibitors I and J, respectively, and α is an interaction term that defines the effect of the binding of one inhibitor on the affinity of the second inhibitor. If the two inhibitors bind in a mutually exclusive fashion, $\alpha = \infty$. If the two bind completely independently, $\alpha = 1$. If the two inhibitors bind nonexclusively but influence each other's affinity for the enzyme, then α will be finite, but less than or greater than 1. When α is less than 1, the binding of one inhibitor increases the affinity of the enzyme for the second inhibitor, and the binding of the two is said to be *synergistic* (i.e., exhibiting positive cooperativity). When α is greater than 1, the binding of one inhibitor decreases the affinity of the enzyme for the second inhibitor, and in this case the binding is said to be *antagonistic* (i.e., exhibiting negative cooperativity).

Loewe (1957) has described the isobologram method for determining exclusivity of binding. In this analysis different concentrations of I and J are combined to yield the same fractional activity (v_i/v_0). The different concentrations of I in these combinations are plotted on the y axis, and the corresponding concentrations of J are plotted on the x axis. If the binding of the two inhibitors is mutually exclusive, the data points on such a plot fall on a straight line. If, however, the two inhibitors bind nonexclusively, the data points will form an outwardly concave curve on the isobologram, the curvature depending on the value of α. A number of other graphic methods have been described for this type of analysis (see, e.g., Chou and Talalay, 1977); of all these methods, the most popular is that of Yonetani and Theorell (1964).

In the Yonetani–Theorell method, the data are arranged as Dixon plots, where $1/v_{ij}$ is plotted as a function of [I] at varying fixed concentrations of J. Consideration of Equation 8.26 will reveal that when α is infinity, the data points will form a series of parallel lines when plotted by the method of Yonetani and Theorell (Figure 8.12A). This is an indication that the two inhibitors bind in a mutually exclusive fashion, competing with one another for the same enzyme form. If α is 1, the two inhibitors bind independently, and the lines in the Yonetani–Theorell plot intersect on the x axis (Figure 8.12B). If α exceeds 1, the two inhibitors antagonize each other's binding, and the lines on the plot intersect below the x axis. Alternatively, if the two inhibitors are synergistic with one another, α is less than 1 and the lines intersect above the x axis. For any Yonetani–Theorell plot in which the lines intersect (i.e., $\alpha \neq \infty$), the x-axis value at the point of intersection provides an estimate of $-\alpha K_i$ when [I] is plotted on the x axis, or $-\alpha K_j$ when [J] is the variable inhibitor concentration. If the values of K_i and K_j are known from independent measurements, the value of α is then easily calculated.

A common motivation for performing the analysis described in this section is to determine whether two structurally distinct inhibitors share a common binding site on the enzyme molecule. If two inhibitors are found to bind in a mutually exclusive fashion, through either kinetic analysis or direct binding measurements, it is tempting to conclude that they bind to the same site on the

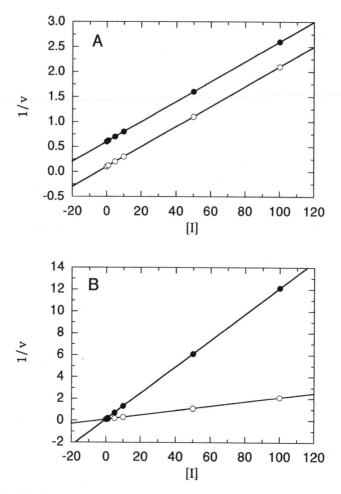

Figure 8.12 (A) Yonetani–Theorell plot for two inhibitors I and J that bind in a mutually exclusive fashion ($\alpha = \infty$) to a common enzyme. (B) Yonetani–Theorell plot for two nonexclusive inhibitors for which $\alpha = 1$. Open circles are data points for [J] = 0; solid circles are data points for [J] = K_J.

enzyme. While this is often true, the caveat described for competitive inhibition with substrate (Section 8.2.1) holds here as well: mutually exclusive binding is observed when the two inhibitors bind to a common site on the enzyme, but it can potentially be observed if the two inhibitors bind at independent sites that strongly affect each other through conformational communication, so that ligand binding at one site precludes ligand binding at the second site. Hence, some caution is required in the interpretation of the results of studies of these types.

8.6 STRUCTURE–ACTIVITY RELATIONSHIPS AND INHIBITOR DESIGN

Modern attempts to identify inhibitors of specific enzymes have largely focused on elucidating the stereochemical and physicochemical features of inhibitory molecules that allow them to bind well to the enzyme (Suckling, 1991). Measures of inhibitor potency, such as K_i or IC_{50}, reflect the change in free energy that accompanies transfer of the inhibitor from the solvated aqueous state to the bound state in the enzyme binding pocket (i.e., the ΔG of binding). We have already discussed the types of physicochemical force that are important in protein structure and ligand binding: hydrophobic interactions, hydrogen bonding, electrostatic interactions, and van der Waals forces. The same forces determine the strength of interaction between an inhibitor and an enzyme.

Likewise, we have seen that the shape or topology of a substrate will determine its ability to fit well into the binding pocket of an enzyme, based on the structural complementarity between the enzyme binding pocket and the substrate molecule. It stands to reason that the same structural complementarity should be important in inhibitor binding as well, and this is what is observed empirically. Thus if one can somehow identify a reasonably potent inhibitor of an enzyme, one can begin to make analogues of that molecule with varied structural and physicochemical properties to determine what effect these changes have of inhibitor potency. Attempts to correlate these structural changes with inhibitor potency are referred to as *structure–activity relationship* (SAR) studies.

Today SAR studies can be divided into two major strategic categories: SAR in the absence of structural information on the target enzyme and SAR that utilizes structural information about the enzyme that is obtained from x-ray crystallographic or multidimensional NMR studies. This latter category is also referred to as rational or structure-based inhibitor design. Section 8.6.1 and 8.6.2 introduce some of the techniques used for both these strategies.

8.6.1 SAR in the Absence of Enzyme Structural Information

Any SAR study begins with the identification of a lead compound that shows some potency for inhibiting the target enzyme. This lead compound might be identified by random screening of a compound library, such as a natural products library, or it might be based on the known structures of the substrate or product of a particular enzymatic reaction. With a lead compound in hand, analysts subject the substance to small structural perturbations and test these analogues for inhibitor potency. This most basic form of SAR study has been conducted in one way or another since the nineteenth century. The goals of these studies are to determine what structural changes will lead to improved inhibitor potency and to identify the *pharmacophore*, the minimal structure required for inhibition. Once identified, the pharmacophore serves as a template for further inhibitor design.

Consider the enzyme dihydrofolate reductase (DHFR), which catalyzes a key step in the biosynthesis of deoxythymidine. Inhibition of this enzyme blocks DNA replication and thus acts to inhibit cell growth and proliferation. DHFR inhibitors are therefore potentially useful therapeutic agents for the control of aberrant cell growth in cancer, and as antibiotics for the control of bacterial growth. Early attempts to identify inhibitors of this enzyme were based on synthesizing analogues of the substrate dihydrofolate. Figure 8.13 illustrates the chemical structures of dihydrofolate and two classes of DHFR inhibitors; the pteridines, exemplified by methotrexate, and the 5-substituted 2,4-diaminopyrimidines. Methotrexate was identified as a potent inhibitor of DHFR because of its striking structural similarity to the substrate dihydrofolate.

Next, the question of what portions of the methotrexate molecule were critical for DHFR inhibition was addressed by synthesizing various structural analogues of methotrexate. From these studies it was determined that the critical pharmacophore (i.e., the minimal structural component required for inhibition) was the 2,4-diaminopyrimidine ring system. This discovery led to the development of the second class of inhibitors illustrated in Figure 8.13, the 5-substituted 2,4-diaminopyrimidines, of which trimethoprim is a well-known example. Methotrexate is now a prescribed drug for the treatment of human cancers and certain immune-based diseases. Trimethoprim also is a prescribed drug, but its use is in the control of bacterial infections.

An unexpected outcome of the studies on these inhibitors was the finding that the pteridines, such as methotrexate, are potent inhibitors of both mammalian and bacterial DHFR, while trimethoprim and its analogues are much better inhibitors of the bacterial enzymes. The K_i values for trimethoprim for *E. coli*, and human lymphoblast DHFR are 1.35 and 170,000 nM, respectively (Li and Poe, 1988), a selectivity for the bacterial enzyme of about 126,000-fold! The reason for this spectacular species selectivity was not clear until the crystal structures of mammalian and bacterial DHFR were solved. (See Matthews et al., 1985, for a very clear and interesting account of these crystallographic studies and their interpretation.)

Today the enzymologist attempts to develop higher potency inhibitors not simply by random replacement of structural components on a molecule, but rather by systematic and rational changes in stereochemical and physicochemical properties of the substituents. Some properties to be changed are obvious from the structure of the lead compound. If, for example, the lead inhibitor contains a carboxylic acid group, one immediately wonders whether acid–base-type interactions with a group on the enzyme are involved in binding. One might substitute the carboxylate moiety with an ester, for example, to determine the importance of the carboxylate in binding. More general properties of chemical substituents can be examined as well. These studies call for quantitative measures of the different physicochemical properties to be considered. Among the relevant general properties of chemical substituents, steric bulk, hydrophobicity, and electrophilicity/nucleophilicity are generally agreed

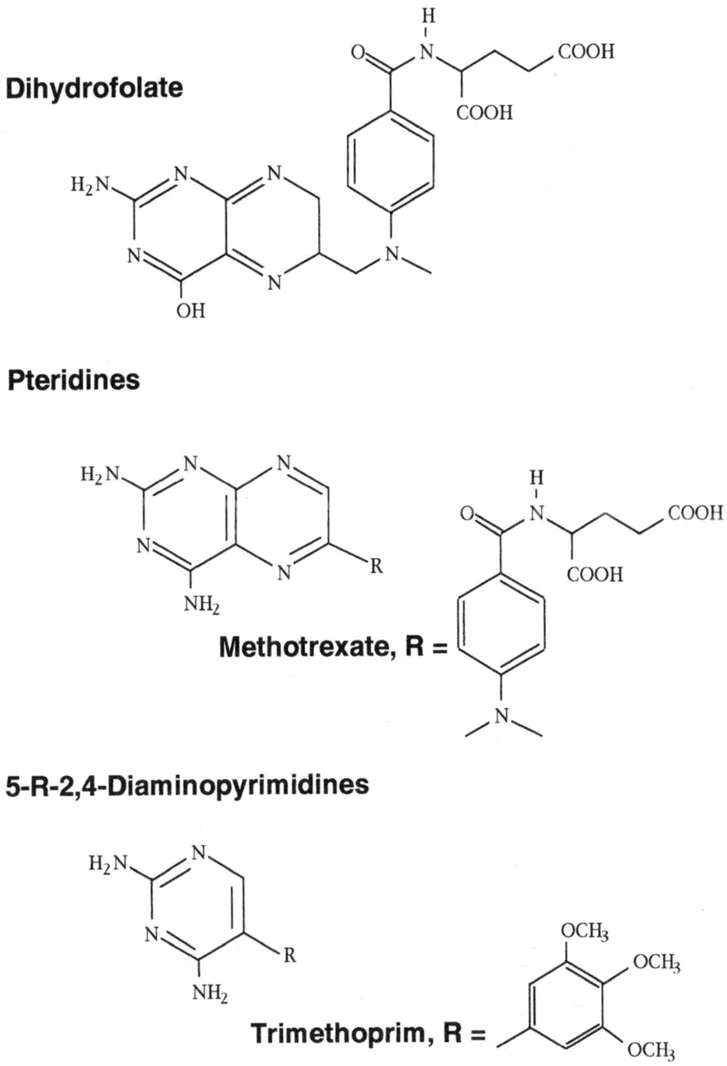

Figure 8.13 Chemical structures of the substrate (dihydrofolate) and two types of inhibitor of the enzyme dihydrofolate reductase (DHFR).

to be important factors, and chemists have therefore developed quantitative measures of these parameters.

Several measures have been suggested to quantify steric bulk or molecular volume. One of the earliest attempts at this was the Taft steric parameter E_s, which was defined as the logarithm of the rate of acid-catalyzed hydrolysis of a carboxymethyl-substituted molecule relative to the rate for the methyl acetate analogue (Taft, 1953; Nogrady, 1985):

$$E_s = \log(k_{XCOOCH_3}) - \log(k_{CH_3COOCH_3}) \tag{8.27}$$

A more geometric measure of steric bulk is provided by the Verloop steric parameter, which basically measures the bond angles and bond lengths of the substituent group (Nogrady, 1985). Chemists also have used the molar refractivity as a measure of molecular volume of substituents (Pauling and Pressman, 1945; Hansch and Klein, 1986). The molar refractivity, MR, is defined as follows:

$$MR = \frac{n^2 - 1}{n^2 + 1} \frac{MW}{d} \tag{8.28}$$

where n is the index of refraction, MW is the molecular weight, and d is the density of the substituent under consideration. Since n does not vary widely among organic molecules, MR is mainly a measure of molecular volume.

The relative hydrophobicity of molecular substituents is most commonly measured by their partition coefficient between a polar and nonpolar solvent. For this purpose, chemists have made water and octanol the solvents of choice. The molecule is dissolved in a 1:1 mixture of the two solvents, and its concentration in each solvent is measured at equilibrium. The partition coefficient is then calculated as the equilibrium constant:

$$P = \frac{[I]_{octanol}}{[I]_{water}} \tag{8.29}$$

In measuring relative hydrophobicity, the effect of different substituents, on the partition coefficient of benzene in octanol/water is used as a standard. The hydrophobic parameter π is used for this purpose, and is defined as follows (Hansch and Klein, 1986):

$$\pi = \log(P_x) - \log(P_H) \tag{8.30}$$

where P_x is the partition coefficient for a monosubstituted benzene with substituent x, and P_H is the partition coefficient of benzene itself.

The most widely used index of electronic effects in inhibitor design is the Hammett σ constant. Originally developed to correlate quantitatively the relationship between the electron-donating or -accepting nature of a parasubstituent on the ionization constant for benzoic acid in water (Hammett, 1970; Nogrady, 1985); this index is defined as follows:

$$\sigma = \log(K_x) - \log(K_H) \tag{8.31}$$

where K_x is the ionization constant for the parasubstituted benzoic acid with substituent x and K_H is the ionization constant for benzoic acid. Groups that are electron acceptors (e.g., COOH, NO_2, NR_3^+) withdraw electron density from the ring system, hence stabilize the ionized form of the acid; such groups have positive values of σ. Electron-donating groups (e.g., OH, OCH$_3$, NH$_2$) have the opposite effect on the ionization constant and thus have negative values of σ. Values of σ for a very large number of organic substituents have been tabulated by several authors. One of the most comprehensive list of σ values can be found in the text by Martin (1978).

The ability to quantify these various physicochemical properties has led to attempts to express the inhibitor potency of molecules as a mathematical function of these parameters. This strategy of *quantitative structure–activity relationships* (QSAR) was first championed by Hansch and his coworkers (Hansch, 1969; Hansch and Leo, 1979). In a typical QSAR study, a series of analogues of a lead inhibitor is prepared with substituents that systematically vary the parameters described earlier. The experimentally determined potencies of these compounds are then fit to varying linear and nonlinear weighted sums of the parameter indices to obtain the best correlation by regression analysis. Equations 8.32–8.34 illustrate forms typically used in QSAR work.

$$\log\left(\frac{1}{K_i}\right) = a\pi + b\sigma + cMR + d \tag{8.32}$$

$$\log\left(\frac{1}{K_i}\right) = a\pi^2 + b\sigma + cMR + d \tag{8.33}$$

$$\log\left(\frac{1}{K_i}\right) = a\pi^2 + b\sigma + c\log(\beta MR) + d \tag{8.34}$$

Equation 8.32 is a simple linear relationship, while Equations 8.33 and 8.34 have nonlinear components. In these equations the values a, b, c, d, and β are proportionality constants, determined from the regression analysis. In developing a mathematical expression for the correlation relationship here, one hopes to predict the inhibitory potency of further compounds prior to their synthesis, based on the equation established from the QSAR. In practice, the predictive power of these QSAR equations varies dramatically. When such predictions fail, it is usually because additional factors that influence inhibitor potency were not quantitatively included in the functional expression. In some cases these additional factors are neither well understood nor easily quantified.

As a simple example of QSAR, let us again consider the inhibition of bacterial DHFR by pteridines and 5-substituted 2,4-diaminopyrimidines. Coats et al. (1984) studied the QSARs of both classes of compounds for their ability to inhibit the DHFR from the bacterium *Lactobacillus casei*. They measured the IC$_{50}$ values for 25 pteridine analogues with different R substituents and also for 33 5-substituted 2,4-diaminopyrimidine analogues with different R groups (Figure 8.13). From these data they determined the QSAR equations given by Equations 8.35 and 8.36 for the pteridines and 5-R-2,4-

diaminopyrimidines, respectively:

$$\log\left(\frac{1}{IC_{50}}\right) = 0.23\pi - 0.004\pi^2 + 0.77I + 3.39 \qquad (8.35)$$

$$\log\left(\frac{1}{IC_{50}}\right) = 0.38\pi - 0.007\pi^2 + 0.66I + 2.15 \qquad (8.36)$$

In these equations, the π parameter refers to the hydrophobicity of the R group, and the index I is an empirical parameter related to the presence of a —N—C— or —C—N— bridge between the parent ring system and an aromatic ring on the substituent (Coats et al., 1984). The relationships between the IC_{50} values calculated from these equations and the experimentally determined IC_{50} values are illustrated in Figure 8.14. Again, one must keep in mind that the correlations illustrated are for the molecules used to establish the QSAR equations. The value of these equations in predicting the inhibitor potency of other molecules will depend on how significantly other unaccounted-for factors influence potency. Nevertheless, QSAR provides a means of rationalizing the observed potencies of structurally related compounds in terms of familiar physicochemical properties. An up-to-date volume by Kubinyi (1993) provides a detailed and practical introduction to the field of QSAR. This text should be consulted as a starting point for acquiring a more in-depth treatment of the subject.

Another approach to designing potent inhibitors of enzymes is to consider the probable structure of the transition state of the chemical reaction catalyzed by the enzyme. As described in Chapter 6, the catalytic efficiency of enzymes is due largely to their ability to achieve transition state stabilization. If this stabilization is equated with binding energy, a stable analogue that mimics the structure of the transition state should bind to an enzyme some 10^{10}–10^{15} times greater than the corresponding ground state substrate molecule (Wolfenden, 1972). Since typical substrate K_m values are in the millimolar-to-nanomolar range, a true transition state mimic may bind to its target enzyme with a K_i value between 10^{-13} and 10^{-24} M! With such incredibly tight binding affinities, such inhibitors would behave practically as irreversible enzyme inactivators.

The foregoing approach to inhibitor design has been hindered by the great difficulty of obtaining information on transition state structure by traditional physical methods. Because they are so short-lived, the transition state species of most enzymatic reactions are present under steady state conditions at very low concentrations (i.e., femtomolar or less). Hence, attempts to obtain structural information on these species from spectroscopic or crystallographic methods have been largely unsuccessful. Information on transition state structure can, however, be gleaned from analysis of kinetic isotope effects on enzyme catalysis, as recently reviewed by Schramm et al. (1994). As discussed in Chapter 7, kinetic isotope effects are observed because of the changes in

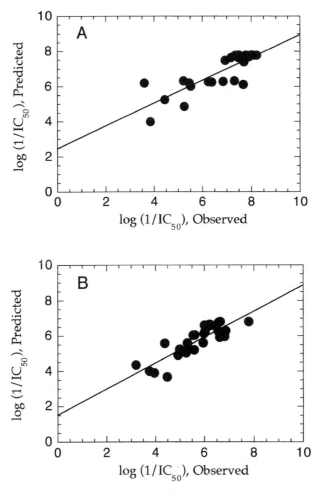

Figure 8.14 QSAR correlation plots for the potencies of pteridines (A) and 5-substituted 2,4-diaminopyrimidines (B) as inhibitors of the dihydrofolate reductase from *L. casei*. [Data from Coats et al. (1984).]

vibrational frequencies for the reactant and transition state species that accompany heavy isotope incorporation. By synthesizing substrate analogues with heavy isotopes at specific locations, one can determine the kinetic isotope effects imparted by each replacement. From this type of information, one can use vibrational normal mode calculations to identify the vibrational modes that are most strongly perturbed in the transformation from reactant to transition state of the substrate, hence to map out the structural changes that have occurred in the molecule. The information thus obtained can then be used to design molecules that mimic the structure of the reaction transition state.

This approach has been applied to the design of transition state analogues of renin, an aspartyl protease (Blundell et al., 1987).

In vivo, renin is responsible for the proteolytic processing of angiotensinogen to angiotensin I, the progenitor of the vasoconstrictor peptide angiotensin II. The substrate is hydrolyzed at a Leu-Val peptide bond, and the hydrolysis reaction is proposed to utilize an active site water molecule as the attacking nucleophile to produce the tetrahedral transition state illustrated in Figure 8.15A. The peptide sequence of the renin substrate angiotensinogen is shown in Figure 8.15C. In their first attempt at an inhibitor of renin, Blundell and coworkers replaced the P1 carbonyl group by a methylene linkage, yielding the reduced isostere (—CH$_2$—NH—) containing peptide inhibitor **1** (Figure 8.15C). The investigators next noted that the pepstatins are naturally occurring protease inhibitors that contain the unusual amino acid statine (Figure 8.15B), which in turn contains a —CH(OH)— moiety that resembles the proposed transition state of renin. Pepstatin is a poor renin inhibitor; but, reasoning that the statine group was a better transition state mimic than the reduced peptide isostere (—CH$_2$—NH—), Blundell et al. incorporated this structure into their peptide inhibitor to produce **2** (Figure 8.15C). The closest analogue to the true transition state of the reaction would be one incorporating

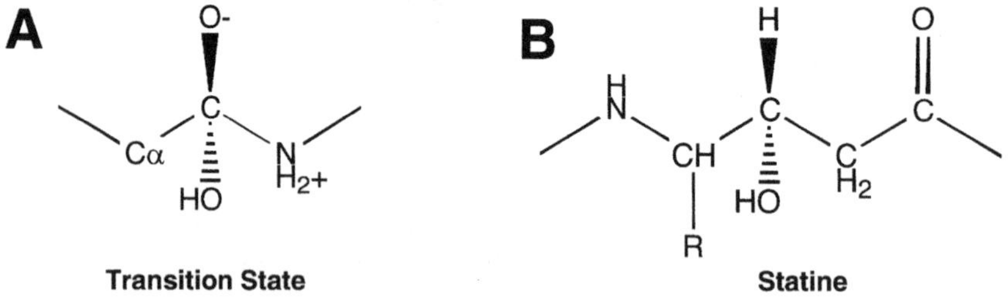

Transition State

Statine

C

Compound	P1		P1'	IC$_{50}$ (µM)
Substrate	Ile His Pro Phe His Leu-	CO - NH-	Val Ile His Asn	-
1	Pro His Pro Phe His Leu-	CH$_2$ -NH-	Val Ile His Lys	0.01
2	Boc His Pro Phe His Leu- CH(OH)-CH$_2$-CO-NH-Leu Phe			0.01
3	Boc His Pro Phe His Leu- CH(OH)-CH$_2$-		Val Ile His	0.0007

Figure 8.15   (A) Proposed structure of the tetrahedral transition state of the renin proteolysis reaction. (B) Chemical structure of the statine moiety of natural protease inhibitors, such as pepstatin. (C) Structures and IC$_{50}$ values for the peptide substrate and inhibitors of renin, incorporating various forms of transition state analogues. [Data taken from Blundell et al. (1987).]

a —CH(OH)—NH— group at P1—P1′ of the peptide. Since, however, synthesis of this transition state analogue was hampered by the instability of the resulting compound, Blundell's group instead synthesized a closely related analogue containing a hydroxyethylene moiety [(—CH(OH)—CH$_2$—), **3**, which proved to be an extremely potent inhibitor of the enzyme (Figure 8.15C).

To date, the de novo design of transition state analogues as enzyme inhibitors has been applied only to a limited number of enzymes by a handful of laboratories. With improvements in the computational methods associated with this strategy, however, more widespread use of this approach is likely to be seen in the future.

8.6.2 Inhibitor Design Based on Enzyme Structure.

In the search for potent enzyme inhibitors, knowledge of the three dimensional structure of the inhibitor binding site on the enzyme provides the ultimate guide to designing new compounds. The structures of enzyme active sites can be obtained in atomic detail from x-ray crystallography and multidimensional NMR spectroscopy. A detailed discussion of these methods is beyond the scope of the present text. Our discussion will focus instead on the use of the structural details obtained from these techniques. The reader interested in learning about protein crystallography and NMR spectroscopy can find many excellent review articles and texts (see McRee, 1993, and Fesik, 1991, for good introductions to protein crystallography and NMR spectroscopy, respectively).

The crystal or NMR structure of an enzyme with an inhibitor bound provides structural details at the atomic level on the interactions between the inhibitor and the enzyme that promote binding. Hydrogen bonding, salt bridge formation, other electrostatic interactions, and hydrophobic interactions can be readily inferred from inspection of a high resolution structure. Figure 8.16 provides simplified schematic representations of the binding interactions between DHFR and its substrate dihydrofolate and inhibitor methotrexate, illustrating the involvement of common amino acid residues in the binding of both ligands. These structural diagrams also indicate that the orientation and hydrogen bonding patterns are not identical for the substrate and the inhibitor. Nevertheless, the major forces involved in binding of both ligands to the enzyme are hydrogen bonds between amino acid residues of the active site and the 2,4-diaminopyrimidine ring of the ligands. Visual inspection by means of molecular graphics methods suggested that this ring constituted the critical pharmacophore and led to the design of trimethoprim, the prototypical 5-substituted 2,4-diaminopyrimidine (Marshall and Cramer, 1988). As we have seen, these structural inferences are consistent with the SAR and QSAR studies of DHFR inhibitors.

The example of trimethoprim suggests a straightforward, if tedious, means of utilizing structural information in the design of new enzyme inhibitors: namely, the iterative design, synthesis, and crystallization of inhibitor–enzyme complexes. In this approach, one starts with the crystal structure of the free

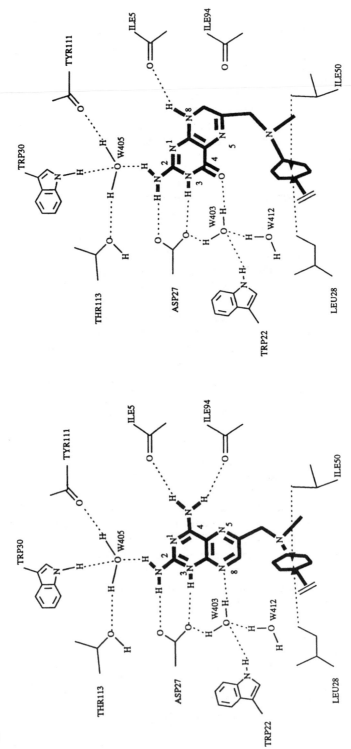

Figure 8.16 Interactions of the dihydrofolate reductase active site with the inhibitor methotrexate (left) and the substrate dihydrofolate (right). [Reprinted from Klebe (1994) with permission from Academic Press Limited.]

enzyme or of the enzyme–lead inhibitor complex. Based on inspection of the crystal structure, one suggests, changes in chemical structure of the inhibitor to better engage the enzyme active site. The new compound is then synthesized and tested for inhibitory potency. Next, to determine whether the predicted interactions in fact occur, a crystal structure of the enzyme with this new inhibitor bound is obtained. This new structure is then used to search for additional changes to the inhibitor structure that might further improve potency, and the process is continued until an inhibitor of sufficient potency is obtained. This iterative structure-based inhibitor design method was used in the design and synthesis of inhibitors of thymidylate synthase reported by Appelt et al. (1991); this paper provides a good illustration of the method.

Thus, the first step to structure-based inhibitor design is to obtain a crystal or NMR structure of the target enzyme, with or without a lead inhibitor bound to it. In some cases, the determination of a crystal or NMR structure of the target protein proves problematic because of the technical difficulties associated with crystallographic and NMR methods. If the structure of a closely related enzyme has been reported, however, one can still attempt to model the three-dimensional structure of the target enzyme by means of homology modeling (Lesk and Boswell, 1992). In homology modeling one attempts to build a model of the target enzyme by superimposing the amino acid residues of the target onto the three-dimensional structure of the homologous protein whose structure has been solved. For these method to work, the target enzyme and its homologue must share at least 30% amino acid sequence identity. The accuracy of the model obtained in this way is directly related to the degree of sequence identity between the two proteins: the greater the sequence identity, the more accurate the modeled structure.

With the modeled or actual structure of the target enzyme active site in hand, the next step is to assess the active site structure in a meaningful way, to permit the use of this information to predict inhibitor binding motifs. The simple visual inspection of such structures can be augmented today with computer programs that allow the analyst to map the electrostatic potential surface of the active site, identify and localize specific types of functional group within the active site (potential acid–base groups, hydrogen-bonding acceptors or donors, etc.), and the like. When the active site has been well described, one attempts to design inhibitors with stereochemical and functional complementarity to the active site structure. Again, these activities are greatly aided by high powered computer programs that make possible the probing of complementarity between a potential inhibitor and the enzyme active site. Assessment of the stereochemical complementarity of a potential inhibitor is aided by the use of molecular dynamics simulation programs by means of which the most energetically favorable conformations of inhibitory molecules can be assessed to determine whether they will adapt a conformation that is complementary to the enzyme active site.

New programs allow one to perform free energy perturbation calculations in which a bound inhibitor is slowly mutated and the difference in calculated free energy of binding between the starting and final structures determined (Marshall and Cramer, 1988). In this way, one can search for structural perturbations that will increase the affinity of an inhibitor for the enzyme active site. The complementarity of functional groups can be probed by computational methods as well. For example, the computer program GRID (Goodford, 1985) can be used to search the structure of an enzyme active site for areas that are likely to interact strongly with a particular functional group probe.

A recent example of the use of such programs comes from the studies by von Itzstein et al. (1993) aimed at designing potent inhibitors of the sialidase enzyme from influenza virus. This group started with the enzyme active site obtained from a series of crystals for the enzyme to which various sialic acid analogues were bound. Visual inspection of the cocrystal structure for the enzyme bound to the unsaturated sialic acid analogue Neu5Ac2en suggested that replacement of the 4-hydroxyl group of the substrate by an amino group might be useful. A GRID calculation was performed with a protonated primary amine group as the probe, and a "hot spot," or area of likely strong interaction, was identified within the enzyme active site.

The results of this process of visual inspection and calculation suggested that replacement of the 4-hydroxyl group with an amino group would lead to much tighter binding because a salt bridge would form between the amino group and the side chain carboxylate of Glu 119 of the enzyme. Further evaluation of the computational data suggested that replacement of the 4-hydroxyl group with a guanidinyl group would even further enhance inhibitor binding by engaging both Glu 119 and Glu 227 through lateral binding of the two terminal nitrogens on this functional group. Based on these results, the 4-amino and 4-guanidino derivatives of Neu5Ac2en were synthesized and, as expected, found to be potent inhibitors of the enzyme, with K_i values of 50 and 0.2 nM, respectively. When the crystal structures of the enzyme bound to each of these new inhibitors was determined, the predicted modes of inhibitor interactions with the enzyme were by and large confirmed.

The design of new enzyme inhibitors, both by structure-based design methods and in the absence of enzyme structural information, is a large and growing field. We have only briefly introduced this complex and exciting area. There are many excellent sources for additional information on strategies for inhibitor design. These include several texts devoted entirely to this subject (e.g., Sandler and Smith, 1994; Gringauz, 1996). Also, most modern medicinal chemistry textbooks contain sections on SAR and inhibitor design (see, e.g., Nogrady, 1985; Dean, 1987). Finally, a number of primary journals commonly feature papers in the field of inhibitor design and SAR. These include *Journal of Medicinal Chemistry* (ACS), *Journal of Enzyme Inhibitors, Bioorganic and Medicinal Chemistry Letters*, and *Journal of Computer-Aided Molecular Design*. These sources, and the specific references at the end of this chapter, will provide good starting points for the reader interested in exploring these subjects in greater depth.

8.7 SUMMARY

In this chapter we described the modes by which an inhibitor can bind to an enzyme molecule and thus render it inactive. Graphical methods were introduced for the diagnosis of the mode of inhibitor interaction with the enzyme on the basis of the effects of that inhibitor on the apparent values of the kinetic constants K_m and V_{max}. Having thus identified the inhibitor modality, we described methods for quantifying the inhibitor potency in terms of K_i, the dissociation constant for the enzyme–inhibitor complex.

Also in this chapter, we introduced some of the physicochemical determinants of enzyme–inhibitor interactions and saw how these could be systematically varied for the design of more potent inhibitors. Finally we introduced the concept of structure-based inhibitor design in which the crystal or NMR structure of the target enzyme is used to aid the design of new inhibitor molecules in an iterative process of enzyme–inhibitor structure determination, new inhibitor design and synthesis, and quantitation of new inhibitor potency.

REFERENCES AND FURTHER READING

Appelt, K., Bacquet, R. J., Bartlett, C. A., Booth. C. L. J., Freer, S. T., Fuhry, M. A. M., et al. (1991) *J. Med. Chem.* **34**, 1925.

Blundell, T. L., Cooper, J., Foundling, S. I., Jones, D. M., Atrash, B., and Szelke, M. (1987) *Biochemistry* **26**, 5586.

Chaiken, I., Rose, S., and Karlsson, R. (1991) *Anal. Biochem.* **201**, 197.

Cheng, Y.-C., and Prusoff, W. H. (1973) *Biochem. Pharmacol.* **22**, 3099.

Chou, T.-C., and Talalay, P. (1977) *J. Biol. Chem.* **252**, 6438.

Cleland, W. W. (1979) *Methods Enzymol.* **63**, 103.

Coats, E. A., Genther, C. S., and Smith, C. C. (1984) In *QSAR in Design of Bioactive Compounds*, M. Kuchar, Ed., J. R. Prous Science, Barcelona, Spain, pp. 71–85.

Dean, P. M. (1987) *Molecular Foundations of Drug–Receptor Interactions*, Cambridge University Press, New York.

Dixon, M. (1953) *Biochem. J.* **55** 170.

Dougas, H., and Penney, C. (1981) *Bioorganic Chemistry: A Chemical Approach to Enzyme Action*, Springer-Verlag, New York.

Fesik, S. W. (1991) *J. Med. Chem.* **34**, 2937.

Furfine, E. S., D'Souza, E., Ingold, K. J., Leban, J. J., Spectro, T., and Porter, D. J. T. (1992) *Biochemistry*, **31**, 7886.

Goodford, P. J. (1985) *J. Med. Chem.* **28**, 849.

Gringauz, A. (1996) *Medicinal Chemistry: How Drugs Act and Why*, Wiley, New York.

Hammett, L. P. (1970) *Physical Organic Chemistry*, McGraw-Hill, New York.

Hansch, C. (1969) *Acc. Chem. Res.* **2**, 232.

Hansch, C., and Klein, T. E. (1986) *Acc. Chem. Res.* **19**, 392.

Hansch, C., and Leo, A. (1979) *Substituent Constants for Correlation Analysis in Chemistry and Biology*, Wiley, New York.

Karlsson, R. (1994) *Anal. Biochem.* **221**, 142.

Klebe, G. (1994) *J. Mol. Biol.* **237**, 212.

Kubinyi, H. (1993) *QSAR: Hansch Analysis and Related Approaches*, VCH, New York.

Lesk, A. M., and Boswell, D. R. (1992) *Curr. Opin. Struct. Biol.* **2**, 242.

Li, R.-L., and Poe, M. (1988) *J. Med. Chem.* **31**, 366.

Loewe, S. (1957) *Pharmacol. Rev.* **9**, 237.

Ma, H., Yang, H. Q., Takano, E., Hatanaka, M., and Maki, M. (1994) *J. Biol. Chem.* **269**, 24430.

Marshall, G. R., and Cramer, R. D., III (1988) *Trends Pharmacol. Sci.* **9**, 285.

Martin, Y. C. (1978) *Quantitative Drug Design*, Dekker, New York.

Martinez-Irujo, J. J., Villahermosa, M. L., Mercapide, J., Cabodevilla, J. F., and Santiago, E. (1998) *Biochem. J.* **329**, 689.

Matthews, D. A., Bolin, J. T., Burridge, J. M., Filman, D. J., Volz, K. W., and Kraut, *J.* (1985) *J. Biol. Chem.* **260**, 392.

McRee, D. E. (1993) *Practical Protein Crystallography*, Academic Press, San Diego, CA.

Nogrady, T. (1985) *Medicinal Chemistry, A Biochemical Approach*, Oxford University Press, New York.

Pauling, L., and Pressman, D. (1945) *J. Am. Chem. Soc.* **75**, 4538.

Sandler, M., and Smith, H. J. (1994) *Design of Enzyme Inhibitors as Drugs*, Vols. 1 and 2, Oxford University Press, New York.

Schramm, V. L., Horenstein, B. A., and Kline, P. C. (1994) *J. Biol. Chem.* **269**, 18259.

Segel, I. H. (1975) *Enzyme Kinetics*, Wiley, New York.

Segel, I. H. (1976) *Biochemical Calculations*, 2nd ed., Wiley, New York.

Suckling, C. J. (1991) *Experimentia*, **47**, 1139.

Taft, R. W. (1953) *J. Am. Chem. Soc.* **75**, 4538.

Von Itzstein, M., Wu, W.-Y., Kok, G. B., Pegg, M. S., Dyason, J. C., Jin, B., Phan, T. V., Smythe, M. L., White, H. F., Oliver, S. W., Colman, P. M., Varghese, J. N., Ryan, D. M., Woods, J. M., Bethell, R. C., Hotham, V. J., Cameron, J. M., and Penn, C. R. (1993) Nature, **363**, 418.

Wolfenden, R. (1972) *Acc. Chem. Res.* **5**, 10.

Yonetani, T., and Theorell, H. (1964) *Arch. Biochem. Biophys.* **106**, 243.

9

TIGHT BINDING INHIBITORS

In Chapter 8 we discussed reversible inhibitors of enzymes that bind and are released at rates that are rapid in comparison to the rate of enzyme turnover and have overall dissociation constants that are large in comparison to the total concentration of enzyme present. We are able to analyze the interactions of these inhibitors with their target enzymes by means of equations of the Henri–Michaelis–Menten type discussed in Chapters 5 and 8 because we can generally assume that the free inhibitor concentration is well modeled by the total concentration of added inhibitor: that is, since [E] is much smaller than K_i, the concentration of the EI complex is held to be very small compared to [I]. This assumption, however, is not valid for all inhibitors. Some inhibitors bind to their target enzyme with such high affinity that the population of free inhibitor molecules is significantly depleted by formation of the enzyme-inhibitor complex. For these *tight binding inhibitors*, the steady state approximations used thus far are no longer valid; in fact, it has been suggested that these assumptions should be abandoned whenever the K_i of an inhibitor is less than 1000-fold greater than the total enzyme concentration (Goldstein, 1944; Dixon and Webb, 1979). In this chapter we shall describe alternative methods for data analysis in the case of tight binding inhibitors that allow us to characterize the type of inhibition mechanism involved and to quantify correctly the dissociation constant for the enzyme–inhibitor complex.

9.1 IDENTIFYING TIGHT BINDING INHIBITION

In this chapter we shall consider the steady state approach to studying tight binding inhibitors. Such work requires assay conditions that permit all the equilibria involving the inhibitor, substrate, and enzyme to be reached rapidly with respect to the measurement of the steady state velocities. Many tight binding inhibitors, however, are restricted in their action by a slow onset of inhibition—that is, by a time-dependent component to their inhibition. We deal with time-dependent inhibition explicitly in Chapter 10. For our present discussion, we shall assume either that the establishment of equilibrium is rapid or that sufficient time has been allowed before the initiation of reaction by the substrate for the inhibitor and enzyme to establish equilibrium (i.e., a preincubation of the enzyme with the inhibitor has been incorporated into the experimental design).

The simplest determination that tight binding inhibition is occurring comes from measurement of the dose–response curve for inhibition (see Chapter 8). An IC_{50} value obtained from this treatment of the data that is similar to the concentration of total enzyme in the sample (i.e., within a factor of 10) is a good indication that the inhibitor is of the tight binding type. A more defining feature of tight binding inhibitors is the variation of the IC_{50} value observed for these inhibitors with total enzyme concentration at a fixed substrate concentration. This is true because a tight binding inhibitor interacts with the enzyme in nearly stoichiometric fashion. Hence, the higher the concentration of enzyme present, the higher the concentration of inhibitor required to reach half-maximal saturation of the inhibitor binding sites (Figure 9.1A). Several authors have derived equations similar in form to Equation 9.1, which demonstrates that the IC_{50} value of a tight binding inhibitor will track linearly with the total concentration of enzyme, [E] (Myers, 1952; Cha et al., 1975; Williams and Morrison, 1979; Greco and Hakala, 1979):

$$IC_{50} = \tfrac{1}{2}[E] + K_i^{app} \tag{9.1}$$

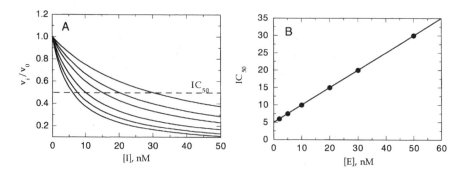

Figure 9.1 (A) Dose–response plot of fractional velocity as a function of tight binding inhibitor concentration at different enzyme concentrations. Note that this plot is the same as the dose–response plots introduced in Chapter 8, except that here the x axis is plotted on a linear, rather than a logarithmic, scale. (B) Plot of IC_{50} value obtained from the curves in (A) as a function of enzyme concentration.

Thus, a plot of IC_{50} as a function of $[E]$ (at a single, fixed substrate concentration) is expected to yield a straight line with slope of 0.5 and y intercept equal to K_i^{app}. The value K_i^{app} is related to the true K_i by factors involving the substrate concentration and K_m, depending on the mode of interaction between the inhibitor and the enzyme.

9.2 DISTINGUISHING INHIBITOR TYPE FOR TIGHT BINDING INHIBITORS

Morrison (Morrison, 1969; Williams and Morrison, 1979) has provided in-depth mathematical treatments of the effects of tight binding inhibitors on the initial velocities of enzymatic reactions. These studies revealed, among other things, that the classical double-reciprocal plots used to distinguish inhibitor type for simple enzyme inhibitors fail in the case of tight binding inhibitors. For example, based on the work just cited by Morrison and coworkers, the double-reciprocal plot for a tight binding competitive inhibitor would give the pattern of lines illustrated in Figure 9.2. The data at very high substrate concentrations curve downward in this plot, and the curves at different inhibitor concentrations converge at the y axis. Note, however, that this curvature is apparent only at very high substrate concentrations and in the presence of high inhibitor concentrations. This subtlety in the data analysis is easy to miss if care is not taken to include such extreme conditions, or if these conditions are not experimentally attainable. Hence, if the few data points in the very high substrate region are ignored, it is tempting to fit the data in Figure 9.2 to a series of linear functions, as has been done in this illustration. The pattern of lines that emerges from this treatment of the data is a series of

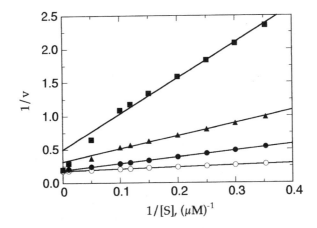

Figure 9.2 Double-reciprocal plot for a tight binding competitive inhibitor: the pattern of lines is similar to that expected for a classical noncompetitive inhibitor (see Chapter 8).

lines that intersect at or near the x axis, to the left of the y axis. This is the expected result for a classical noncompetitive inhibitor (see Chapter 8), and we can generally state that *regardless of their true mode of interaction with the enzyme, tight binding inhibitors display double-reciprocal plots that appear similar to the classical pattern for noncompetitive inhibitors.*

As one might imagine, this point has led to a number of misinterpretations of kinetic data for inhibitors in the literature. For example, the naturally occurring inhibitors of ribonuclease are nanomolar inhibitors of the enzyme. Initial evaluation of the inhibitor type by double-reciprocal plots indicated that these inhibitors acted through classical noncompetitive inhibition. It was not until Turner et al. (1983) performed a careful examination of these inhibitors, over a broad range of inhibitor and substrate concentrations, and properly evaluated the data (as discussed below) that these proteins were recognized to be tight binding *competitive* inhibitors.

How then can one determine the true mode of interaction between an enzyme and a tight binding inhibitor? Several graphical approaches have been suggested. One of the most straightforward is to determine the IC_{50} values for the inhibitor at a fixed enzyme concentration, but at a number of different substrate concentrations. As with simple reversible inhibitors, the IC_{50} of a tight binding inhibitor depends on the K_i of the inhibitor, the substrate concentration, and the substrate K_m in different ways, depending on the mode of inhibition. For tight binding inhibitors we must additionally take into consideration the enzyme concentration in the sample, since this will affect the measured IC_{50}, as discussed earlier. The appropriate relationships between these factors and the IC_{50} for different types of tight binding inhibitor have been derived several times in the literature (Cha, 1975; Williams and Morrison, 1979; Copeland et al., 1995). Rather than working through these derivations again, we shall simply present the final form of the relationships.

For tight binding competitive inhibitors:

$$IC_{50} = K_i\left(1 + \frac{[S]}{K_m}\right) + \frac{[E]}{2} \tag{9.2}$$

For tight binding noncompetitive inhibitors:

$$IC_{50} = \frac{[S] + K_m}{\dfrac{K_m}{K_i} + \dfrac{[S]}{\alpha K_i}} + \frac{[E]}{2} \tag{9.3}$$

when $\alpha = 1$:

$$IC_{50} = K_i + \frac{[E]}{2} \tag{9.4}$$

For tight binding uncompetitive inhibitors:

$$IC_{50} = K_i \left(1 + \frac{K_m}{[S]} \right) + \frac{[E]}{2} \tag{9.5}$$

From the form of these equations, we see that a plot of the IC_{50} value as a function of substrate concentration will yield quite different patterns, depending on the inhibitor type. For a tight binding competitive inhibitor, the IC_{50} value will increase linearly with increasing substrate concentration (Figure 9.3A). For an uncompetitive inhibitor, a plot of IC_{50} value as a function of substrate concentration will curve downward sharply (Figure 9.3A), while for a noncompetitive inhibitor the IC_{50} will curve upward or downward, or be independent of [S], depending on whether α is greater than, less than, or equal to 1.0 (Figure 9.3A and B).

In an alternative graphical method for determining the inhibitor type, and obtaining an estimate of the inhibitor K_i, the fractional velocity of the enzyme reaction is plotted as a function of inhibitor concentration at some fixed substrate concentration (Dixon, 1972). The data can be fit to Equation 8.20 to yield a curvilinear fit as shown in Figure 9.4A. (Note that this is the same as the dose–response plots discussed in Chapter 8, except here the x axis is plotted on a linear, rather than a logarithmic, scale). A line is drawn from the v/v_0 value at $[I] = 0$ (referred to here as the starting point) through the point on the curve where $v = v_0/2$ ($n = 2$) and extended to the x axis. A second line is drawn from the starting point through the point on the curve where $v = v_0/3$ ($n = 3$), and, in a similar fashion, additional lines are drawn from the starting point through other points on the curve where $v = v_0/n$ (where n is an integer). The nest of lines thus drawn will intersect the x axis at a constant spacing, which is defined as K.

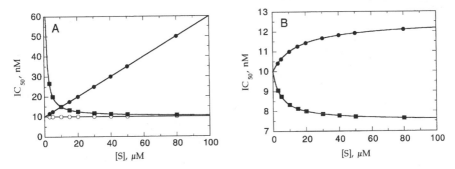

Figure 9.3 (A) The effects of substrate concentration on the IC_{50} values of competitive (solid circles), noncompetitive when $\alpha = 1$ (open circles), and uncompetitive (solid squares) tight binding inhibitors. (B) The effects of substrate concentration on the IC_{50} values of noncompetitive tight binding inhibitors when $\alpha > 1$ (squares) and when $\alpha < 1$ (circles).

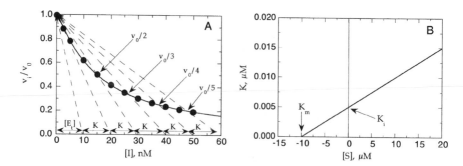

Figure 9.4 (A) Determination of "K" by the graphical method of Dixon (1972): dashed lines connect the starting point ($v_i/v_0 = 1$, [I] = 0) with points on the curve where $v_i/v_0 = v_0/n$ ($n = 2, 3, 4,$ and 5). Additional lines are drawn for apparent $n = 1$ and apparent $n = 0$, based on the x-axis spacing value "K," determine from the $n = 2$–5 lines (see text for further details). (B) Secondary plot of the "K" as a function of substrate concentration for a tight binding competitive inhibitor. Graphical determinations of K_i and K_m are obtained from the values of the y and x intercepts of the plot, respectively, as shown.

Knowing the value of K from a nest of these lines, one can draw additional lines from the x axis to the origin at spacing of K on the x axis, for apparent values of $n = 1$ and $n = 0$. From this treatment, the line corresponding to $n = 0$ will intersect the x axis at a displacement from the origin that is equal to the total enzyme concentration, [E]. Dixon goes on to show that in the case of a noncompetitive inhibitor ($\alpha = 1$), the spacing value K is equal to the inhibitor K_i, and a plot of K as a function of substrate concentration will be a horizontal line; that is, the value of K for a noncompetitive inhibitor is independent of substrate concentration. For a competitive inhibitor, however, the measured value of K will increase with increasing substrate concentration. A replot of K as a function of substrate concentration yields estimates of the K_i of the inhibitor and the K_m of the substrate from the y and x intercepts, respectively (Figure 9.4B).

9.3 DETERMINING K_i FOR TIGHT BINDING INHIBITORS

The literature describes several methods for determining the K_i value of a tight binding enzyme inhibitor. We have already discussed the graphical method of Dixon (1972), which allows one to simultaneously distinguish inhibitor type and calculate the K_i. A more mathematical treatment of tight binding inhibitors, presented by Morrison (1969), led to a generalized equation to describe the fractional velocity of an enzymatic reaction as a function of inhibitor concentration, at fixed concentrations of enzyme and substrate. This equation, commonly referred to as the Morrison equation, is derived in a manner similar to Equation 4.38, except that here the equation is cast in terms of fractional enzymatic activity in the presence of the inhibitor (i.e., in terms of the fraction of free enzyme instead of the fraction of inhibitor-bound enzyme).

$$\frac{v_i}{v_0} = 1 - \frac{([E] + [I] + K_i^{app}) - \sqrt{([E] + [I] + K_i^{app})^2 - 4[E][I]}}{2[E]} \quad (9.6)$$

The form of K_i^{app} in Equation 9.6 varies with inhibitor type. The following explicit forms of this parameter for the different inhibitor types are similar to those presented in Equations 9.2–9.5 for the IC_{50} values.

For competitive inhibitors:

$$K_i^{app} = K_i \left(1 + \frac{[S]}{K_m} \right) \quad (9.7)$$

For Noncompetitive Inhibitors:

$$K_i^{app} = \frac{[S] + K_m}{\dfrac{K_m}{K_i} + \dfrac{[S]}{\alpha K_i}} \quad (9.8)$$

when $\alpha = 1$:

$$K_i^{app} = K_i \quad (9.9)$$

For uncompetitive inhibitors:

$$K_i^{app} = K_i \left(1 + \frac{K_m}{[S]} \right) \quad (9.10)$$

Prior to the widespread use of computer-based routines for curve fitting, the direct use of the Morrison equation was inconvenient for extracting inhibitor constants from experimental data. To overcome this limitation, Henderson (1972) presented the derivation of a linearized form of the Morrison equation that allowed graphical determination of K_i and $[E]$ from measurements of the fractional velocity as a function of inhibitor concentration at a fixed substrate concentration. The generalized form of the Henderson equation is as follows:

$$\frac{[I]}{1 - \dfrac{v_i}{v_0}} = K_i^{app} \left(\frac{v_0}{v_i} \right) + [E] \quad (9.11)$$

where K_i^{app} has the same forms as presented in Equations 9.7–9.10 for the various inhibitor types.

Inspection reveals that Equation 9.11 is a linear equation. Hence, if one were to plot $[I]/(1 - v_i/v_0)$ as a function of v_0/v_i (i.e., the reciprocal of the fractional

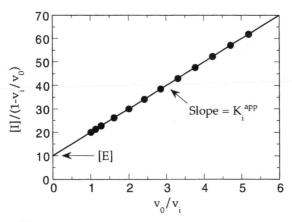

Figure 9.5   Henderson plot for a tight binding inhibitor.

velocity), the data could be fit to a straight line with slope equal to K_i^{app} and y intercept equal to [E], as illustrated in Figure 9.5. Note that the Henderson method yields a straight-line plot regardless of the inhibitor type. The slope of the lines for such plots will, however, vary with substrate concentration in different ways depending on the inhibitor type. The variation observed is similar to that presented in Figure 9.3 for the variation in IC_{50} value for different tight binding inhibitors as a function of substrate concentration. Thus, the Henderson plots also can be used to distinguish among the varying inhibitor binding mechanisms.

While linearized Henderson plots are convenient in the absence of a computer curve-fitting program, the data treatment does introduce some degree of systematic error (see Henderson, 1973, for a discussion of the statistical treatment of such data). Today, with the availability of robust curve-fitting routines on laboratory computers, it is no longer necessary to resort to linearized treatments of data such as the Henderson plots. The direct fitting of fraction velocity versus inhibitor concentration data to the Morrison equation (Equation 9.6) is thus much more desirable, and is strongly recommended.

Figure 9.6 illustrates the direct fitting of fractional velocity versus inhibitor concentration data to Equation 9.6. Such data would call for predetermination of the K_m value for the substrate (as described in Chapter 5) and knowledge of the substrate concentration in the assays. Then the data, such as the points in Figure 9.6, would be fit to the Morrison equation, allowing both K_i^{app} and [E] to be simultaneously determined as fitting parameters. Measurements of this type at several different substrate concentrations would allow determination of the mode of inhibition, and thus the experimentally measured K_i^{app} values could be converted to true K_i values.

In the case of competitive tight binding inhibitors, an alternative method for determining inhibitor K_i is to measure the iniital velocity under conditions of

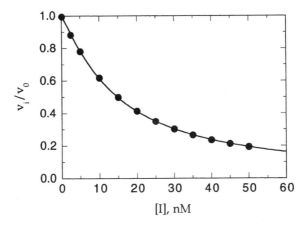

Figure 9.6 Plot of fractional velocity as a function of inhibitor concentration for a tight binding inhibitor. The solid curve drawn through the data points represents the best fit to the Morrison equation (Equation 9.6).

extremely high substrate concentration (Tornheim, 1994). Reflecting on Equation 9.2, we see that if the ratio $[S]/K_m$ is very large, the IC_{50} will be much greater than the enzyme concentration, even though the K_i is similar in magnitude to $[E]$. Thus, if a high enough substrate concentration can be experimentally achieved, the tight binding nature of the inhibitor can be overcome, and the K_i can be determined from the measured IC_{50} by application of a rearranged form of Equation 9.2. Tornheim recommends adjusting $[S]$ so that the ratios $[S]/K_m$ and $[I]/K_i$ are about equal for these measurements. Not all enzymatic reactions are amenable to this approach, however, because of the experimental limitations on substrate concentration imposed by the solubility of the substrate and the analyst's ability to measure a linear initial velocity under such extreme conditions. In favorable cases, however, this approach can be used with excellent results.

9.4 USE OF TIGHT BINDING INHIBITORS TO DETERMINE ACTIVE ENZYME CONCENTRATION

In many experimental strategies one wishes to know the concentration of enzyme in a sample for subsequent data analysis. This approach applies not only to kinetic data, but also to other types of biochemical and biophysical studies with enzymes. The literature gives numerous methods for determining total protein concentration in a sample, on the basis of spectroscopic, colorimetric, and other analytical techniques (see Copeland, 1994, for some examples). All these methods, however, measure bulk protein concentration rather than the concentration of the target enzyme in particular. Also, these

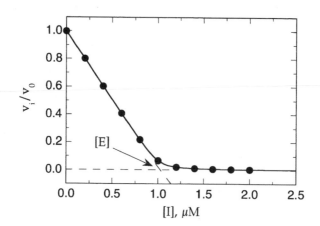

Figure 9.7 Determination of active enzyme concentration by titration with a tight binding inhibitor. $[E] = 1.0$ μM, $K_i = 5$ nM (i.e., $[E]/K_i = 200$). The solid curve drawn through the data is the best fit to the Morrison equation (Equation 9.6). The dashed lines were drawn by linear least-squares fits of the data at inhibitor concentrations that were low (0–0.6 μM) and high (1.4–2.0 μM), respectively. The active enzyme concentration is determined from the x-axis value at the intersection of the two straight lines.

methods do not necessarily distinguish between active enzyme molecules, and molecules of denatured enzyme. In many of the applications one is likely to encounter, it is the concentration of active enzyme molecules that is most relevant. The availability of a tight binding inhibitor of the target enzyme provides a convenient means of accurately determining the concentration of active enzyme in the sample, even in the presence of denatured enzyme or other nonenzymatic proteins.

Referring back to Equation 9.6, if we set up an experiment in which both $[E]$ and $[I]$ are much greater than K_i^{app}, we can largely ignore the K_i^{app} term in this equation. Under these conditions, the fractional velocity of the enzymatic reaction will fall off quasi-linearly with increasing inhibitor concentration until $[I] = [E]$. At this point the fractional velocity will approach zero and remain there at higher inhibitor concentrations. In this case, a plot of fractional velocity as a function of inhibitor concentration will look like Figure 9.7 when fit to the Morrison equation. The data in figure 9.7 were generated for a hypothetical situation: K_i of inhibitor, 5 nM; active enzyme concentration of the sample, 1.0 µM (i.e., $[E]/K_i = 200$). The data at lower inhibitor concentration can be fit to a straight line that is extended to the x axis (dashed line in Figure 9.7), and the data points at higher inhibitor concentrations can be fit to a straight horizontal line at $v_i/v_0 = 0$ (longer dashed line in Figure 9.7). The two lines thus drawn will intersect at a point on the x axis where $[I] = [E]$. Note, however, that this treatment works only when $[E]$ is much greater than K_i. When $[E]$ is less than about $200K_i$, the data are not well described by two intersecting straight lines. In such cases the data can be fit directly to Equation 9.6 to determine [E], as described earlier.

This type of treatment is quite convenient for determining the active enzyme concentration of a stock enzyme solution (i.e., at high enzyme concentration) that will be diluted into a final reaction mixture for experimentation. For example, one might wish to store an enzyme sample at a nominal enzyme concentration of 100 μM in a solution containing 1 mg/mL gelatin for stability purposes (see discussion in Chapter 7). The presence of the gelatin would preclude accurate determination of enzyme concentration by one of the traditional colorimetric protein assays; moreover, *active* enzyme concentration could not be determined by means of such assays. Given a nanomolar inhibitor of the target enzyme, one could dilute a sample of the stock enzyme to some convenient concentration for an enzymatic assay that was still much greater than the K_i (e.g., 1 μM). Treatment of the fractional velocity versus inhibitor concentration as described here would thus lead to determination of the true concentration of active enzyme in the working solution, and from this one could back-calculate to arrive at the true concentration of active enzyme in the enzyme stock. This is a routine strategy in many enzymology laboratories, and numerous examples of its application can be found in the literature.

A comparable assessment of active enzyme concentration can be obtained by the reverse experiment in which the inhibitor concentration is fixed at some value much greater than the K_i (about $200 K_i$ or more), and the amount of enzyme added to the reaction mixture is varied. The results of such an experiment are illustrated in Figure 9.8. The initial velocity remains zero until equal concentrations of enzyme and inhibitor are present in solution. As the enzyme concentration is titrated beyond this point, the stoichiometric inhibition is overcome, and a linear increase in initial velocity is then observed. Again, from the point of intersection of the two dashed lines drawn through the data as in Figure 9.8, the true concentration of active enzyme can be determined (Williams and Morrison, 1979). An advantage of this second approach to active enzyme concentration determination is that it typically uses up less of the enzyme stock to complete the titration. Hence, when the enzyme is in limited supply, this alternative is recommended.

9.5 SUMMARY

In this Chapter we have described a special case of enzyme inhibition, in which the dissociation constant of the inhibitor is similar to the total concentration of enzyme in the sample. These inhibitor offer a special challenge to the enzymologist, because they cannot be analyzed by the traditional methods described in Chapter 8. We have seen that tight binding inhibitors yield double-reciprocal plots that appear to suggest noncompetitive inhibition regardless of the true mode of interaction between the enzyme and the inhibitor. Thus, whenever noncompetitive inhibition is diagnosed through the use of double reciprocal

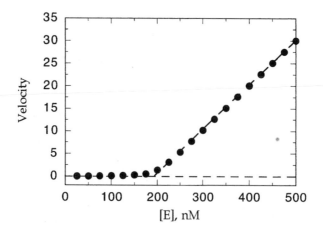

Figure 9.8 Determination of active enzyme concentration by titration of a fixed concentration of a tight binding inhibitor with enzyme: [I] = 200 nM, K_i = 1 nM (i.e., [I]/K_i = 200). The data analysis is similar to that described for Figure 9.7 and in the text. Velocity is in arbitrary units.

plots, the data should be reevaluated to ensure that tight binding inhibition is not occurring. Methods for determining the true mode of inhibition and the K_i for these tight binding inhibitors were described in this chapter.

Tight binding inhibitors are an important class of molecules in many industrial enzyme applications. Many contemporary therapeutic enzyme inhibitors, for example, act as tight binders. Recent examples include inhibitors of dihydrofolate reductase (as anticancer drugs), inhibitors of the HIV aspartyl protease, (as anti-AIDS drugs), and inhibitors of metalloproteases (as potential cartilage protectants). Many of the naturally occurring enzyme inhibitors, which play a role in metabolic homeostasis, are tight binding inhibitors of their target enzymes. Thus tight binding inhibitors are an important and commonly encountered class of enzyme inhibitor. The need for special treatment of enzyme kinetics in the presence of these inhibitors must not be overlooked.

REFERENCES AND FURTHER READING

Bieth, J. (1974) In *Proteinase Inhibitors, Bayer-Symposium V,* Springer-Verlag, New York, pp. 463–469.

Cha, S. (1975) *Biochem. Pharmacol.* **24**, 2177.

Cha, S. (1976) *Biochem. Pharmacol.* **25**, 2695.

Cha, S., Agarwal, R. P., and Parks, R. E., Jr. (1975) *Biochem. Pharmacol.* **24**, 2187.

Copeland, R. A. (1994) *Methods of Protein Analysis, A Practical Guide to Laboratory Protocols,* Chapman & Hall, New York.

Copeland, R. A., Lombardo, D., Giannaras, J., and DeCicco, C. P. (1995) *Bioorg. Med. Chem. Lett.* **5**, 1947.

Dixon, M. (1972) *Biochem. J.* **129**, 197.

Dixon, M., and Webb, E. C. (1979) *Enzymes*, 3rd ed., Academic Press, New York.

Goldstein, A. (1944) *J. Gen. Physiol.* **27**, 529.

Greco, W. R., and Hakala, M. T. (1979) *J. Biol. Chem.* **254**, 12104.

Henderson, P. J. F. (1972) *Biochem. J.* **127**, 321.

Henderson, P. J. F. (1973) *Biochem. J.* **135**, 101.

Morrison, J. F. (1969) *Biochim. Biophys. Acta,* **185**, 269.

Myers, D. K. (1952) *Biochem. J.* **51**, 303.

Szedlacsek, S. E., and Duggleby, R. G. (1995) *Methods Enzymol.* **249**, 144.

Tornheim, K. (1994) *Anal. Biochem.* **221**, 53.

Turner, P. M., Lerea, K. M., and Kull, F. J. (1983) *Biochem. Biophys. Res. Commun.* **114**, 1154.

Williams, J. W., and Morrison, J. F. (1979) *Methods Enzymol.* **63**, 437.

Williams, J. W., Morrison, J. F., and Duggleby, R. G. (1979) *Biochemistry,* **18**, 2567.

10

TIME-DEPENDENT INHIBITION

All the inhibitors we have encountered thus far have established their binding equilibrium with the enzyme on a time scale that is rapid with respect to the turnover rate of the enzyme-catalyzed reaction. In Chapter 9 we noted that many tight binding inhibitors establish this equilibrium on a slower time scale, but in our discussion we eliminated this complication by pretreating the enzyme with the inhibitor long enough to ensure that equilibrium had been fully reached before steady state turnover was initiated by addition of substrate. In this chapter we shall explicitly deal with inhibitors that bind slowly to the enzyme on the time scale of enzymatic turnover, and thus display a change in initial velocity with time. These inhibitors, that is, act as *slow binding* or *time-dependent* inhibitors of the enzyme.

We can distinguish four different modes of interaction between an inhibitor and an enzyme that would result in slow binding kinetics. The equilibria involved in these processes are represented in Figure 10.1. Figure 10.1A shows the equilibrium associated with the uninhibited turnover of the enzyme, as we discussed in Chapter 5: k_1, the rate constant associated with substrate binding to the enzyme to form the ES complex, is sometimes refered to as k_{on} (for substrate coming *on* to the enzyme). The constant k_2 in Figure 10.1A is the dissociation or off rate constant for the ES complex dissociating back to free enzyme and free substrate, and k_{cat} is the catalytic rate constant as defined in Chapter 5.

In the remaining schemes of Figure 10.1 (B–D), the equilibrium described by Scheme A occurs as a competing reaction (as we saw in connection with simple reversible enzyme inhibitors in Chapter 8).

Scheme B illustrates the case of the inhibitor binding to the enzyme in a simple bimolecular reaction, similar to what we discussed in Chapters 8 and 9.

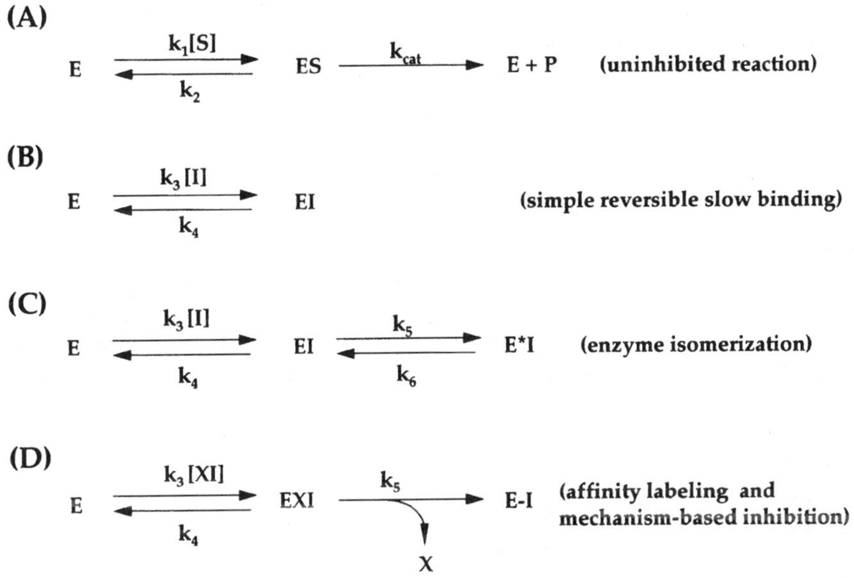

(A)

$$E \underset{k_2}{\overset{k_1[S]}{\rightleftharpoons}} ES \xrightarrow{k_{cat}} E + P \quad \textbf{(uninhibited reaction)}$$

(B)

$$E \underset{k_4}{\overset{k_3[I]}{\rightleftharpoons}} EI \quad \textbf{(simple reversible slow binding)}$$

(C)

$$E \underset{k_4}{\overset{k_3[I]}{\rightleftharpoons}} EI \underset{k_6}{\overset{k_5}{\rightleftharpoons}} E^*I \quad \textbf{(enzyme isomerization)}$$

(D)

$$E \underset{k_4}{\overset{k_3[XI]}{\rightleftharpoons}} EXI \xrightarrow{k_5} E\text{-}I \quad \begin{array}{l}\textbf{(affinity labeling and}\\ \textbf{mechanism-based inhibition)}\end{array}$$

$$X$$

Figure 10.1 Schemes for time-dependent enzyme inhibition. Scheme A, which describes the turnover of the enzyme in the absence of inhibitor, is a competing reaction for all the other schemes. Scheme B illustrates the equilibrium for a simple reversible inhibition process that leads to time-dependent inhibition because of the low values of k_3 and k_4 relative to enzyme turnover. In Scheme C, an initial binding of the inhibitor to the enzyme leads to formation of the EI complex, which undergoes an isomerization of the enzyme to form the new complex E*I. Scheme D represents the reactions associated with irreversible enzyme inactivation due to covalent bond formation between the enzyme and some reactive group on the inhibitor, leading to the covalent adduct E—I. Inhibitors that conform to Scheme D may act as affinity labels of the enzyme, or they may be mechanism-based inhibitors.

Here, however, the association and dissociation rate constants (k_3 and k_4, respectively) are such that the equilibrium is established slowly. As with rapid binding inhibitors, the equilibrium dissociation constant K_i is given here by:

$$K_i = \frac{k_4}{k_3} = \frac{[E][I]}{[EI]} \tag{10.1}$$

Morrison and Walsh (1988) have pointed out that even when k_3 is diffusion limited, if K_i is low and [I] is varied in the region of K_i, both $k_3[I]$ and k_4 will be low in value. Hence, under these circumstances onset of inhibition would be slow even though the magnitude of k_3 is that expected for a rapid reaction. This is why most tight binding inhibitors display time-dependent inhibition. If the observed time dependence is due to an inherently slow rate of binding, the inhibitor is said to be a *slow binding* inhibitor, and its dissociation constant is given by Equation 10.1. If, on the other hand, the inhibitor is also a tight binder, it is said to be a *slow, tight binding* inhibitor, and the depletion of the free enzyme and free inhibitor concentrations due to formation of the EI complex also must be taken into account:

$$K_i = \frac{([E_t] - [EI])([I] - [EI])}{EI} \tag{10.2}$$

where $[E_t]$ represents the concentration of total enzyme (i.e., in all forms) present in solution.

In Scheme C, the enzyme encounters the inhibitor and establishes a binding equilibrium that is defined by the on and off rate constants k_3 and k_4, just as in Scheme B. In Scheme C, however, the binding of the inhibitor induces in the enzyme a conformational transition, or isomerization, that leads to a new enzyme–inhibitor complex E*I; the forward and reverse rate constants for the equilibrium between these two inhibitor-bound conformations of the enzyme are given by k_5 and k_6, respectively. The dissociation constant for the initial EI complex is still given by K_i (i.e., k_4/k_3), but a second dissociation constant for the second enzyme conformation K_i^* must be considered as well. This second dissociation constant is given by:

$$K_i^* = \frac{K_i k_6}{k_5 + k_6} = \frac{[E][I]}{[EI] + [E^*I]} \tag{10.3}$$

To observe a slow onset of inhibition, K_i^* must be much less than K_i. Hence, in this situation, the isomerization of the enzyme leads to much tighter binding between the enzyme and the inhibitor. As with Scheme B, if the inhibitor is of the slow, tight binding variety, the diminution of free enzyme and free inhibitor must be explicitly accounted for in the expressions for both K_i and K_i^* (see Morrison and Walsh, 1988).

Note that to observe slow binding kinetics it is not sufficient for the conversion of EI to E*I alone to be slow. The reverse reaction must be slow as well. In fact, for the slow binding to be detected, the reverse rate constant (k_6) must be less than the forward isomerization rate (k_5). In the extreme case ($k_6 \ll k_5$), one does not observe a measurable return to the EI conformation and the enzyme isomerization step will appear to lead to irreversible inhibition. Under these conditions, k_6 can be considered to be insignificant, and the isomerization can be treated practically as an irreversible step dominated by the rate constant k_5.

Finally, in Scheme D we consider two modes of interaction of the inhibitor with the enzyme for which k_6 is truly equal to zero; that is, we are dealing with irreversible enzyme inactivation. We must make the distinction here between reversible and irreversible inhibition. In all the inhibitory schemes we have considered thus far, even in the case of slow tight binding inhibition, k_6 has been nonzero. This rate constant may be very small, and the inhibitors may act, for all practical purposes, as irreversible. With enough dilution of the EI complex and enough time, however, one can eventually recover an active free enzyme population. In the case of an irreversible inhibitor, the enzyme molecule that has bound the inhibitor is permanently incapacitated. No amount of time or dilution will result in a reactivation of the enzyme that has encountered inhibitors of these types. Such inhibitors hence are often referred to as *enzyme inactivators.*

The first example of irreversible inhibition is the process known as *affinity labeling* or *covalent modification* of the enzyme. In this case, the inhibitory compound binds to the enzyme and covalently modifies a catalytically essential residue or residues on the enzyme. The covalent modification involves some chemical alteration of the inhibitory molecule, but the process is based on chemistry that occurs at the modification site in the absence of any enzyme-catalyzed reaction. Affinity labels are useful not only as inhibitors of enzyme activity; they also have become valuable research tools. Some of these compounds are very selective for specific amino acid residues and can thus be used to identify key residues involved in the catalytic cycle of the enzyme. See Section 10.5.3 and Lundblad (1991), and Copeland (1994).

In the second form of irreversible inactivation we shall consider, *mechanism-based inhibition*, the inhibitory molecule binds to the enzyme active site and is recognized by the enzyme as a substrate analogue. The inhibitor is therefore chemically transformed through the catalytic mechanism of the enzyme to form an E–I complex that can no longer function catalytically. Many of these inhibitors inactivate the enzyme by forming an irreversible covalent E–I adduct. In other cases, the inhibitory molecule is subsequently released from the enzyme (a process referred to as noncovalent inactivation), but the enzyme has been permanently trapped in a form that can no longer support catalysis. Because they are chemically altered via the mechanism of enzymatic catalysis at the active site, mechanism-based inhibitors always act as competitive enzyme inactivators. These inhibitors have been referred to by a variety of names in the literature: suicide substrates, suicide enzyme inactivators, k_{cat} inhibitors, enzyme-activated irreversible inhibitors, Trojan horse inactivators, enzyme-induced inactivators, dynamic affinity labels, trap substrates, and so on (Silverman, 1988a).

In the discussion that follows we shall describe experimental methods for detecting the time dependence of slow binding inhibitors, and data analysis methods that allow us to distinguish among the different potential modes of interaction with the enzyme. We shall also discuss the appropriate determination of the inhibitor constants K_i and K_i^* for these inhibitors.

10.1 PROGRESS CURVES FOR SLOW BINDING INHIBITORS

The progress curves for an enzyme reaction in the presence of a slow binding inhibitor will not display the simple linear product-versus-time relationship we have seen for simple reversible inhibitors. Rather, product formation over time will be a curvilinear function because of the slow onset of inhibition for these

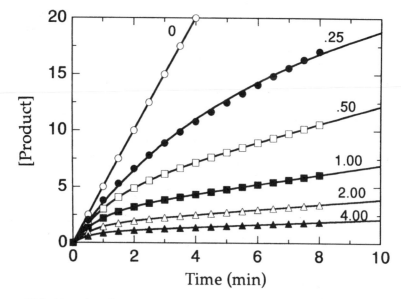

Figure 10.2 Examples of progress curves in the presence of varying concentrations of a time-dependent enzyme inhibitor for a reaction initiated by adding enzyme to a mixture containing substrate and inhibitor. Curves are numbered to indicate the relative concentrations of inhibitor present. Note that over the entire 10-minute time window, the uninhibited enzyme displays a linear progress curve.

compounds. Figure 10.2 illustrates typical progress curves for a slow binding inhibitor when the enzymatic reaction is initiated by addition of enzyme. Over a time period in which the uninhibited enzyme displays a simple linear progress curve, the data in the presence of the slow binding inhibitor will display a quasi-linear relationship with time in the early part of the curve, converting later to a different (slower) linear relationship between product and time. Note that it is critical to establish a time window covering the linear portion of the uninhibited reaction progress curve, during which one can observe the change in slope that occurs with inhibition. If the onset of inhibition is very slow, a long time window may be required to observe the changes illustrated in Figure 10.2. With long time windows, however, one runs the risk of reaching significant substrate depletion, which would invalidate the subsequent data analysis. Thus it may be necessary to evaluate several combinations of enzyme, substrate, and inhibitor concentrations to find an appropriate range of each for conducting time-dependent measurements. With these cautions addressed, the progress curves at different inhibitor concentrations can be described by Equation 10.4:

$$[P] = v_s t + \frac{v_i - v_s}{k_{obs}} [1 - \exp(-k_{obs}t)] \tag{10.4}$$

where v_i and v_s are the initial and steady state (i.e., final) velocities of the reaction in the presence of inhibitor, k_{obs} is the apparent first-order rate constant for the interconversion between v_i and v_s, and t is time.

Morrison and Walsh (1988) have provided explicit mathematical expressions for v_i and v_s in the case of a competitive slow binding inhibitor, illustrating that v_i and v_s are functions (similar to Equation 8.10) of V_{max}, [S], K_m, and either K_i or K_i^* (for inhibitors that act according to Scheme C in Figure 10.1), respectively. For our purposes, it is sufficient to treat Equation 10.4 as an empirical equation that makes possible the extraction from the experimental data of values for v_i, v_s, and most importantly, k_{obs}. Note that v_i may or may not vary with inhibitor concentration, depending on the relative values of K_i and K_i^*, and the ratio of [I] to K_i (Morrison and Walsh, 1988). The value of v_s will be a finite, nonzero value as long as the inhibitor is not an irreversible enzyme inactivator. In the latter case, the value of v_s will eventually reach zero.

A second strategy for measuring progress curves for slow binding inhibitors is to preincubate the enzyme with the inhibitor for a long time period relative to the rate of inhibitor binding, and to then initiate the reaction by diluting the enzyme–inhibitor solution with a solution containing the substrate for the enzyme. During the preincubation period the equilibria between enzyme and inhibitor are established, and addition of substrate perturbs this equilibrium. Because of the slow off rate of the inhibitor, the progress curve will display an initial shallow slope, which eventually turns over to the steady state velocity, as illustrated in Figure 10.3. The progress curves seen here also are well described by Equation 10.4, except that now the initial velocity is lower than the steady state velocity, whereas for data obtained by initiating the reaction with enzyme, the initial velocity is greater than the steady state velocity. To highlight this difference, some authors replace the term v_i in Equation 10.4 with v_r in the case of reactions initiated with substrate. Morrison and Walsh (1988) again provide an explicit mathematical form for v_r, which depends on the V_{max}, [S], K_m, [I], K_i, K_i^*, and the volume ratio between the preincubation enzyme–inhibitor solution and the final volume of the total reaction mixture. Again, for our purposes we can use Equation 10.4 as an empirical equation, allowing v_i (or v_r), v_s, and k_{obs} to be adjustable parameters whose values are determined by nonlinear curve-fitting analysis.

Inhibitors that are very tight binding, as well as time dependent, almost always conform to Scheme C of Figure 10.1 (Morrison and Walsh, 1988). In this case the progress curves also will be influenced by the depletion of the free enzyme and free inhibitor populations that occurs. To account for these diminished populations, Equation 10.4 must be modified as follows:

$$[P] = v_s t + \frac{(v_i - v_s)(1 - \gamma)}{k_{obs}\gamma} \ln \left\{ \frac{[1 - \gamma \exp(-k_{obs}t)]}{1 - \gamma} \right\} \qquad (10.5)$$

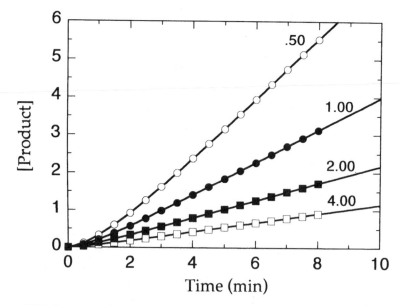

Figure 10.3 Examples of progress curves in the presence of varying concentrations of a time-dependent enzyme inhibitor for a reaction initiated by diluting an enzyme—inhibitor complex into the reaction buffer containing substrate. Curves are numbered to indicate the relative concentrations of inhibitor.

where γ is given by

$$\gamma = \frac{K_i^{*app} + [E_t] + [I_t] - Q}{K_i^{*app} + [E_t] + [I_t] + Q} = \frac{[E_t]}{[I_t]}\left(1 - \frac{v_s}{v_0}\right)^2 \qquad (10.6)$$

where v_0 is the initial velocity in the absence of inhibitor, and

$$Q = [(K_i^{*app} + [I_t] - [E_t])^2 + 4(K_i^{*app}[E_t])]^{1/2} - (K_i^{*app} + [I_t] - [E_t]) \qquad (10.7)$$

Throughout Equations 10.5–10.7, $[E_t]$ and $[I_t]$ refer to the total concentrations (i.e., all forms) of enzyme and inhibitor, respectively. Further discussion of the data analysis for slow, very tight binding inhibitors can be found in the review by Morrison and Walsh (1988).

If inhibitor binding (or release) is very slow compared to the rate of uninhibited enzyme turnover, another convenient experimental strategy can be employed to determine k_{obs}. Essentially, the enzyme is preincubated with the inhibitor for different lengths of time before the steady state velocity of the reaction is measured. For example, if the steady state velocity of the reaction can be measured over a 30-second time window, but the inhibitor binding event occurs over the course of tens of minutes, the enzyme could be preincubated with the inhibitor between 0 and 120 minutes in 5-minute intervals, and the velocity of the reaction measured after each of the different preincubation times. Figure 10.4 illustrates the type of data this treatment would produce. For a fixed inhibitor concentration, the fractional velocity remaining after a given preincubation time will fall off according to Equation 10.8:

$$\frac{v}{v_0} = \exp(-k_{obs}t) \tag{10.8}$$

Therefore, at a fixed inhibitor concentration, the fractional velocity will decay exponentially with preincubation time, as in Figure 10.4A. For convenience, we can recast Equation 10.8 by taking the logarithm of each side to obtain a linear function:

$$2.303 \log_{10}\left(\frac{v}{v_0}\right) = -k_{obs}t \tag{10.9}$$

Thus the value of k_{obs} at a fixed inhibitor concentration can be determined directly from the slope of a semilog plot of fractional velocity as a function of preincubation time, as in Figure 10.4B.

10.2   DISTINGUISHING BETWEEN SLOW BINDING SCHEMES

To distinguish among the schemes illustrated in Figure 10.1, one must determine the effect of inhibitor concentration on the apparent first-order rate constant k_{obs}. We shall present the relationships between k_{obs} and [I] for these various schemes without deriving them explicitly. A full treatment of the derivation of these equations can be found in Morrison and Walsh (1988) and references therein.

10.2.1   Scheme B

For an inhibitor that binds according to Scheme B of Figure 10.1, the relationship between k_{obs} and [I] is given by Equation 10.10:

$$k_{obs} = k_4\left(1 + \frac{[I]}{K_i^{app}}\right) \tag{10.10}$$

where K_i^{app} is the apparent K_i, which is related to the true K_i by different functions depending on the mode of inhibitor interaction with the enzyme (i.e., competitive, noncompetitive, uncompetitive, etc.; see Section 10.3). From Equation 10.10 we see that a plot of k_{obs} as a function of [I] should yield a straight line with slope equal to k_4/K_i^{app} and y intercept equal to k_4 (Figure 10.5). Thus from linear regression analysis of such data, one can simultaneously

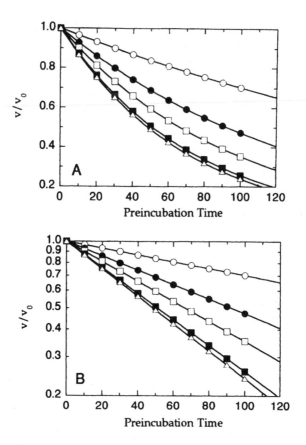

Figure 10.4 Preincubation time dependence of the fractional velocity of an enzyme-catalyzed reaction in the presence of varying concentrations of a slow binding inhibitor: data on a linear scale (A) and on a semilog scale (B).

determine the values of k_4 and K_i^{app}. If the inhibitor modality is known, K_i^{app} can be converted into K_i (Section 10.3), and from this the value of k_3 can be determined by means of Equation 10.1.

10.2.2 Scheme C

For inhibitors corresponding to Scheme C of Figure 10.1, k_{obs} is related to [I] as follows:

$$k_{obs} = k_6 + \left[\frac{k_5[I]}{K_i^{app} + [I]} \right]$$ (10.11)

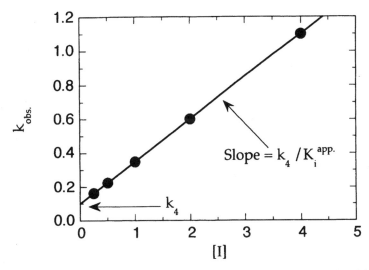

Figure 10.5 Plot of k_{obs} as a function of inhibitor concentration for a slow binding inhibitor that conforms to Scheme B of Figure 10.1.

which can be recast thus:

$$k_{obs} = k_6 \left[\frac{1 + \dfrac{[I]}{K_i^{*app}}}{1 + \dfrac{[I]}{K_i^{app}}} \right] \qquad (10.12)$$

The form of Equations 10.11 and 10.12 predicts that k_{obs} will vary as a hyperbolic function of [I], as illustrated in Figure 10.6. The y intercept of the curve in this figure provides an estimate of the rate constant k_6, while the maximum value of k_{obs} expected at infinite inhibitor concentration according to Equation 10.11, is $k_6 + k_5$. Hence, by nonlinear curve fitting of the data to Equation 10.11 one can simultaneously determine the values of k_6, K_i^{app}, and K_i^{*app}.

Note that if K_i were much greater than K_i^*, the concentrations of inhibitor required for slow binding inhibition would be much less than K_i. Under these circumstances, the steady state concentration of [EI] would be kinetically insignificant, and Equation 10.11 would thus reduce to:

$$k_{obs} = k_6 \left[1 + \frac{[I]}{K_i^{*app}} \right] \qquad (10.13)$$

Thus for this situation a plot of k_{obs} as function of [I] would again yield a straight-line relationship, as we saw for inhibitors associated with Scheme B. In fact, when a straight-line relationship is observed in the plot of k_{obs} versus [I], one cannot readily distinguish between these two situations.

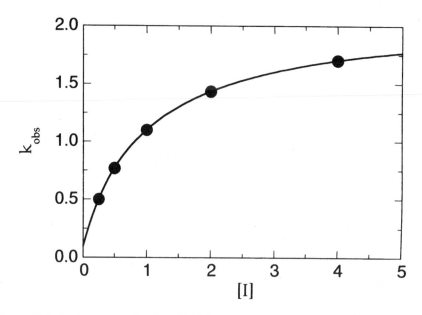

Figure 10.6   Plot of k_{obs} as a function of inhibitor concentration for a slow binding inhibitor that conforms to Scheme C of Figure 10.1.

10.2.3 Scheme D

If the kinetic constant k_6 is very small in Scheme C, or zero as in Scheme D, the inhibitor acts, for all practical purposes, as an irreversible inactivator of the enzyme. In such cases, Equation 10.11 reduces to:

$$k_{obs} = \frac{k_5[I]}{K_i^{app} + [I]} \tag{10.14}$$

Here again, a plot of k_{obs} as a function of [I] will yield a hyperbolic curve (Figure 10.7A), but now the y intercept will be zero (reflecting the zero, or near-zero, value of k_6).

For irreversible inhibitors, the return to free E and free I from the EI complex is greatly perturbed by the irreversibility of the subsequent inactivation event (represented by k_5). For this reason, Tipton (1973) and Kitz and Wilson (1962) make the point that for irreversible inactivators, the term K_i no longer represents the simple dissociation constant for the EI complex. Rather, the term K_i^{app} in Equation 10.14 is defined as the apparent concentration of inhibitor required to reach half-maximal rate of inactivation of the enzyme. Kitz and Wilson (1962) also replace k_5 in Equation 10.14 with k_{inact}, which they define as the maximal rate of enzyme inactivation. With these definitions, the parameters k_{inact} and K_i^{app} are reminiscent of the parameters V_{max} and K_m, respectively, from the Henri–Michaelis–Menten equation (Chapter 5). Just as

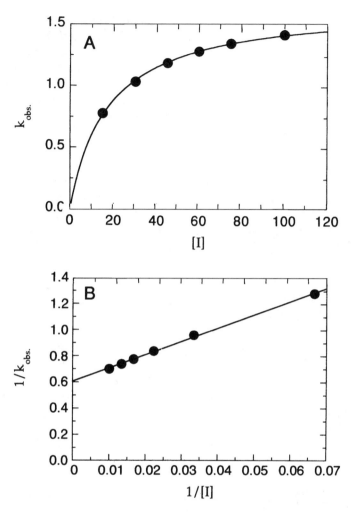

Figure 10.7 (A) Plot of k_{obs} as a function of inhibitor concentration for a slow binding inhibitor that conforms to Scheme D of Figure 10.1. (B) The data as in (A) presented as a double-reciprocal plot. The nonzero intercept indicates that the inactivation proceeds through a two-step mechanism: an initial binding step followed by a slower inactivation event.

the ratio k_{cat}/K_m is the best measure of the catalytic efficiency of an enzyme-catalyzed reaction, the best measure of inhibitory potency for an irreversible inhibitor is the second-order rate constant obtained from the ratio k_{inact}/K_i.

Similar to the Lineweaver–Burk plots encountered in Chapter 5, a double-reciprocal plot of $1/k_{obs}$ as a function of $1/[I]$ yields a straight-line relationship. Most irreversible inhibitors bind to the enzyme active site in a reversible manner (represented by K_i^{app}) before the slower inactivation event (represented by k_5) proceeds. Thus, as illustrated by Scheme D in Figure 10.1, the inactivation of the enzyme requires two sequential steps: a binding event and an inactivation event. Irreversible inhibitors that behave in this fashion display a linear relationship between $1/k_{obs}$ and $1/[I]$ that intersects the y axis at a value greater than zero (Figure 10.7B). If, however, the formation of the reversible EI complex is kinetically insignificant relative to the rate of inactivation, the double-reciprocal plot will pass through the origin, reflecting a single-step inactivation process (Kitz and Wilson, 1962):

$$E + I \xrightarrow{k_{inact}} E - I$$

Although not as common as the two-step inactivation scheme shown in Figure 10.1D, this type of behavior is sometimes seen for small molecule affinity labels of enzymes. For example, Kitz and Wilson (1962) showed that the compound methylsulfonyl fluoride inactivates acetylcholinesterase by irreversible formation of a sulfonyl–enzyme adduct that appears to form in a single inactivation step (Figure 10.8).

10.3 DISTINGUISHING BETWEEN MODES OF INHIBITOR INTERACTION WITH ENZYME

Morrison states that almost all slow binding enzyme inhibitors act as competitive inhibitors, binding at the enzyme active site (Morrison, 1982; Morrison and Walsh, 1988). Nevertheless it is possible, in principle at least, for slow binding inhibitors to interact with the enzyme by competitive, noncompetitive, or uncompetitive inhibition patterns. In the preceding equations, the relationships between K_i^{app} and K_i, and between K_i^{*app} and K_i^*, are the same as those presented in Chapters 8 and 9 for the relationships between K_i^{app} and K_i for the different modes of inhibition.

To distinguish the mode of inhibition that is taking place, hence to ensure the use of the appropriate relationships for K_i and K_i^* in the equations, one must determine the effects of varying substrate concentration on the value of k_{obs} at a fixed concentration of inhibitor. Tian and Tsou (1982, and references therein) have presented derivations of the relationships between k_{obs} and substrate concentration for competitive, noncompetitive, and uncompetitive irreversible inhibitors. (Similar patterns will be observed for slow binding inhibitors that conform to Scheme C as well.) More generalized forms of these relationships are given in Equations 10.15–10.17.

For competitive inhibition:

$$k_{obs} = \frac{k}{1 + [S]/K_m} \tag{10.15}$$

For noncompetitive inhibition ($\alpha = 1$):

$$k_{obs} = k \tag{10.16}$$

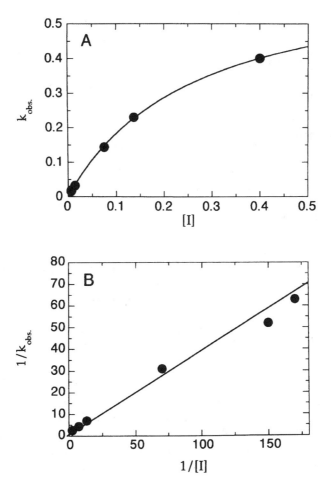

Figure 10.8 (A) Plot of k_{obs} as a function of inhibitor concentration for inhibition of acetyl-cholinesterase by methylsulfonyl fluoride. (B) The data in (A) as a double-reciprocal plot. [Data adapted from Kitz and Wilson (1962).]

For uncompetitive inhibiton:

$$k_{obs} = \frac{k}{1 + K_m/[S]} \tag{10.17}$$

The constant k in these equations can be treated as an empirical variable for curve-fitting purposes (see Tian and Tsou, 1982, for the explicit form of k for irreversible inhibitors).

From the forms of Equations 10.15–10.17, we see that a competitive slow binding inhibitor will display a diminution in k_{obs} as the substrate concentra-

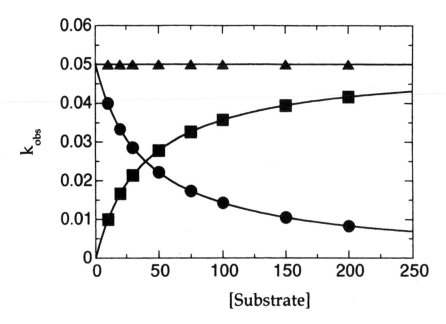

Figure 10.9 Expected dependence of k_{obs} on substrate concentration for time-dependent irreversible inhibitors that conform to competitive (circles), noncompetitive (triangles), and uncompetitive (squares) modes of interaction with the enzyme.

tion is raised. For a noncompetitive inhibitor, on the other hand, k_{obs} will not vary with substrate concentration (when $\alpha = 1$), while for an uncompetitive inhibitor the value of k_{obs} will increase with increasing substrate concentration. These relationships between k_{obs} and substrate concentration are illustrated in Figure 10.9.

10.4 DETERMINING REVERSIBILITY

It has been established that when an inhibitor conforming to Scheme C of Figure 10.1 has a very low value of k_6, it is difficult to differentiate this mode of inhibition from true irreversible inactivation according to Scheme D. To distinguish between these two possibilities, one must determine whether enzyme activity can be rescued by removal of unbound inhibitor from the enzyme solution. This is typically accomplished by large dilution, dialysis, filter binding, or size exclusion chromatography (see Chapters 4 and 7 and references therein for details about these methods). Suppose, for example, that a slow binding inhibitor reduces the steady state velocity of an enzyme reaction by 50% at a concentration of 100 nM. If we prepare a 1 mL sample of enzyme with inhibitor present at 100 nM and dialyze this sample extensively against a liter of buffer, the final concentration of enzyme in the dialysis tubing will be essentially unchanged, but the concentration of free inhibitor will be reduced 1000-fold. If the inhibitor were binding reversibly to the enzyme, we would have observed a postdialysis return of enzyme activity to close to the original uninhibited activity. For a very low value of k_6, it might take some time— hours or days—for the new equilibrium between free and bound inhibitor to establish itself after dialysis. If k_6 is nonzero, however, the expected reversal of inhibition eventually will occur. Of course, one must ensure that the enzyme itself is stable during these manipulations. Otherwise, it will be impossible to distinguish residual inhibition (due to the inhibitor) from enzyme inactivation (due to protein denaturation).

To distinguish covalent inactivation from noncovalent inhibition, one can look for the release of the original inhibitor molecule upon denaturation of the enzyme sample. Suppose that a slow binding inhibitor actually acted as a covalent affinity label of the target enzyme. If we were to denature the enzyme after inhibition and then separate the denatured protein from the rest of the solution (see Chapter 7), the inhibitory molecule would remain with the denatured protein, as a result of the covalent linkage between the inhibitor and the enzyme. If, on the other hand, the inhibitor were noncovalently associated with the enzyme, it would be released into the solution upon denaturation of the enzyme.

An illustration of this type of experiment comes from work performed in our laboratory on an inhibitor of the inducible isoform of the enzyme prostaglandin G/H synthase (PGHS2). A compound we were investigating, DuP697, displayed the kinetic features of a competitive, slow binding, irreversible enzyme inactivator (Copeland et al., 1994). A plot of k_{obs} as a function of DuP697 concentration displayed a hyperbolic fit that passed through the origin of the plot. Extensive dialysis of the inhibited enzyme against buffer did not result in a return of enzymatic activity, suggesting either that the inhibitor was covalently associating with the enzyme or that the value of k_6 was extremely small.

To determine which way DuP697 was interacting with the enzyme, we treated a micromolar solution of the enzyme with a substoichiometric concentration of the inhibitor and allowed the resulting solution to equilibrate for a long time period, relative to the rate of enzyme inactivation. The enzyme was then denatured and precipitated by addition of 4 volumes of a 1:1 mixture of methanol/acetonitrile. The denatured protein solution was centrifuged through a 30 kDa cutoff filter, and the filtrate from this treatment was dried under nitrogen and redissolved in a small volume of dimethyl sulfoxide or acetonitrile. The amount of DuP697 released from the enzyme by this treatment was then determined by reversed phase HPLC and by measuring the ability of the redissolved filtrate to effect inhibition of fresh samples of the enzyme (Copeland et al., 1994, 1995). Upon finding that 97% of the DuP697 added to the starting enzyme sample was recovered in this way, we concluded that DuP697 is not a covalent modifier of the enzyme, but rather conforms to Scheme C of Figure 10.1 with an extremely small value for k_6.

10.5 EXAMPLES OF SLOW BINDING ENZYME INHIBITORS

The literature is filled with examples of slow binding, slow tight binding, affinity label, and mechanism-based inhibitors of important enzymes. Extensive examples of slow binding inhibitors were presented in the review by Morrison and Walsh (1988). Silverman has devoted a two-volume text to the subject of mechanism-based enzyme inactivators (1988a), as well as an extensive review article (1988b) of their potential uses in medicine. (See also Trzaskos et al., 1995, for an interesting more recent example of mechanism-based inactivation of lanosterol 14α-methyl demethylase in the design of new cholesterol-lowering therapies.) Rather than providing an exhaustive review of the literature, we shall present two examples of specific enzyme systems that have proved amenable to time-dependent inhibition: the serine proteases and prostaglandin G/H synthase. These examples should suffice to illustrate the importance of this general class of enzyme inhibitors. In Section 10.5.3 we also present a general discussion of irreversible affinity labels as mechanistic probes of enzyme structure and mechanism.

10.5.1 Serine Proteases

As we saw in Chapter 6, the serine proteases hydrolyze peptide bonds through the formation of a tetrahedral transition state involving a peptide carbonyl carbon of the substrate and an active site serine residue as the attacking nucleophile. Several groups have taken advantage of the ability of boron to adopt a tetrahedral ligand sphere in preparing transition state analogues as inhibitors of serine proteases. Kettner and Shervi (1984) have used this strategy to prepare selective inhibitors of the serine proteases chymotrypsin and leukocyte elastase based on α-aminoboronate peptide analogues (Figure 10.10). They found that succinamide methyl esters that incorporate aminoboronate analogues of phenylalanine and valine were highly selective inhibitors of chymotrypsin and leukocyte elastase, respectively. A selective inhibitor of leukocyte elastase could have potential therapeutic value in the treatment of a number of inflammatory diseases of the respiratory system (e.g., cystic fibrosis, asthma). Kinetic studies of R-Pro-boroPhe-OH binding to chymotrypsin and R-Pro-boroVal-OH binding to leukocyte elastase revealed that both inhibitors function as competitive slow binding inhibitors that conform to Scheme C of Figure 10.1. For chymotrypsin inhibition by R-Pro-boroPhe-OH, these workers determined values of K_i and K_i^* of 3.4 and 0.16 nM, respectively. Likewise, for leukocyte elastase inhibition by R-Pro-boroVal-OH, the values of K_i and K_i^* were found to be 15 and 0.57 nM, respectively. Interestingly, Kettner and Shervi also found R-Pro-boroPhe-OH to be a nanomolar inhibitor of the serine protease cathepsin G, but in this case no slow binding behavior was observed. They suggest that the slow binding behavior of these inhibitors reflects the formation of an initial tetrahedral adduct with the active

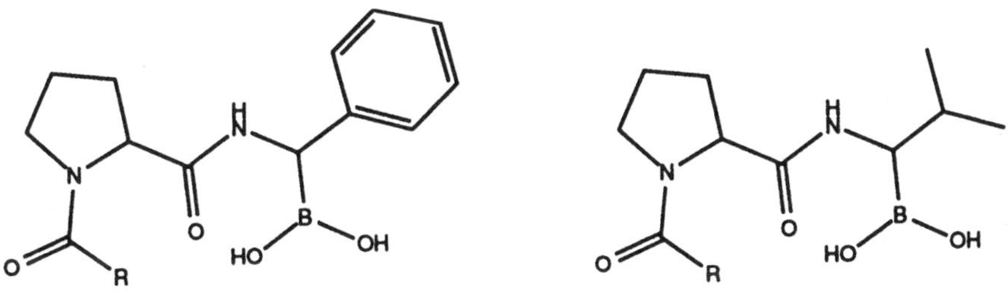

R-Pro-boroPhe-OH   R-Pro-boroVal-OH

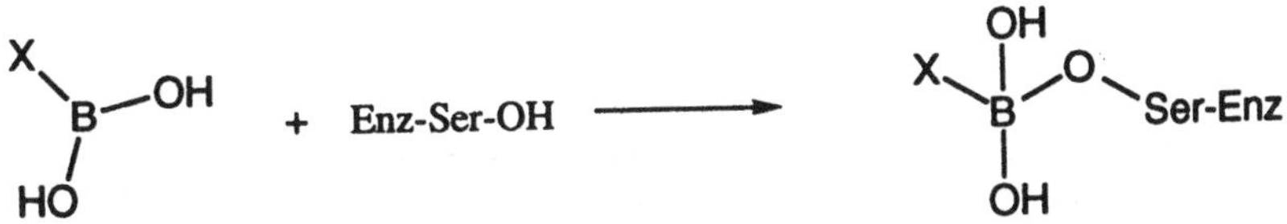

Figure 10.10 Examples of slow binding α-aminoboronate peptide inhibitors of serine proteases. These inhibitors form tetrahedral adducts with the active site serine of the proteases. See Kettner and Shervi (1984) for further details.

site serine, followed by a conformational rearrangement of the enzyme to optimize binding (presumably this conformational readjustment does not occur in the case of cathepsin G).

10.5.2 Prostaglandin G/H Synthase

Prostaglandins are mediators of many of the physiological effects associated with inflammation that lead to such symptoms as pain, swelling, and fever. The biosynthesis of these mediators is rate-limited by the conversion of arachidonic acid to prostaglandin GH_2 by the enzyme prostaglandin G/H synthase (PGHS). One of the most widely used drugs today for the treatment of the pain, swelling, and fever associated with inflammation is aspirin, a compound first isolated from the bark of a certain type of willow tree that had been used for centuries as a folk treatment for pain and fever (Weissman, 1991).

In the early 1970s Vane and his coworkers showed that aspirin elicits its anti-inflammatory effects by inhibition of prostaglandin biosynthesis (Vane,

1971). It was subsequently found that aspirin functions as an affinity label for the enzyme PGHS, covalently inhibiting the enzyme by acetylating an active site serine (Ser 530). The acetylation of this residue irreversibly blocks the binding of arachidonic acid to the enzyme active site. Chronic aspirin use, however, may lead to stomach pain and ulceration, and renal failure, as a result of the breakdown of the mucosal linings of the stomach, intestines, and kidneys. For years, scientists and physicians have searched for anti-inflammatory drugs that could be taken over time without severe side effects. From their efforts a broad class of drugs, known as nonsteroidal anti-inflammatory drugs (NSAIDs) has emerged.

A large and highly prescribed class of NSAIDs are the carboxylic acid containing compounds, typified by the drugs flurbiprofen and indomethacin (Figure 10.11). These compounds have been shown to act as time-dependent inactivators of PGHS, conforming to Scheme C of Figure 10.1. However, the value of k_6 is so low that for all practical purposes, these compounds function as irreversible inactivators and can be treated kinetically as such. Rome and Lands (1975), who have studied the time dependence of these inhibitors in detail, noted a common structural feature, a carboxylic acid group. Reasoning that some acid–base chemistry at the enzyme active site might account for the time-dependent inhibitory effects observed, these workers prepared the methyl esters of eight carboxylate-containing PGHS inhibitors. The results of these studies are summarized in Table 10.1.

Rome and Lands found that for the most part, binding of the inhibitor (reflected in K_i) was not significantly affected by esterification, but in all cases the time dependence (reflected in k_{inact}) was completely lost.

A structural rationale for the foregoing results may now be available. Picot et al. (1994) have reported the crystal structure of PGHS with the carboxylate inhibitor flurbiprofen bound at the active site. They found that the carboxylate moiety of this inhibitor engages in formation of a salt bridge with Arg 120 in the arachidonic acid binding cavity. The formation of this salt bridge, by displacement of nearby amino acid residues, may be the rate-limiting step in the time-dependent inactivation of the enzyme by these inhibitors. Figure 10.11 provides a cartoon version of the proposed interactions between the active site arginine and the carboxylate group of NSAIDs. Note that residue Ser 530 is in close proximity to the bound inhibitor in the crystal structure. This serine is the site of covalent modification and irreversible inactivation by aspirin.

Recently Penning and coworkers (Tang et al., 1995) developed affinity labels of PGHS by preparing bromoacetamido analogues of the existing NSAIDs indomethacin and mefenamic acid. The bromoacetamido group is attacked by an active site nucleophile to form a covalent adduct that leads to irreversible inactivation of the enzyme (Figure 10.12). Under strong acidic conditions, the amine-containing NSAID moiety is cleaved off, leaving behind a carboxymethylated version of the active site nucleophile. These affinity labels can thus be used as mechanistic probes of the enzyme active site. By incorporating a radiolabel into the methylene carbon of the bromoacetamido group, one can

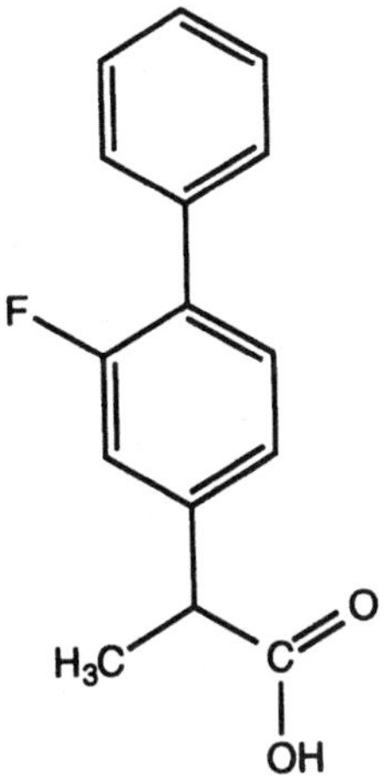

Flurbiprofen 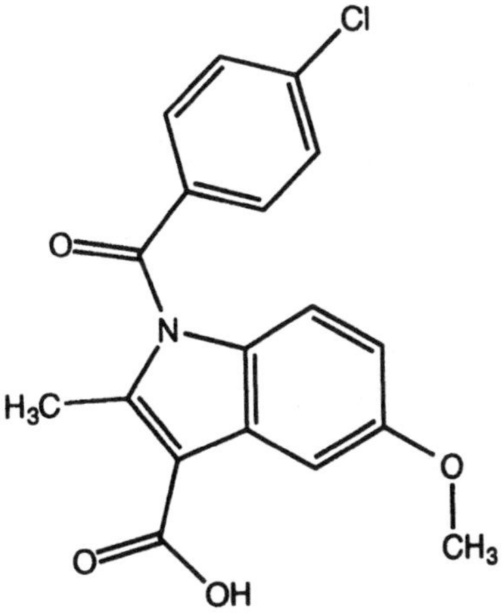   Indomethacin

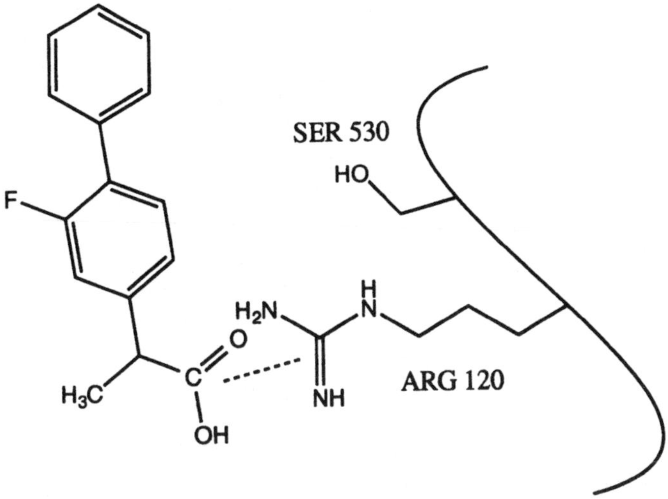

Figure 10.11 Examples of carboxylate-containing NSAIDs that act as slow binding inhibitors of PGHS. The cartoon of the binding of flurbiprofen to the active site of PGHS through salt bridge formation with Arg 120 (bottom) is based on the crystal structure of the PGHS1–flurbiprofen complex reported by Picot et al. (1994).

Table 10.1 Time-dependent kinetic parameters for carboxylate inhibitors of PGHS and their methyl esters

	K_i (μM)		k_{inact} (μM^{-1} min^{-1})	
Inhibitor	Free Acid	Ester	Free Acid	Ester
Indomethacin	100	1	0.04	0
Aspirin	14,000	16,000	0.0003	0
Flurbiprofen	1	0.5	1.1	0
Ibuprofen	3	6	0	0
Meclofenamic acid	4	1	0.4	0
Mefenamic acid	1	3	0	0
BL-2338	1	5	0.08	0
BL-2365	14	9	0	0

Source: Data from Rome and Lands (1975).

obtain selective radiolabel incorporation into the enzyme at the attacking nucleophile.

PGHS performs two catalytic conversions of its substrate, arachidonic acid: a cyclooxygenase step (in which two equivalents of molecular oxygen are added) and a peroxidase step (in which the incorporated peroxide moiety is converted to the final alcohol of prostaglandin GH$_2$). The classical NSAID inhibitors block enzyme turnover by inhibiting selectively the cyclooxygenase step of the reaction. This observation has raised the question of whether the two enzymatic reaction steps involve the same set of active site residues or use distinct catalytic centers for each reaction. Tang et al. (1995) demonstrated that the bromoacetamido affinity labels bind to PGHS in a 2:1 stoichiometry and, unlike their NSAID analogues, abolish both the cyclooxygenase and peroxidase activities of the enzyme. Interestingly, they found that pretreatment of the enzyme with aspirin or mefenamic acid reduces the stoichiometry of the affinity label incorporation to 1:1. Furthermore, if the mefenamic acid saturated enzyme is treated with the affinity label and subsequently dialyzed to remove mefenamic acid, the version of the enzyme that results retains its cyclooxygenase activity but is devoid of peroxidase activity (Tang et al., 1995). These findings support the hypothesis that the catalytic centers for cyclooxygenase and peroxidase activities are distinct in PGHS. These affinity labels, and the peroxidase-deficient enzyme they provide, should prove to be useful tools in dissecting the mechanism of PGHS turnover.

A continuing problem in the treatment of inflammatory diseases with NSAIDs is the gastrointestinal and renal damage observed among patients who receive chronic treatment with these drugs. The side effects are mechanism-based in that inhibition of PGHS not only blocks the symptoms of inflammation but also interferes with the maintenance of the protective mucosal linings of the digestive system. In the early 1990s it was discovered

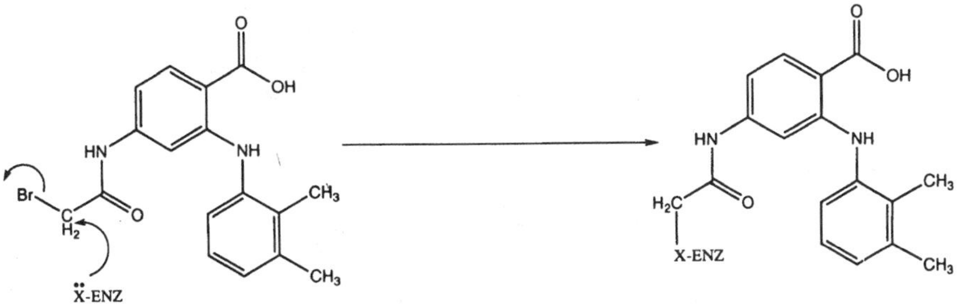

Figure 10.12 Affinity labeling of PGHS by the bromoacetamido analogue of the NSAID 2,3-dimethylanthranilic acid. [Adapted from Tang et al. (1995).]

that humans and other mammals contain two isoforms of this enzyme: PGHS1, which is constitutively expressed in a wide variety of tissues, including gastrointestinal and renal tissue, and PGHS2, which is induced in response to inflammatory stimuli and is primarily localized to cells of the immune system and the brain. This discovery immediately suggested a mechanism for treating inflammatory disease without triggering the side effects of traditional NSAID therapy, namely, selective inhibition of the inducible isoform, PGHS2.

In the hope of developing new and safer anti-inflammatory drugs, a number of laboratories, including ours, set out to identify compounds that would inhibit PGHS2 selectively over PGHS1. One compound that seemed to fit this selectivity profile was DuP697, a methylsulfonyl-containing diaryl thiophene (Figure 10.13). Kinetic studies of DuP697 inhibition of PGHS1 and PGHS2 revealed an unusual basis for the isozyme selectivity of this compound. DuP697 appeared to bind weakly, but with equal affinity, to both isozymes (Copeland et al., 1994). For PGHS1, we demonstrated that DuP697 acted as a classic reversible competitive inhibitor (Copeland et al., 1995). For PGHS2, however, the binding of DuP697 induced an isomerization of the enzyme that led to much tighter association of the inhibitor–enzyme complex, according to Scheme C of Figure 10.1. This isomerization step in fact led to such tight binding that the inhibition could be treated as a two-step irreversible inactivation of the enzyme (Scheme D of Figure 10.1). Plots of k_{obs} as a function of DuP697 concentration showed the hyperbolic behavior expected for inactivation where k_6 was zero or near zero. From this we determined values of K_i and k_{inact} of 2.19 μM and 0.017 s^{-1}, respectively, or a second-order rate constant k_{inact}/K_i of 7.76 × 10^{-3} μM^{-1}·s^{-1}. Thus the isozyme selectivity of this compound resulted from its ability to induce a conformational transition in one isozyme but not the other. The structural basis for this inhibitor-induced conformational transition remains to be fully elucidated.

We then explored analogues of DuP697 in an attempt to identify the minimal structural requirements for selective PGHS2 inhibition and to search

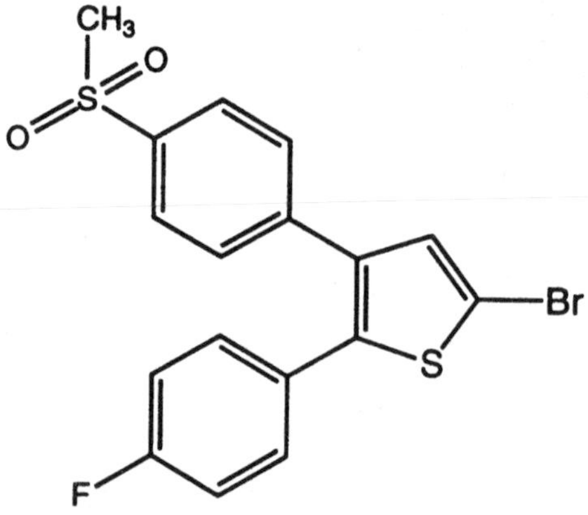

DuP697

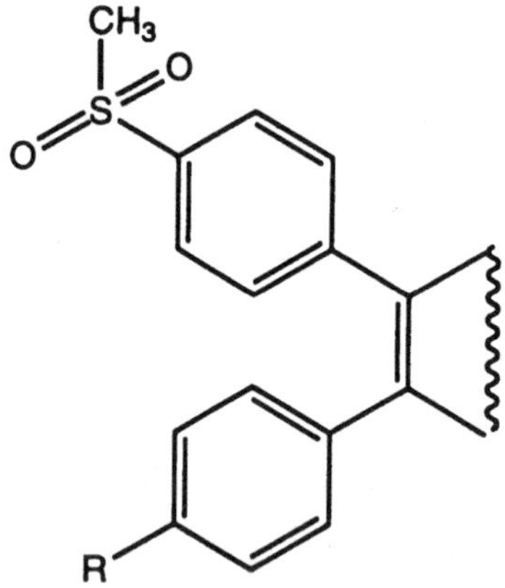

Generic PGHS2 Inhibitor

Figure 10.13 Chemical structures of DuP697 and the generic form of a PGHS2 selective inhibitor. [Based on the data from Copeland et al. (1995).]

for more potent and more selective compounds. The results of these studies identified the structural component labeled "generic PGHS2 inhibitor" in Figure 10.13 as the critical pharmacophore for selective PGHS2 inhibition (Copeland et al., 1995). Within this general class of compounds we were able to prepare inhibitors that showed complete discrimination between the two isozymes: that is, inhibitors that demonstrated potent, time-dependent inhibition of PGHS2, while showing no inhibition of PGHS1 at any concentration up to their solubility limits (Copeland et al., 1995). The information obtained from these studies, and similar studies from other laboratories, provided a clear direction for the development of PGHS2 specific inhibitors. These compounds have proved useful in the design of new NSAIDs that are now in clinical use, with significant benefits to patients suffering from inflammatory diseases.

10.5.3   Chemical Modification as Probes of Enzyme Structure and Mechanism

The use of chemical modifiers has provided a wealth of structural insights for a wide variety of enzymes and receptors. These reagents act as irreversible inactivators, conforming to Scheme D of Figure 10.1, that covalently modify a specific amino acid (or group of amino acids) that is critical to the catalytic function of the enzyme. Quantitative analysis of such inactivation can provide information on the number of residues modified and their structural type. Proteolytic mapping of the covalently modified enzyme can allow the researcher to identify the specific residue(s) modified, and thus obtain some insight into the structural determinants of catalysis.

10.5.3.1 Amino Acid Selective Chemical Modification. A number of chemicals are known to selectively modify specific amino acid side chains within proteins (Glazer et al., 1975; Lundblad, 1991); some of these that are commonly used to study enzyme inactivation are summarized in Table 10.2. These compounds covalently modify the accessible amino acids in a general way, so that treatment of an enzyme with such reagents will lead to modification of both catalytically critical residues and nonessential residues as well. The reagents listed in Table 10.2 either produce chromophoric labels on the modified enzyme or can be obtained in radiolabeled versions, so that by one of these means the total number of covalent labels incorporated into each molecule of enzyme (z) can be quantified. Knowing the concentrations of enzyme and modifying reagent used in such experiments, the researcher can titrate the enzyme with modifying reagent to determine the mole ratio of modifier required to inactivate the enzyme (i.e., the number of moles of modifier required to inactivate one mole of enzyme molecules).

Suppose, for example, that there are n accessible amino acid residues that react equally with a chemical modifying reagent, such as those listed in Table 10.2. Of these, x residues are essential for catalyic activity. If we incubate the enzyme with the modifying reagent for a period of time so an average of z residues on each enzyme molecule are modified, the probability that any particular residue has been modified is z/n, and likewise the probability that any particular residue remains unmodified is $1 - z/n$. For the enzyme to continue to display activity, all the x essential residues must remain unmodified. The probability of this occurrence is given by $(1 - z/n)^x$. Thus, the fractional activity remaining after modification of z groups per molecule is

Table 10.2 Some examples of amino acid–selective chemical modifying agents

Preferred Amino Acid Modified	Modifying Reagent(s)	Comments
Carboxylates	Isoxazolium salts (e.g., trimethyloxonium fluoroborate), carbodiimides	
Cysteine	Iodoacetamide, maleimides, Ellman's reagent, p-hydroxymercuribenzoate	Iodoacetamide can also modify histidine and lysine residues
Histidine	Diethyl pyrocarbonate	Can also react with lysines, cysteines, and tyrosines
Lysine	Acid anhydrides, succinimidyl esters, isothiocyanates, trinitrobenzenesulfonic acid	Reagents react with primary amines. Thus modification of the amino terminus of proteins can also occur
Serine	Halomethyl ketones, peptidic aldehydes	Attack serine nucleophiles, useful for modification of active sites of serine proteinases. Peptide aldehydes also modify active site cysteines of cysteine proteinases
Tryptophan	N-Bromosuccinimide, nitrobenzyl halides	
Tyrosine	Tetranitromethane, chloramine T, NaI, and peroxidases	Chloramine T also modifies histidine and methionine residues

given by:

$$\frac{v_i}{v_0} = \left(1 - \frac{z}{n}\right)^x \qquad (10.18)$$

therefore

$$\left(\frac{v_i}{v_0}\right)^{1/x} = 1 - \frac{z}{n} \qquad (10.19)$$

The value of v_i/v_0 can be measured at each concentration of modifying agent, and the value of z for each experiment can be determined from measuring the amount of spectroscopic or radioactive label associated with the enzyme. A plot of $(v_i/v_0)^{1/x}$ as a function of z yields a straight line according to Equation

10.19. Since, however, we do not know the value of x, we construct a series of plots for v_i/v_0, $(v_i/v_0)^{1/2}$, $(v_i/v_0)^{1/3}$, and so on against z and evaluate them to determine the value of x that gives the best linear fit. Plots of this type are known as Tsou plots (Tsou, 1962), and they provide a good measure of the number of catalytically critical residues that are modified by a specific inactivator.

One complication with the foregoing approach is that not all amino acid residues will necessarily be modified at equal rates by a particular chemical modifier (see Tsou, 1962, for a detailed discussion of this complication). It commonly happens in experiments that some number of nonessential residues are modified at a faster rate than the catalytically essential residues. The effect of this situation is that an initial region of the Tsuo plot will occur where no decrease in enzymatic activity is realized, followed by a region of the expected linear decrease in activity with increasing value of z (Figure 10.14). The number of essential residues modified can still be ascertained from evaluation of the linear portion of such plots, as discussed by Tsou (1962). Norris and Brockel-hurst (1976) extended this approach to the evaluation of multisubunit enzymes, where residues on each subunit are modified.

To clarify this approach, let us walk through an example of the experimental details of such a chemical modification study. Paterson and Knowles (1972) wished to determine the number of carboxylic acid groups required for catalytic activity in the proteolytic enzyme pepsin. To quantify this they treated the enzyme with [^{14}C]trimethyloxonium fluoroborate, a reagent that esterifies carboxylate groups in proteins, hence imparting a ^{14}C label to the protein after

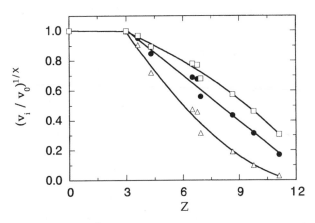

Figure 10.14 Tsou plot of $(v_i/v_0)^{1/x}$ as a function of z for modification of the carboxylate groups of pepsin by trimethyloxonium fluoroborate. The data are plotted for $x = 1$ (triangles), $x = 2$ (solid circles), and $x = 3$ (squares). Below $z = 3$ modification of carboxylates has no effect on enzymatic activity. The data above $z = 3$ are fit to a linear equation for $x = 2$ and to third-order polynomials for $x = 1$ and 3. The linear fit of the data for $x = 2$ suggests that two carboxylates are critical for enzymatic activity. See text for further details. [Data adapted from Paterson and Knowles (1972).]

each esterification reaction. Knowing the specific radioactivity of the modifying reagent (see Chapter 7), the researchers could quantify the number of ^{14}C atoms incorporated into the enzyme after each reaction.

Varying concentrations of $[^{14}C]$trimethyloxonium fluoroborate were added to samples of a solution of pepsin (20 mg/mL), the pH was maintained by addition of NaOH, and the sample was incubated until the modification reaction was complete. The radiolabeled protein was then separated from free reactants and by-products by size exclusion chromatography (see Chapter 7), after which the radioactivity associated with the protein was quantified by scintillation counting. Enzymatic activity of the samples after size exclusion chromatography was assessed by the ability of the enzyme to catalyze the hydrolysis of N-acetyl-L-phenylalanyl-L-phenylalanylglycine, a known substrate for pepsin.

The results of the experiments by Paterson and Knowles (1972) are summarized in Figure 10.14, where $(v_i/v_0)^{1/x}$ is plotted as a function of z (the number of ^{14}C atoms incorporated per mole of enzyme). We see immediately from this figure that a fraction of nonessential carboxylates is rapidly modified without effect of enzymatic activity. From Figure 10.14 we can estimate that approximately three such groups occur in pepsin. After this, the activity of the enzyme decreases with increasing esterification of the carboxylates. To determine the number of carboxylates essential for catalysis, the fractional activity data are plotted in three different ways in Figure 10.14: as v_i/v_0 (i.e., for $x = 1$, triangles); as $(v_i/v_0)^{1/2}$ (i.e., $x = 2$, solid circles); and as $(v_i/v_0)^{1/3}$ (i.e., $x = 3$, squares). Paterson and Knowles fit the data in each form to both linear and polynomial functions, from which they concluded that the best fit to a straight line was obtained for $x = 2$. From this analysis they were able to conclude that two carboxylate residues are essential for catalysis in the enzyme pepsin.

10.5.3.2 Substrate Protection Experiments.

When catalytically essential groups are identified by chemical modification studies, a question that often arises is whether these groups are located within the substrate binding pocket (i.e., active site) of the enzyme. This issue can often be addressed by *substrate protection* experiments, in which one assesses the ability of the substrate, product, or a reversible competitive inhibitor to protect the enzyme against inactivation by the modifying reagent. If an essential amino acid side chain is located in the active site of an enzyme, formation of the reversible binary enzyme–substrate, enzyme–product, or enzyme–inhibitor (for competitive inhibitors) complex may occlude the amino acid so that it is no longer exposed to the chemical modifying reagent during inactivation studies. In this case, removal of free modifying reagent and protectant (i.e. substrate, etc.) by dialysis, size exclusion chromatography, and so on will reveal that enzymatic activity has been retained where the comparable experiment in the absence of protectant resulted in irreversible inactivation.

If the rate of inactivation is followed as described earlier (Scheme D of Figure 10.1, Figure 10.4, and Section 10.2.3) in the presence of varying concentrations of substrate, the observed rate constant for inactivation is found to depend on both substrate and inactivator concentrations as follows:

$$k_{obs} = \frac{k_{inact}[I]}{K_i\left(1 + \dfrac{[S]}{K_m}\right) + [I]}$$ (10.20)

Similar equations can be derived when a product or reversible competitive inhibitor is used as the protectant. Equation 10.20 provides a simple test for whether a catalytically essential group that is chemically modified by a particular reagent is localized to the active site. By measuring the diminution in rate of inactivation with increasing substrate concentration, the researcher can fit the experimental data to Equation 10.20 to determine whether the results are quantitatively consistent with this hypothesis (Figure 10.15). Equation 10.20 additionally provides a means of estimating the value of K_m for a substrate when k_{inact} and K_i are known from previous experiments. This is most useful in the case of multisubstrate enzymes that follow sequential-mechanisms (see Chapter 11). If the first substrate to bind to the enzyme is varied in an inactivation experiment, its binding is in equilibrium with the free enzyme and the inactivator molecule. Hence, the K_m term in Equation 10.20 is replaced by K_S in this case, and the value of K_S for the substrate can be determined directly. An advantage of this approach over more conventional

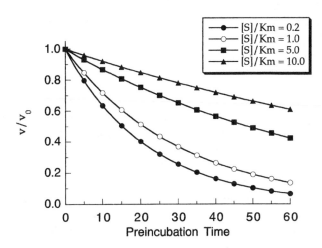

Figure 10.15 Substrate protection against inactivation by chemical modification of an active site amino acid residue. As the substrate concentration is raised, the ability of the active site—directed inactivator to compete for binding and chemical modification of the enzyme is diminished. The symbols represent different $[S]/K_m$ ratios at fixed concentrations of inactivator and enzyme. The lines drawn through the data were obtained by fitting to Equation 10.20 with K_i, [I], and k_{inact} set to 1.0 μM, 1.0 μM, and 0.1 min^{-1}, respectively.

equilibrium methods, such as equilibrium dialysis (Chapter 4) is that here only catalytic amounts of the enzyme are required for the determination of K_S. Thus, when enzyme supplies are limited, this type of experiment can be used to great advantage. See Malcolm and Radda (1970) and Anderton and Rabin (1970) for examples of this approach.

10.5.3.3 Affinity Labels. Covalent modifying groups can often be incorporated into substrate analogues and other ligands (cofactors, inhibitors, activators, etc.) to direct covalent modification to specific functional sites on the enzyme molecule. The work of Tang et al. (1995), discussed in Section 10.5.2, introduced the concept of affinity labeling an enzyme with a covalent modifier as a probe of enzyme structure and mechanism. This approach can help to identify key residues within a ligand binding pocket of an enzyme or receptor through a combination of covalent modification and subsequent peptide mapping studies.

As we just saw in Section 10.5.3.2, certain chemical functionalities will react selectively with specific amino acid side chains, and some of these functionalities can be synthetically incorporated into substrate molecules. Maleimides and succinyl esters, for example, have been incorporated into substrate and/or inhibitor analogues to specifically modify active site cysteine and lysine residues, respectively. Peptidic substrates of serine and cysteine proteinases can have halomethyl ketones and aldehydes incorporated into them to covalently modify the active site nucleophiles of these enzymes specifically. Likewise, metal chelating groups such as carboxylic and hydroxamic acids can be incorporated into the peptidic substrates of metalloproteinases to bind the active site metal in a slowly reversible manner. Alternatively, more permissive crosslinking agents can be incorporated into ligand analogues to determine the identify of amino acids in the ligand binding pocket. A particularly useful strategy is the use of nonselective photoaffinity labels for this purpose.

Photoaffinity labels are molecules that form highly reactive excited states when illuminated with light of an appropriate wavelength, leading to covalent modification of groups within the binding site of the protein. The value of these reagents is that they can be mixed with proteins under varying conditions, and the researcher can control the initiation of crosslinking by illuminating the sample. Two examples of such functionalities are aryl azides and the benzophenone group (Figure 10.16). For both groups, excitation into the π^* excited state results in formation of reactive centers that will combine with nearby methylene groups in the target enzyme or receptor. (In the case of the aryl azides, photolysis leads to formation of an aryl nitrene functionality, which then reacts with carbon–hydrogen or, preferably, oxygen–hydrogen bonds.) By incorporating such functionalities into a ligand molecule, photocrosslinking is targeted to the ligand binding pocket of the target protein. After photolysis, the ligand analogue is covalently attached to a group or groups within the binding pocket. In addition to the photocrosslinker, researches usually incorporate a chromophore, radiolabel, or other affinity label (e.g., biotin) into the

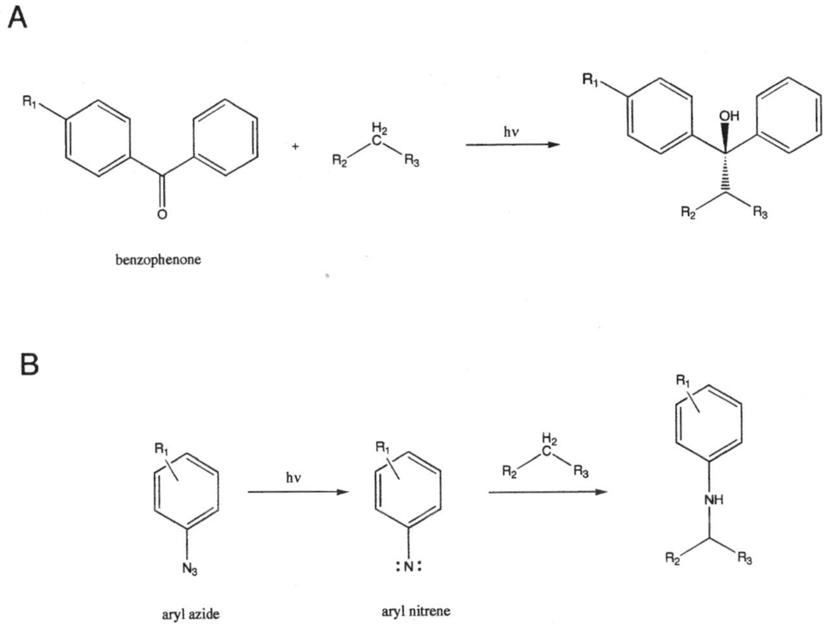

A

benzophenone

B

aryl azide   aryl nitrene

Figure 10.16 Examples of photoaffinity labels that can be incorporated into substrate and inhibitor analogues to covalently modify residues within the ligand binding pocket of proteins: (A) reaction of the benzophenone group and (B) reaction of the aryl azide group.

ligand structure to facilitate detection of the covalently linked species. The crosslinked protein–ligand complex can then be treated with an appropriate proteinase to cleave the target protein into a number of peptide fragments. These fragments are separated by HPLC or electrophoretic methods, and the fragment containing the crosslinked group is collected. The amino acid sequence of the labeled peptide can then be determined by mass spectroscopic methods or by traditional Edman sequencing chemistry. In this way the amino acids located within the ligand binding pocket can be identified.

An example of this approach comes from the work of DeGrado and coworkers (Kauer et al., 1986; O'Neil et al., 1989). This group wished to determine the location of the peptide binding pocket on the protein calmodulin. They had previously identified a peptide of the following sequence that bound tightly (K_d of 400 pM) and specifically to calmodulin:

Lys-Leu-Trp-Lys-Lys-Leu-Leu-Lys-Leu-Leu-Lys-Lys-Leu-Leu-Lys-Leu-Gly

They next synthesized a peptide of similar sequence in which the tryptophan at position 3 was replaced by a benzophenone. Mixing this peptide with calmodulin and subsequent photolysis led to a covalent peptide–calmodulin complex that could be separated from free calmodulin by SDS-PAGE or reversed phase HPLC. The same peptide was also synthesized with a ^{3}H-containing acetyl cap on the N-terminal lysine to impart a radiolabel to the peptide and photolysis product. Cleavage of the photoproduct with cyanogen bromide or *S. aureus* V8 proteinase led to selective cleavage of amide bonds within the calmodulin polypeptide without any cleavage of the peptide ligand. The tritium-containing cleavage product was separated by reversed phase HPLC and subjected to N-terminal amino acid sequence analysis. From these studies DeGrado and coworkers were able to identify Met 144 and Met 71 as the primary sites of photolabeling. These results allowed the researchers to build a model of the three-dimensional structure of the peptide binding pocket in calmodulin.

Affinity labeling of enzymes is a common and powerful tool for studying enzyme structure and mechanism. We have barely scratched the surface in our brief description of these methods. Fortunately there are several excellent in-depth reviews of these methods in the literature. General affinity labeling is covered in a dedicated volume of *Methods in Enzymology* (Jakoby and Wilchek, 1977). General chemical modification of proteins is covered well in the texts by Lundblad (1991) and Glazer et al. (1975). Photoaffinity labeling is covered in the *Methods in Enzymology* volume edited by Jakoby and Wilchek (1977) and also in review articles by Dorman and Prestwich (1994) and by Chowdhry (1979). These references should serve as good starting points for the reader who wishes to explore these tools in greater detail.

10.6 SUMMARY

In this chapter we have described the behavior of enzyme inhibitors that elicit their inhibitory effects slowly on the time scale of enzyme turnover. These slow binding, or time-dependent, inhibitors can operate by any of several distinct mechanisms of interaction with the enzyme. Some of these inhibitors bind reversibly to the enzyme, while others irreversibly inactivate the enzyme molecule. Irreversible enzyme inactivators that function as affinity labels or mechanism-based inactivators can provide useful structural and mechanistic information concerning the types of amino acid residue that are critical for ligand binding and catalysis.

We discussed kinetic methods for properly evaluating slow binding enzyme inhibitors, and data analysis methods for determining the relevant rate constants and dissociation constants for these inhibition processes. Finally, we presented examples of slow binding inhibitors and irreversible inactivators to illustrate the importance of this class of inhibitors in enzymology.

REFERENCES AND FURTHER READING

Anderton, B. H., and Rabin, B. R. (1970) *Eur. J. Biochem.* **15**, 568.

Chowdhry, V. (1979) *Annu. Rev. Biochem.* **48**, 293.

Copeland, R. A. (1994) *Methods for Protein Analysis: A Practical Guide to Laboratory Protocols*, Chapman & Hall, New York, pp. 151–160.

Copeland, R. A., Williams, J. M., Giannaras, J., Nurnberg, S., Covington, M., Pinto, D., Pick, S., and Trzaskos, J. M. (1994) *Proc. Natl. Acad. Sci. USA*, **91**, 11202.

Copeland, R. A., Williams, J. M., Rider, N. L., Van Dyk, D. E., Giannaras, J., Nurnberg, S., Covington, M., Pinto, D., Magolda, R. L., and Trzaskos, J. M. (1995) *Med. Chem. Res.* **5**, 384.

Dorman, G., and Prestwich, G. D. (1994) *Biochemistry*, **33**, 5661.

Glazer, A. N., Delange, R. J., and Sigman, D. S. (1975) *Chemical Modification of Proteins*, Elsevier, New York.

Jakoby, W. B., and Wilchek, M., Eds. (1977) *Methods in Enzymology*, Vol. 46, Academic Press, New York.

Kauer, J. C., Erickson-Viitanen, S., Wolfe, H. R., Jr., and DeGrado, W. F. (1986) *J. Biol. Chem.* **261**, 10695.

Kettner, C., and Shervi, A. (1984) *J. Biol. Chem.* **259**, 15106.

Kitz, R., Wilson, I. B. (1962) *J. Biol. Chem.* **237**, 3245.

Lundblad, R. (1991) *Chemical Reagents for Protein Modification*, CRC Press, Boca Raton, FL.

Malcolm, A. D. B., and Radda, G. K. (1970) *Eur. J. Biochem.* **15**, 555.

Morrison, J. F. (1982) *Trends Biochem. Sci.* **7**, 102.

Morrison, J. F., and Walsh, C. T. (1988) *Adv. Enzymol.* **61**, 201.

Norris, R., and Brocklehurst, K. (1976) *Biochem. J.* **159**, 245.

O'Neil, K. T., Erickson-Viitanen, S., and DeGrado, W. F. (1989) *J. Biol. Chem.* **264**, 14571.

Paterson, A. K., and Knowles, J. R. (1972) *Eur. J. Biochem.* **31**, 510.

Picot, D., Loll, P. J., and Garavito, M. R. (1994) *Nature*, **367**, 243.

Rome, L. H., and Lands, W. E. M. (1975) *Proc. Natl. Acad. Sci. USA*, **72**, 4863.

Silverman, R. B. (1988a) *Mechanism-Based Enzyme Inactivation: Chemistry and Enzymology*, Vols. I and II, CRC Press, Boca Raton, FL.

Silverman, R. B. (1988b) *J. Enzyme Inhib.* **2**, 73.

Tang, M. S., Askonas, L. J., and Penning, T. M. (1995) *Biochemistry*, **34**, 808.

Tian, W.-X., and Tsou, C.-L. (1982) *Biochemistry*, **21**, 1028.

Tipton, K. F. (1973) *Biochem. Pharmacol.* **22**, 2933.

Trzaskos, J. M., Fischer, R. T., Ko, S. S., Magolda, R. L., Stam, S., Johnson, P., and Gaylor, J. L. (1995) *Biochemistry*, **34**, 9677.

Tsou, C.-L. (1962) *Sci. Sin. Ser. B* (English ed.) **11**, 1536.

Vane, J. R. (1971) *Nature New Biol.* **231**, 232.

Weissman, G. (1991) *Sci. Am.* January, p. 84.

11

ENZYME REACTIONS WITH MULTIPLE SUBSTRATES

Until now we have considered only the simplest of enzymatic reactions, those involving a single substrate being transformed into a single product. However, the vast majority of enzymatic reactions one is likely to encounter involve at least two substrates and result in the formation of more than one product. Let us look back at some of the enzymatic reactions we have used as examples. Many of them are multisubstrate and/or multiproduct reactions. For example, the serine proteases selected to illustrate different concepts in earlier chapters use two substrates to form two products. The first, and most obvious, substrate is the peptide that is hydrolyzed to form the two peptide fragment products. The second, less obvious, substrate is a water molecule that indirectly supplies the proton and hydroxyl groups required to complete the hydrolysis. Likewise, when we discussed the phosphorylation of proteins by kinases, we needed a source of phosphate for the reaction, and this phosphate source itself is a substrate of the enzyme. An ATP-dependent kinase, for example, requires the protein and ATP as its two substrates, and it yields the phosphoprotein and ADP as the two products. A bit of reflection will show that many of the enzymatic reactions in biochemistry proceed with the use of multiple substrates and/or produce multiple products. In this chapter we explicitly deal with the steady state kinetic approach to studying enzyme reactions of this type.

11.1 REACTION NOMENCLATURE

A general nomenclature has been devised to describe the number of substrates and products involved in an enzymatic reaction, using the Latin prefixes uni,

Table 11.1   General nomenclature for enzymatic reactions

Reaction	Name
A $\longrightarrow$ P	Uni uni
A + B $\longrightarrow$ P	Bi uni
A + B $\longrightarrow$ P$_1$ + P$_2$	Bi bi
A + B + C $\longrightarrow$ P$_1$ + P$_2$	Ter bi
$\vdots$	$\vdots$

bi, ter, and so on to refer to one, two, three, and more chemical entities. For example, a reaction that utilizes two substrates to produce two products is referred to as a *bi bi* reaction, a reaction using three substrates to form two products is as a *ter bi* reaction, and so on (Table 11.1).

Let us consider in some detail a group transfer reaction that proceeds as a bi bi reaction:

$$E + AX + B \rightleftharpoons E + A + BX$$

The reaction scheme as written leaves several important questions unanswered. Does one substrate bind and leave before the second substrate can bind? Is the order in which the substrates bind random, or must binding occur in a specific sequence? Does group X transfer directly from A to B when both are bound at the active site of the enzyme, or does the reaction proceed by transfer of the group from the donor molecule, A, to a site on the enzyme, whereupon there is a second transfer of the group from the enzyme site to the acceptor molecule B (i.e., a reaction that proceeds through formation of an E–X intermediate)?

These questions raise the potential for at least three distinct mechanisms for the generalized scheme; these are referred to as random ordered, compulsory ordered, and double-displacement or "Ping-Pong" bi bi mechanisms. Often a major goal of steady state kinetic measurements is to differentiate between these varied mechanisms. We shall therefore present a description of each and describe graphical methods for distinguishing among them.

In the treatments that follow we shall use the general steady state rate equations of Alberty (1953), which cast multisubstrate reactions in terms of the equilibrium constants that are familiar from our discussions of the Henri–Michaelis–Menten equation. This approach works well for enzymes that utilize one or two substrates and produce one or two products. For more complex reaction schemes, it is often more informative to view the enzymatic reactions instead in terms of the rate constants for individual steps (Dalziel, 1975). At the end of this chapter we shall briefly introduce the method of King and Altman (1956) by which relevant rate constants for complex reaction schemes can be determined diagrammatically.

11.2 Bi Bi REACTION MECHANISMS

11.2.1 Random Ordered Bi Bi Reactions

In the random ordered bi bi mechanism, either substrate can bind first to the enzyme, and either product can leave first. Regardless of which substrate binds first, the reaction goes through an intermediate ternary complex (E · AX · B), as illustrated:

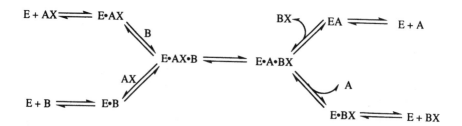

Here the binding of AX to the free enzyme (E) is described by the dissociation constant K^{AX}, and the binding of B to E is likewise described by K^B. Note that the binding of one substrate may very well affect the affinity of the enzyme for the second substrate. Hence, we may find that the binding of AX to the preformed E · B complex is described by the constant αK^{AX}. Likewise, since the overall equilibrium between E · AX · B and E must be path independent, the binding of B to the preformed E · AX complex is described by αK^B. When B is saturating, the value of αK^{AX} is equal to the Michaelis constant for AX (i.e., K_m^{AX}). Likewise, when AX is saturating, $\alpha K^B = K_m^B$. The velocity of such an enzymatic reaction is given by Equation 11.1:

$$v = k_{cat}[E \cdot AX \cdot B] = \frac{k_{cat}[E_t][E \cdot AX \cdot B]}{[E] + [E \cdot AX] + [E \cdot B] + [E \cdot AX \cdot B]} \quad (11.1)$$

If we express the concentrations of the various species in terms of the free enzyme concentration [E], we obtain:

$$v = \frac{V_{max}[AX][B]}{\alpha K^{AX}K^B + \alpha K^B[AX] + \alpha K^{AX}[B] + [AX][B]} \quad (11.2)$$

If we fix the concentration of one of the substrates, we can rearrange and simplify Equation 11.2 significantly. For example, when [B] is fixed and [AX] varies, we obtain:

$$v = \frac{V_{max}[AX]}{\alpha K^{AX}\left(1 + \dfrac{K^B}{[B]}\right) + [AX]\left(1 + \dfrac{\alpha K^B}{[B]}\right)} \quad (11.3)$$

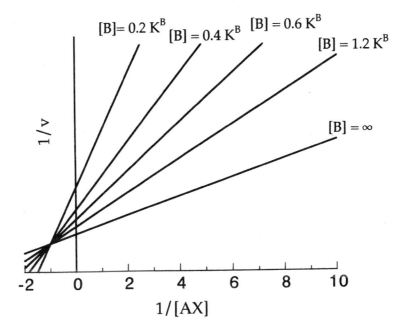

Figure 11.1 Double-reciprocal plot for a random ordered bi bi enzymatic reaction.

At high, fixed concentrations of B, the terms $K^B/[B]$ and $\alpha K^B/[B]$ go to zero. Thus, at saturating concentrations of B we find:

$$v = \frac{V_{\max}^{\mathrm{app}}[\mathrm{AX}]}{K_m^{\mathrm{AX,\,app}} + [\mathrm{AX}]} \qquad (11.4)$$

and likewise, at fixed, saturating [AX]:

$$v = \frac{V_{\max}^{\mathrm{app}}[\mathrm{B}]}{K_m^{\mathrm{B,\,app}} + [\mathrm{B}]} \qquad (11.5)$$

If we measure the reaction velocity over a range of AX concentrations at several, fixed concentrations of B, the reciprocal plots will display a nest of lines that converge to the left of the y axis, as illustrated in Figure 11.1. The data from Figure 11.1 can be replotted as the slopes of the lines as a function of $1/[B]$, and the y intercepts (i.e., $1/V_{\max}^{\mathrm{app}}$) as a function of $1/[B]$ (Figure 11.2). The y intercept of the plot of slope versus $1/[B]$ yields an estimate of $\alpha K^{\mathrm{AX}}/V_{\max}$, and the x intercept of this plot yields an estimate of $-1/K^B$. The y and x intercepts of the plot of $1/V_{\max}^{\mathrm{app}}$ versus $1/[B]$ yield estimates of $1/V_{\max}$ and $-1/\alpha K^B$, respectively. Thus from the data contained in the two replots, one can calculate the values of K^{AX}, K^B, and $V_{\max}$ simultaneously.

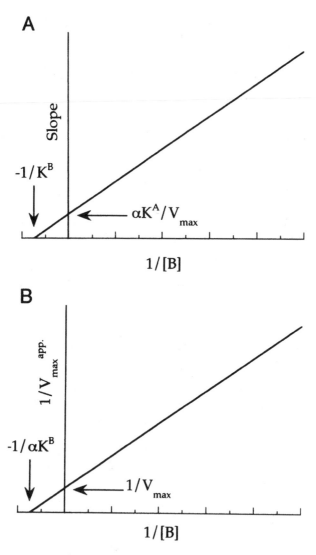

Figure 11.2 (A) Slope and (B) *y*-intercept replots of the data from Figure 11.1, illustrating the graphical determination of K^{AX}, K^B, and V_{max} for a random ordered bi bi enzymatic reaction.

11.2.2 Compulsory Ordered Bi Bi Reactions

In compulsory ordered bi bi reactions, one substrate, say AX, must bind to the enzyme before the other substrate (B) can bind. As with random ordered reactions, the mechanism proceeds through formation of a ternary intermediate. In this case the reaction scheme is as follows:

$$E + AX \rightleftharpoons E \cdot AX \xrightarrow{B} E \cdot AX \cdot B \rightleftharpoons E \cdot A \cdot BX \xrightarrow{-BX} E \cdot A \rightleftharpoons E + A$$

If conversion of the $E \cdot AX \cdot B$ complex to $E \cdot A \cdot BX$ is the rate-limiting step in catalysis, then E, AX, B, and $E \cdot AX \cdot B$ are all in equilibrium, and the velocity of the reaction will be given by:

$$v = \frac{V_{max}[AX][B]}{K^{AX}K^B + K^B[AX] + [AX][B]} \tag{11.6}$$

If, however, the conversion of $E \cdot AX \cdot B$ to $E \cdot A \cdot BX$ is as rapid as the other steps in catalysis, steady state assumptions must be used in the derivation of the velocity equation. For a compulsory ordered bi bi reaction, the steady state treatment yields Equation 11.7:

$$v = \frac{V_{max}[AX][B]}{K^{AX}K_m^B + K_m^B[AX] + K_m^{AX}[B] + [AX][B]} \tag{11.7}$$

As we have described before, the term K^{AX} in Equation 11.7 is the dissocation constant for the $E \cdot AX$ complex, and K_m^{AX} is the concentration of AX that yields a velocity of half V_{max} at fixed, saturating [B].

The pattern of reciprocal plots observed for varied [AX] at different fixed values of [B] is identical to that seen in Figure 11.1 for a random ordered bi bi reaction (note the similarity between Equations 11.2 and 11.7). Hence, *one cannot distinguish between random and compulsory ordered bi bi mechanisms on the basis of reciprocal plots alone.* It is necessary to resort to the use of isotope incorporation studies, or studies using product-based inhibitors.

11.2.3 Double Displacement or Ping-Pong Bi Bi Reactions

The double displacement, or Ping-Pong, reaction mechanism involves binding of AX to the enzyme and transfer of the group, X, to some site on the enzyme. The product, A, can then leave and the second substrate, B, binds to the E–X form of the enzyme (in this mechanism, B cannot bind to the free enzyme form). The group, X, is then transferred (i.e., the second displacement reaction) to the bound substrate, B, prior to the release from the enzyme of the final product, BX. This mechanism is diagrammed as follows:

$$E + AX \rightleftharpoons E \cdot AX \rightleftharpoons EX \cdot A \xrightarrow{-A} EX \xrightarrow{B} EX \cdot B \rightleftharpoons E \cdot BX \rightleftharpoons E + BX$$

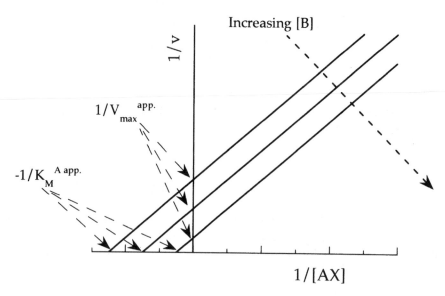

Figure 11.3 Double-reciprocal plot for a double-displacement (Ping-Pong) bi bi enzymatic reaction.

Using steady state assumptions, the velocity equation for a double-displacement reaction can be obtained:

$$v = \frac{V_{max}[AX][B]}{K_m^B[AX] + K_m^{AX}[B] + [AX][B]} \tag{11.8}$$

If we fix the value of [B], then Equation 11.8 for variable [AX] becomes:

$$v = \frac{V_{max}[AX]}{K_m^{AX} + [AX]\left(1 + \dfrac{K_m^B}{[B]}\right)} \tag{11.9}$$

Reciprocal plots of a reaction that conforms to the double-displacement mechanism for varying concentrations of AX at several fixed concentrations of B will yield a nest of parallel lines, as seen in Figure 11.3. For each concentration of substrate B, the values of $1/V_{max}^{app}$ and $-1/K_m^{AX,app}$ can be determined from the y and x intercepts, respectively, of the double-reciprocal plot. The data contained in Figure 11.3 can be replotted in terms of $1/V_{max}^{app}$ as a function of $1/[B]$, and $1/K_m^{AX,app}$ as illustrated in Figure 11.4. The value of $-1/K_m^B$ can be determined from the x intercepts of either replot in Figure 11.4. The y intercepts of the two replots yield estimates of $1/V_{max}$ (for the $1/V_{max}^{app}$ versus $1/[B]$ replot) and $1/K_m^{AX}$ (for the $1/K_m^{app}$ versus $1/[B]$ replot) for the reaction, as seen in Figure 11.4.

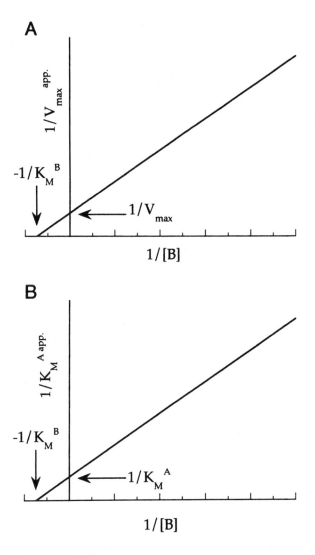

Figure 11.4 Replots of the data from Figure 11.3 as (A) $1/V_{max}^{app}$ versus $1/[B]$ and (B) $1/K_m^{AX,app}$ versus $1/[B]$, illustrating the graphical determination of K_m^{AX}, K_m^B, and V_{max} for a double-displacement (Ping-Pong) bi bi enzymatic reaction.

11.3 DISTINGUISHING BETWEEN RANDOM AND COMPULSORY ORDERED MECHANISMS BY INHIBITION PATTERN

It should be clear from Figures 11.1 and 11.3, and the foregoing discussion, that the qualitative form of the double-reciprocal plots makes it easy to distinguish between a double-displacement mechanism and a mechanism involving ternary complex formation. But again, it is not possible to further distinguish between random and compulsory ordered mechanisms on the basis of reciprocal plots alone. If, however, there is available an inhibitor that binds to the same site on the enzyme as one of the substrates (i.e., is a competitive inhibitor with respect to one of the substrates), addition of this compound will slow the overall forward rate of the enzymatic reaction and can allow one to kinetically distinguish between random and compulsory ordered reaction mechanisms. Because of their structural relationship to the substrate, the product molecules of enzymatic reactions themselves are often competitive inhibitors of the substrate binding site; this situation is referred to as *product inhibition.*

Recall from Chaepter 8 that competitive inhibition is observed when the inhibitor binds to the same enzyme form as the substrate that is being varied in the experiment, or alternatively, binds to an enzyme form that is connected by reversible steps to the form that binds the varied substrate. The pattern of reciprocal lines observed with different inhibitor concentrations is a nest of lines that converge at the y intercept (see Chapter 8). For an enzyme that requires two substrates, a competitive inhibitor of one of the substrate binding sites will display the behavior of a competitive, noncompetitive, or even uncompetitive inhibitor, depending on which substrate is varied, whether the inhibitor is a reversible dead-end (i.e., an inhibitor that does not permit product formation to occur when it is bound to the enzyme, corresponding to

Table 11.2 Patterns of dead-end inhibition observed for the Bi Bi reaction E + AX + B → E + A + BX for differing reaction mechanisms

Mechanism	Competitive Inhibitor for Substrate	Inhibitor Pattern Observed[a]	
		For Varied [AX]	For Varied [B]
Compulsory ordered with [AX] binding first	AX	Competitive	Noncompetitive
Compulsory ordered with [AX] binding first	B	Uncompetitive	Competitive
Compulsory ordered with [B] binding first	AX	Competitive	Uncompetitive
Compulsory ordered with [B] binding first	B	Noncompetitive	Competitive
Random ordered	AX	Competitive	Noncompetitive
Random ordered	B	Noncompetitive	Competitive
Double displacement	AX	Competitive	Uncompetitive
Double displacement	B	Uncompetitive	Competitive

[a]At nonsaturating ([S] $\sim K_m$) concentration of the fixed substrate.

**Table 11.3 Pattern of product inhibition observed for the Bi Bi reaction
E + AX + B → E + A + BX for differing reaction mechanisms**

		Inhibitor Pattern Observed[a]			
		For Varied [AX]		For Varied [B]	
Mechanism	Product Used As Inhibitor	At Unsaturated [B]	At Saturated [B]	At Unsaturated [AX]	At Saturated [AX]
Compulsory ordered with [AX] binding first	A	N	U	N	N
Compulsory ordered with [AX] binding first	BX	C	C	N	—
Compulsory ordered with [B] binding first	A	N	—	C	C
Compulsory ordered with [B] binding first	BX	N	N	N	U
Random ordered	A	C	—	C	—
Random ordered	BX	C	—	C	—
Double displacement	A	N	—	C	C
Double displacement	BX	C	C	N	—

[a]C, competitive; N, noncompetitive; U, uncompetitive; —, no inhibition.

$\beta = 0$ for the scheme in Figure 8.1) or product inhibitor, and the mechanism of substrate interaction with the enzyme. For a bi bi reaction, one observes specific inhibitor patterns for the different mechanisms we have discussed when a competitive dead-end inhibitor or a product of the reaction is used as the inhibitor. The patterns for both dead-end and product inhibition additionally depend on whether the fixed substrate is at a saturating or nonsaturating (typically at [S] $\sim K_m$) concentration with respect to its apparent K_m.

The relationships leading to these differing patterns of dead-end and product inhibition for bi bi reactions have been derived elsewhere (see, e.g., Segel, 1975). Rather than rederiving these relationships, we present them as diagnostic tools for determining the mechanism of reaction. The patterns are summarized in Tables 11.2 and 11.3 for dead-end and product inhibition, respectively. By measuring the initial velocity of the reaction in the presence of several concentrations of inhibitor, and varying separately the concentrations of AX and B, one can identify the reaction mechanism from the pattern of double-reciprocal plots and reference to these tables.

11.4   ISOTOPE EXCHANGE STUDIES FOR DISTINGUISHING REACTION MECHANISMS

An alternative means of distinguishing among reaction mechanisms is to look at the rate of exchange between a radiolabeled substrate and a product molecule under equilibrium conditions (Boyer, 1959; Segel, 1975).

The first, and simplest mechanistic test using isotope exchange is to ask whether exchange of label can occur between a substrate and product in the presence of enzyme, but in the absence of the second substrate. Looking over the various reaction schemes presented in this chapter, it became obvious that such an exchange could take place only for a double-displacement reaction:

$$E + A^*X \rightleftharpoons E \cdot A^*X \rightleftharpoons EX \cdot A^* \overset{-A^*}{\underset{}{\rightleftharpoons}} EX \overset{B}{\rightleftharpoons} EX \cdot B \rightleftharpoons E \cdot BX \rightleftharpoons E + BX$$

For random or compulsory ordered reactions, the need to proceed through the ternary complex before initial product release would prevent the incorporation of radiolabel into one product in the absence of the second substrate.

Next, let us consider what happens when the rate of isotope exchange is measured under equilibrium conditions for a general group transfer reaction:

$$AX + B \rightleftharpoons A + BX$$

Under these conditions the forward and reverse reaction rates are equivalent, and the equilibrium constant is given by:

$$K_{eq} = \frac{[BX][A]}{[AX][B]} \tag{11.10}$$

If under these conditions radiolabeled substrate B is introduced in an amount so small that it is insufficient to significantly perturb the equilibrium, the rate of formation of labeled BX can be measured. The measurement is repeated at increasing concentrations of A and AX, to keep the ratio [A]/[AX] constant (i.e., to avoid a shift in the position of the equilibrium). As the amounts of A and AX are changed, the rate of radiolabel incorporation into BX will be affected.

Suppose that the reaction proceeds through a compulsory ordered mechanism in which B is the first substrate to bind to the enzyme and BX is the last product to be released. If this is the case, the rate of radiolabel incorporation into BX will initially increase as the concentrations of A and AX are increased. As the concentrations of A and AX increase further, however, the formation of the ternary complexes $E \cdot AX \cdot B$ and $E \cdot A \cdot BX$ will be favored, while dissociation of the EB and EBX complexes will be disfavored. This will have the effect of lowering the rate of isotope exchange between B and BX. Hence, a plot of the rate of isotope exchange as a function of [AX] will display substrate inhibition at high [AX], as illustrated in Figure 11.5A.

The effect of increasing [AX] and [A] on the rate of radiolabel exchange between B and BX will be quite different, however, in a compulsory ordered reaction that requires initial binding of AX to the enzyme. In this case, increasing concentrations of AX and A will disfavor the free enzyme in favor

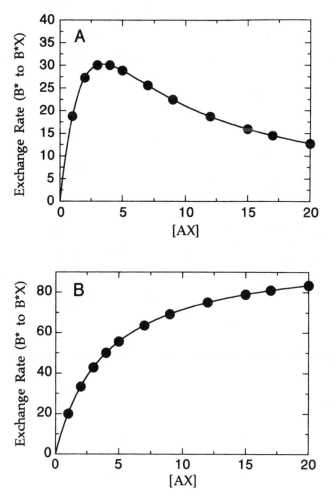

Figure 11.5 Plots of the equilibrium rate of radioisotope exchange between B and BX as a function of [AX] for (A) a compulsory ordered bi bi reaction in which B is the first substrate to bind to the enzyme and BX is the last product to be released, and (B) either a compulsory ordered bi bi reaction in which AX binds first or a random ordered bi bi reaction.

of the EAX and EA forms. The EAX form will react with B, leading to formation of BX, while the EA form will not. Hence, the rate of radiolabel incorporation into BX will increase with increasing [AX] as a hyperbolic function (Figure 11.5B). The same hyperbolic relationship would also be observed for a reaction that proceeded through a random ordered mechanism. In this latter case, however, the hyperbolic relationship also would be seen for experiments performed with labeled AX and varying [B].

Thus isotope exchange in the absence of the second substrate is diagnostic of a double-displacement reaction, while compulsory ordered and random ordered reactions can be distinguished on the basis of the relation of the rate of radiolabel exchange between one substrate and product of the reaction to the concentration of the other substrate and product under equilibrium conditions. (See Segel, 1975, for a more comprehensive treatment of isotope exchange studies for multisubstrate enzymes.)

11.5 USING THE KING–ALTMAN METHOD TO DETERMINE VELOCITY EQUATIONS

The velocity equations for bi bi reactions can be easily related to the Henri–Michaelis–Menten equation described in Chapter 5. However, for more complex reaction schemes, such as those involving multiple intermediate species, it is often difficult to derive the velocity equation in simple terms. An alternative method, devised by King and Altman (1956), allows the derivation of a velocity equation for essentially any enzyme mechanism in terms of the individual rate constants of the various steps in catalysis. On the basis of the methods of matrix algebra, King and Altman derived empirical rules for writing down the functional forms of these rate constant relationships. We provide a couple of illustrative examples of their use and encourage interested readers to explore this method further.

To begin with, we shall consider a simple uni uni reaction as first encountered in Chapter 5:

$$E + S \rightleftharpoons ES \longrightarrow E + P$$

In the King and Altman approach we consider the reaction to be a cyclic process and illustrate it in a way that displays all the interconversions among the various enzyme forms involved:

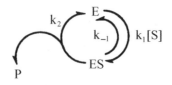

For each step in the reaction we can define a term κ (kappa) which is the product of the rate constant for that step and the concentration of free substrate involved in the step. Next, we determine every pathway by which a particular enzyme species might be formed in the reaction scheme. For the simple uni uni reaction under consideration we have:

Enzyme Form	Pathways to That Form	Σ of Kappa Products
E	$E \xleftarrow{k_{-1}}$ $E \xleftarrow{k_2}$	$k_{-1} + k_2$
ES	$\xrightarrow{k_1[S]} ES$	$k_1[S]$

For any particular enzyme species, the following relationship holds:

$$\frac{[\text{form}]}{[E]} = \frac{\Sigma \kappa_{\text{form}}}{\Sigma \kappa} \tag{11.11}$$

where [form] is the concentration of the particular enzyme form under consideration, $\Sigma \kappa_{\text{form}}$ is the sum of the kappa products for that enzyme form, and $\Sigma \kappa$ is the sum of the kappa products for all species. Applying this to our uni uni reaction we obtain:

$$\frac{[E]}{[E_t]} = \frac{k_{-1} + k_2}{k_{-1} + k_2 + k_1[S]} \tag{11.12}$$

and

$$\frac{[ES]}{[E_t]} = \frac{k_1[S]}{k_{-1} + k_2 + k_1[S]} \tag{11.13}$$

The overall velocity equation can be written as follows:

$$v = k_2[ES] \tag{11.14}$$

Substituting the equalities given in Equations 11.12 and 11.13 into Equation 11.14, we obtain:

$$v = \frac{k_2 k_1[S][E_t]}{k_{-1} + k_2 + k_1[S]} = \frac{k_2[E_t][S]}{\left(\dfrac{k_{-1} + k_2}{k_1}\right) + [S]} \tag{11.15}$$

Inspecting Equation 11.15, we immediately see that k_2 is equivalent to k_{cat}, and $(k_{-1} + k_2/k_1)$ is equivalent to the Michaelis constant, K_m. If we invoke the further equality that $V_{\text{max}} = k_{\text{cat}}[E]$, we see that the King–Altman approach results in the same velocity equation we had derived as Equation 5.24.

Now let us consider the more complex case of a double-displacement bi bi reaction using the King–Altman approach. Note here that the initial concentrations of the two products A and BX are zero, and the release of these products from the enzyme is essentially irreversible. Hence, the cyclic form of the reaction scheme is:

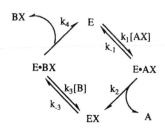

Consideration of this reaction yields the relationships given in Table 11.4. The overall rate equation for a double-displacement reaction is:

$$v = k_2[\text{EAX}] = k_4[\text{EBX}] \tag{11.16}$$

From the preceding relationships, we see that:

$$\frac{[\text{EAX}]}{[\text{E}_t]} = \frac{k_1 k_3 k_4 [\text{AX}][\text{B}]}{k_1 k_3 [\text{AX}][\text{B}](k_2 + k_4) + k_3 k_4 [\text{B}](k_{-1} + k_2) + k_1 k_2 [\text{AX}](k_{-3} + k_4)} \tag{11.7}$$

Combining Equations 11.16 and 11.17, and performing a few rearrangements we obtain:

$$v = \frac{\left(\dfrac{k_2 k_4}{k_2 + k_4}\right)[\text{E}_t][\text{AX}][\text{B}]}{\dfrac{k_2}{k_3}\left(\dfrac{k_{-3} + k_4}{k_2 + k_4}\right)[\text{AX}] + \dfrac{k_4}{k_1}\left(\dfrac{k_{-1} + k_2}{k_2 + k_4}\right)[\text{B}] + [\text{AX}][\text{B}]} \tag{11.18}$$

With the appropriate substitutions, Equation 11.18 can be recast, using the approach of Alberty, to yield the more familiar form first presented as Equation 11.8.

With similar considerations, the velocity equations for random ordered and compulsory ordered bi bi mechanisms can likewise be derived. With some practice, this seemingly cumbersome approach provides a clear and intuitive means of deriving the appropriate velocity equation for complex enzymatic systems. A more thorough treatment of the King–Altman approach can be found in the text by Segel (1975) as well as in the original contribution by King and Altman (1956).

11.6 SUMMARY

In this chapter we have briefly introduced the concept of multisubstrate enzyme reactions and have presented steady state equations to describe the

Table 11.4   King–Altman relationships for a double displacement Bi Bi reaction

Enzyme Form	Pathways to Form	Σ of Kappa Products
E	k_{-1}; $k_3[B]$; E; k_4; k_2; $k_3[B]$ and E; k_4	$k_{-1}k_3k_4[B] + k_2k_3k_4[B] = k_3k_4[B](k_{-1} + k_2)$
E·AX	$k_1[AX]$; $E \bullet AX$; k_4; $k_3[B]$	$k_1k_3k_4[AX][B]$
EX	$k_1[AX]$; k_2; EX; k_4; $k_1[AX]$; k_2; EX; k_{-3} and k_2	$k_1k_2k_4[AX] + k_1k_2k_{-3}[AX] = k_1k_2[AX](k_{-3} + k_4)$
E·BX	$k_1[AX]$; k_2; $E \bullet BX$; $k_3[B]$	$k_1k_2k_3[AX][B]$

velocities for these reactions. We have seen that enzyme reactions involving two substrates and two products can proceed by at least three distinct mechanisms: random ordered, compulsory ordered, and double-displacement reactions. Experimental methods were presented to allow the investigator to distinguish among these mechanisms on the basis of kinetic measurements, product inhibition studies, and radioisotope exchange studies. We briefly described the method of King and Altman for deriving the velocity equation of complex enzymatic reaction, such as those involving multiple substrates.

The importance of multisubstrate enzymatic reactions can hardly be overstated. In fact, the vast majority of enzymatic reactions in nature proceed through the utilization of more than one substrate to yield more than one product.

REFERENCES AND FURTHER READING

Alberty, R. A. (1953) *J. Am. Chem. Soc.* **75**, 1928.

Boyer, P. D. (1959) *Arch. Biochem. Biophys.* **82**, 387.

Cleland, W. W. (1963) *Biochim. Biophys. Acta*, **67**, 188.

Cornish-Bowden, A., and Wharton, C. W. (1988) *Enzyme Kinetics*, IRL Press, Oxford, pp. 25–33.

Dalziel, K. (1975) Kinetics and mechanism of nicotinamide-dinucleotide-linked dehydrogenases, in *The Enzymes*, 3rd ed., P. D. Boyer, Ed., Academic Press, San Diego, CA, pp. 1–60.

King, E. L., and Altman, C. (1956) *J. Phys. Chem.* **60**, 1375.

Palmer, T. (1981) *Understanding Enzymes*, Wiley, New York, pp. 170–189.

Segel, I. H. (1975) *Enzyme Kinetics*, Wiley, New York, pp. 506–883.

12

COOPERATIVITY IN ENZYME CATALYSIS

As we described in Chapter 3, some enzymes function as oligomeric complexes of multiple protein subunits, each subunit being composed of copies of the same or different polypeptide chains. In some oligomeric enzymes, each subunit contains an active site center for ligand binding and catalysis. In the simplest case, the active sites on these different subunits act independently, as if each represented a separate catalytic unit. In other cases, however, the binding of ligands at one active site of the enzyme can increase or decrease the affinity of the active sites on other subunits for ligand binding. When the ligand binding affinity of one active site is affected by ligand occupancy at another active site, the active sites are said to be acting *cooperatively*. In *positive cooperativity* ligand binding at one site *increases* the affinity of the other sites, and in *negative cooperativity* the affinity of other sites is *decreased* by ligand binding to the first site.

For cooperative interaction to occur between two active sites some distance apart (e.g., on separate subunits of the enzyme complex), ligand binding at one site must induce a structural change in the surrounding protein that is transmitted, via the polypeptide chain, to the distal active site(s). This concept of transmitted structural changes in the protein, resulting in long-distance communication between sites, has been termed "allostery," and enzymes that display these effects are known as *allosteric enzymes*. (The word "allosteric," which derives from two Greek words—*allos* meaning different, and *stereos*, meaning structure or solid—was coined to emphasize that the structural *change* within the protein mediates the cooperative interactions among different sites.)

Allosteric effects can occur between separate binding sites for the *same* ligand.within a given enzyme, as just discussed, in *homotropic cooperativity*.

Also, ligand binding at the active site of the enzyme can be affected by binding of a structurally unrelated ligand at a distant separate site; this effect is known as *heterotropic cooperativity*. Thus small molecules can bind to sites other than the enzyme active site and, as a result of their binding, induce a conformational change in the enzyme that regulates the affinity of the active site for its substrate (or other ligands). Such molecules are referred to as *allosteric effectors*, and they can operate to enhance active site substrate affinity (i.e., serving as allosteric activators) or to diminish affinity (i.e., serving as allosteric repressors). Both types of allosteric effector are seen in biology, and they form the basis of metabolic control mechanisms, such as feedback loops.

In this chapter we shall describe some examples of cooperative and allosteric proteins that not only illustrate these concepts but also have historic significance in the development of the theoretical basis for understanding these effects. We shall then briefly describe two theoretical frameworks for describing the two effects. Finally, we shall discuss the experimental consequences of cooperativity and allostery, and appropriate methods for analyzing the kinetics of such enzymes.

The treatment to follow discusses the effects of cooperativity in terms of substrate binding to the enzyme. The reader should note, however, that ligands other than substrate also can display cooperativity in their binding. In fact, in some cases enzymes display cooperative inhibitor binding, but no cooperativity is observed for substrate binding to these enzymes. Such special cases are beyond the scope of the present text, but the reader should be aware of their existence. A relatively comprehensive treatment of such cases can be found in the text by Segel (1975).

12.1 HISTORIC EXAMPLES OF COOPERATIVITY AND ALLOSTERY IN PROTEINS

The proteins hemoglobin and the Trp repressor provide good examples of the concepts of ligand cooperativity and allosteric regulation, respectively. Hemoglobin is often considered to be the paradigm for cooperative proteins. This primacy is in part due to the wealth of information on the structural determinants of cooperativity in this protein that is available as a result of detailed crystallographic studies on the ligand-replete and ligand-free states of hemoglobin. Likewise, much of our knowledge of the regulation of Trp repressor activity comes from detailed crystallographic studies.

Hemoglobin, as described in Chapter 3, is a heterotetramer composed of two copies of the α subunit and two copies of the β subunit. These subunits fold independently into similar tertiary structures that provide a binding site for a heme cofactor (i.e., an iron-containing porphyrin cofactor: see Figure 3.19). The heme in each subunit is associated with the protein by a coordinate bond between the nitrogen of a histidine residue and the central iron atom of the heme. Iron typically takes up an octahedral coordination geometry composed of six ligand coordination sites. In the heme groups of hemoglobin, four of these coordination sites are occupied by nitrogens of the porphyrin ring system and a fifth is occupied by the coordinating histidine, leaving the sixth coordination site open for ligand binding. This last coordination site forms the O_2 binding center for each subunit of hemoglobin.

A very similar pattern of tertiary structure and heme binding motif is observed in the structurally related monomeric protein myoglobin, which also binds and releases molecular oxygen at its heme iron center. Based on the similarities in structure, one would expect each of the four hemes in the hemoglobin tetramer to bind oxygen independently, and with an affinity similar to that of myoglobin. In fact, however, when O_2 binding curves for these two proteins are measured, the results are dramatically different, as illustrated in Figure 12.1. Myoglobin displays the type of hyperbolic saturation curve one would expect for a simple protein–ligand interaction. Hemoglobin, on the other hand, shows not a simple hyperbolic saturation curve but, instead, a sigmoidal dependence of O_2 binding to the protein as a function of O_2 concentration. This is the classic signature for cooperatively interacting binding sites. That is, the four heme groups in hemoglobin are not acting as independent oxygen binding sites, but instead display positive cooperativity in their binding affinities. The degree of cooperativity among these distant sites is such that the data for oxygen binding to hemoglobin are best described by a two-state model in which all the molecules of hemoglobin contain either 4 or

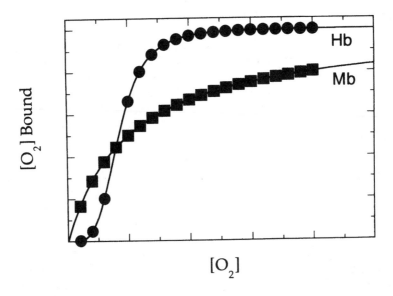

Figure 12.1 Plot of bound molecular oxygen as a function of oxygen concentration for the proteins hemoglobin (Hb) and myoglobin (Mb), illustrating the cooperativity of oxygen binding for hemoglobin.

0 moles of bound O_2; under equilibrium conditions, no significant population of hemoglobin molecules exist with intermediate (i.e., 2 or 3) stoichiometries of O_2 binding.

The crystal structures of oxy- (with four O_2 molecules bound) and deoxy- (with no O_2 bound) hemoglobin provide a clear structural basis for this cooperativity. We know from Chapter 3 that hemoglobin can adopt two distinct quaternary structures; these are referred to as the R (for relaxed) and T (for tense) states (see Section 12.2). The differences between the R and T quaternary structures are relative rotations of two of the subunits, as described in Figure 3.18. These changes in quaternary structure are mediated by changes in intersubunit hydrogen bonding at the subunit interfaces. The crystal structures of oxy- and deoxyhemoglobin reveal that loss of oxygen at the heme of one subunit induces a change in the strength of the iron–histidine bond that occupies the fifth coordination site on the heme iron. This change in bond strength results in a puckering of the porphyrin macrocycle and a displacement of position for the coordinated histidine (Figure 12.2). The coordinated histidine is located in a segment of α-helical secondary structure in the hemoglobin subunit, and the motion of the histidine in response to O_2 binding or release results in a propagated motion of the entire α helix. Ultimately, this propagated motion produces alterations of the intersubunit hydrogen-bonding pattern at the α_1/β_2 subunit interface that acts as a quaternary structure "switch." The accompanying movements of the other subunits leads to alterations of the oxygen affinities for their associated heme cofactors.

The availability of detailed structural information for both the oxy and deoxy structures of hemoglobin has made this molecule the classic model of cooperativity in proteins, illustrating how distant binding sites can interact to control the overall affinity for a single ligand. Likewise, the structural information available for the Trp repressor protein has made this molecule an excellent example of allosteric regulation in biology. As its name implies, the Trp repressor protein acts to inhibit the function of the Trp operon, a segment of DNA that is ultimately responsible for the synthesis of the amino acid tryptophan. The protein accomplishes this task by binding within the major groove of the DNA in its tryptophan-bound form and, when not bound by tryptophan, releasing the DNA. The activity of the Trp repressor is an example of a negative feedback loop, in which the synthesis of an essential molecule of the cell is controlled by the concentration of that molecule itself. At low tryptophan concentrations, the synthesis of tryptophan is required by the cell. Under these conditions the Trp operon must be functional, and thus the Trp repressor must not bind to the DNA.

The crystal structures of the tryptophan-depleted protein shows that the α-helical segments of the protein are arranged in a way that precludes effective DNA binding (Figure 12.3A). Thus, when the tryptophan concentration is low, the protein is found in a conformation that does not allow for DNA binding, and the operon is functional, leading to tryptophan synthesis. When the tryptophan concentration in the cell exceeds some critical concentration,

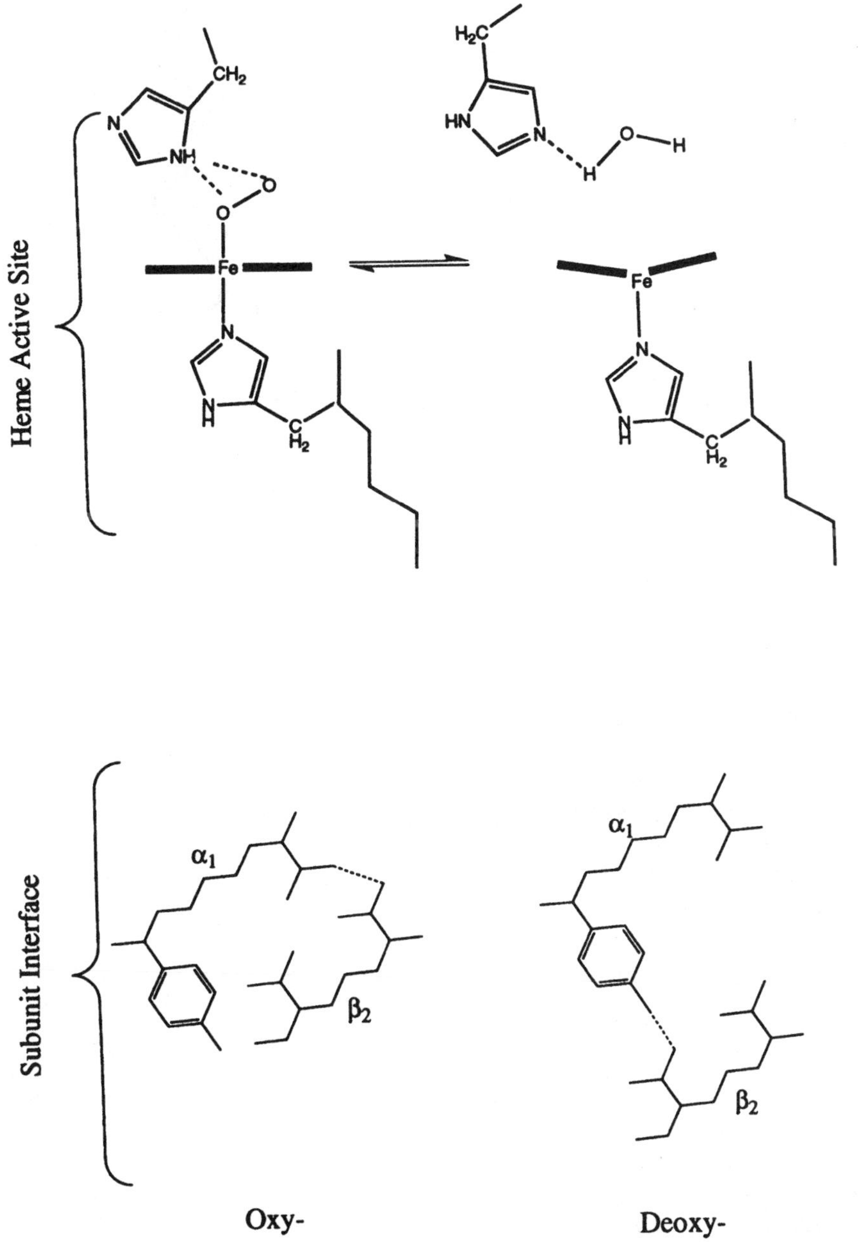

Figure 12.2 Changes in structure of the active site heme that accompany O_2 binding to hemoglobin, and the associated changes in protein structure at the α_1/β_2 subunit interface.

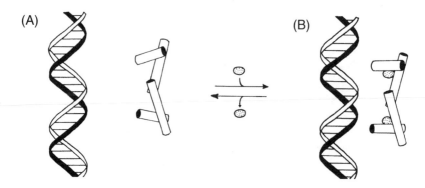

Figure 12.3 Cartoons of the interactions of the Trp repressor protein with Trp operon DNA in the absence (A) and presence (B) of bound tryptophan. This tryptophan-binding-induced conformational transition is the basis for the negative feedback regulation of tryptophan synthesis.

however, the Trp repressor binds tryptophan and, as a result, changes its conformation. The tryptophan-replete form of the protein now has an α-helical arrangement in which two helices are positioned for effective binding to the Trp operon, via interactions between the helices and the double-stranded DNA helical structure (Figure 12.3B). When the Trp repressor binds to the operon, it effectively shuts down the action of this DNA, thus leading to inhibition of further tryptophan synthesis. This simple method of conformationally controlling the activity of the Trp repressor, by binding of tryptophan, provides an elegant mechanism for the metabolic control of the production of an essential amino acid.

Again, we have used hemoglobin and the Trp repressor to illustrate the concepts of cooperativity and allosteric control in structural terms because of the wealth of structural information available for these two proteins. The reader should be aware, however, that the same mechanisms are common in enzymatic systems as well. Numerous examples of cooperativity and allosteric control of enzymatic activity can be found in biology, and these control mechanisms serve vital metabolic roles. For example, many enzymes involved in de novo biosynthetic cascades display the phenomenon of feedback inhibition. Here a metabolite that is the ultimate or penultimate product of the cascade will act as a heterotropic inhibitor of one of the enzymes that occurs early in the biosynthetic cascade, much as tryptophan controls its own rate of synthesis by binding to the Trp repressor.

One of the first examples of this phenomenon came from studies of threonine deaminase from the bacterium *E. coli*. Abelson (1954) observed that addition of isoleucine to cultures of the bacterium inhibited the further biosynthesis of isoleucine. Later workers showed that this effect is due to specific inhibition by isoleucine of threonine deaminase, the first enzyme in the biosynthetic route to isoleucine. Further studies with purified threonine deaminase revealed that the substrate threonine and the inhibitor isoleucine bind to the enzyme at different, nonidentical sites; thus isoleucine is an example of a heterotropic allosteric inhibitor of threonine deaminase.

Another classic example of feedback inhibition comes from aspartate carbamoyltransferase. This enzyme catalyzes the formation of carbamoyl-aspartate from aspartate and carbamoylphosphate, which is the first step in the de novo biosynthesis of pyrimidines. The enzyme shows a sigmoidal dependence of reaction rate on the concentration of the substrate, aspartate, demonstrating cooperativity between the active sites of this oligomeric enzyme. Additionally, the enzyme is inhibited by the pyrimidine analogues cytidine, cytidine 5-phosphate, and cytosine triphosphate (CTP), and is activated by adenosine triphosphate (ATP). Structural studies also have revealed that the substrate binding sites and the CTP inhibitory binding sites are separate and distinct; binding of CTP at its exclusive site, however, influences the affinity of the active site for aspartate, via heterotropic allosteric regulation.

Threonine deaminase and aspartate carbamoyltransferase are examples of what is now known to be a ubiquitous means of metabolic control, namely, feedback inhibition. To understand fully this important biological control mechanism requires a theoretical framework for describing how distant sites within an enzyme can interact to affect one another's affinity for similar or different ligands. We now turn to a brief description of two such theoretical frameworks that have proved useful in the study of allosteric enzymes.

12.2 MODELS OF ALLOSTERIC BEHAVIOR

When an oligomeric enzyme contains multiple substrate binding sites that all behave identically (i.e., display the same K_m and k_{cat} for substrate) and independently, the velocity equation can be shown to be identical to that for a single active site enzyme (Segel, 1975). Regardless of whether the binding sites interact, an oligomeric enzyme will have different distributions of ligand occupancy among its different subunits at different levels of substrate saturation. For instance, six possible combinations of subunit occupancies can be envisaged for a tetrameric enzyme with two substrate molecules bound. This is illustrated for the example of a tetrameric enzyme in Figure 12.4.

When the ligand binding sites of an oligomeric enzyme interact cooperatively, we need to modify the existing kinetic equations to account for this intersite interaction. Two theoretical models have been put forth to explain allostery in enzymes and other ligand binding proteins. The first, the *simple sequential interaction* model, proposed by Koshland and coworkers (Koshland et al., 1966), is in some ways an extension of the induced-fit model introduced in Chapter 6. The second, the *concerted transition* or *symmetry* model, is based on the work of Monod, Wyman, and Changeux (1965) and has been widely applied to proteins such as hemoglobin, to explain ligand binding cooperativity.

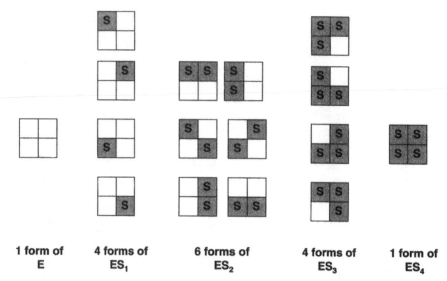

Figure 12.4 Schematic illustration of the number of possible forms of ligand binding to an enzyme that is a homotetramer: open squares, subunits with empty binding site; shaded squares with S, subunits to which a molecule of substrate has bound to the active site.

The simple sequential interaction model assumes that a large conformational change attends each ligand binding event at one of the enzyme active sites. It is this conformational transition that affects the affinity of the enzyme for the next ligand molecule. Let us consider the simplest case of an allosteric enzyme with two substrate binding sites that display positive cooperativity. The equilibria involved in substrate binding and their associated equilibrium constants are schematically illustrated in Figure 12.5 (Segel, 1975). The dissociation constant for the first substrate molecule is give by K_S. When one of the substrate binding sites is occupied, however, the dissociation constant for the second site is modified by the factor a, which for positive cooperativity has a value less than 1. The overall velocity equation for such an enzyme is given by:

$$v = \frac{V_{max}\left(\dfrac{[S]}{K_S} + \dfrac{[S]^2}{aK_S^2}\right)}{1 + \dfrac{2[S]}{K_S} + \dfrac{[S]^2}{aK_S^2}} \tag{12.1}$$

Now let us extend the model to a tetrameric enzyme (Figure 12.6). In this case the binding of the first substrate molecule modifies the dissociation constant of all three other binding sites by the factor a. If a second substrate molecule now binds, the two remaining vacant binding sites will have their dissociation constants modified further by the factor b, and their dissociation

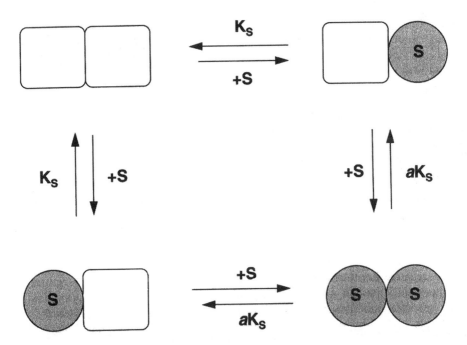

Figure 12.5 Schematic representation of the equilibria involved in substrate binding to a homodimeric enzyme where substrate binding is accompanied by a conformational transition of the subunit to which it binds, according to the model of Koshland et al. (1966).

constants will be abK_S. When a third substrate molecule binds, the final empty binding site will be modified still further by the factor c, and the dissociation constant will be $abcK_S$. Taking into account all these factors, and the occupancy weighing factors from Figure 12.4, we can write overall velocity of the enzymatic reaction as follows:

$$v = \frac{V_{max}\left(\dfrac{[S]}{K_S} + \dfrac{3[S]^2}{aK_S^2} + \dfrac{3[S]^3}{a^2bK_S^3} + \dfrac{[S]^4}{a^3b^2cK_S^4}\right)}{1 + \dfrac{4[S]}{K_S} + \dfrac{6[S]^2}{aK_S^2} + \dfrac{4[S]^3}{a^2bK_S^3} + \dfrac{[S]^4}{a^3b^2cK_S^4}} \tag{12.2}$$

Equation 12.2 is a velocity equation that can account for either positive or negative homotropic cooperativity, depending on the numerical values of the coefficients a, b, and c. In this model each binding event is associated with a separate conformational readjustment of the enzyme. Since, however, the effects are additive and progressive, once a single ligand has bound, the subsequent steps are strongly favored.

The second model for homotrophic cooperativity is the concerted transition or symmetry model, which is also known as the MWC model in honor of its

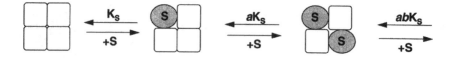

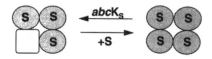

Figure 12.6 Extension of the Koshland model, from Figure 12.5, to a tetrameric enzyme.

original proponents: Monod, Wyman, and Changeux. This model assumes that allosteric enzymes are oligomers made up of identical minimal units (subunits or "protomers") arranged symmetrically with respect to one another and that each unit contains a single ligand binding site. The overall oligomer can exist in either of two conformational states, reflecting either a change in quaternary structure or tertiary structure changes within the individual protomer units, and these two conformations are in equilibrium. Another feature of the MWC model is that the transition between the two conformational states occurs with a retention of symmetry. For this to be so, all the protomer units must change in concert—one cannot have an oligomer in a mixed conformational state (i.e., some protomers in one conformation and some in the other).

Hence, in contrast to the Koshland model, in the MWC model the transition between the two conformational states is highly concerted, and there are no hybrid or intermediate states involved. The affinity of the ligand binding site on a protomer depends on the conformational state of that protomer unit. In other words, the ligand of interest will bind preferentially to one of the two conformational states of the protomer. Thus, binding of a ligand to one binding site will shift the equilibrium between the conformational states in favor of the preferred ligand binding conformation. Since the protomeric units of the oligomer shift conformation in concert, ligand binding to one site has the effect of switching all the ligand binding sites to the higher affinity form. Thus the MWC model explains strong positive cooperativity in terms of the observation that occupancy at a single ligand binding site induces all the other binding sites of the oligomeric protein to take on their high affinity conformation.

The original MWC model assumes that the conformational state with low ligand affinity is a strained structure and that the strain is relieved by ligand binding and the associated conformational transition. For this reason, the state of low binding affinity is often referred to as the "T" state (for tense), and the high affinity conformation is referred to as the "R" state (for relaxed). While the transitions between these states are concerted, as described, for bookkeeping purposes diagrams of the MWC model designate different ligand occupancy states of the two conformations as R_x and T_x, where x indicates the number of ligand bound to the oligomer. Therefore, a tetrameric enzyme could in principle occur in states R_0 through R_4, and T_0 through T_4. The states R_0 and T_0 thus refer to the two conformational states with no ligands bound to the enzyme. The equilibrium constant between these two "empty" states, the *allosteric constant*, is symbolized by L:

$$L = \frac{[T_0]}{[R_0]} \tag{12.3}$$

This dissociation constant of a binding site for ligand, S, on a protomer in the T state is termed K_{ST}, and for the protomer in the R state this dissociation constant is K_{SR}. The ratio K_{SR}/K_{ST} is referred to as the *nonexclusive binding coefficient* and is symbolized by c. Both L and c influence the degree of cooperativity displayed by the enzyme. As L becomes larger, the velocity curve for the enzymatic reaction displays greater sigmoidal character, because the R_0–T_0 equilibrium favors the T_0 state more. As c increases, the affinity of the T state for ligand increases relative to the R state. Hence, high cooperativity is associated with small values of c.

The simplest example of the MWC model is that for a dimeric enzyme in which the T state is assumed to have no affinity at all for the substrate (i.e., $c = 0$). Figure 12.7 schematically represents the equilibria involved in such a system, where substrate binds exclusively to the R state of the dimer. (Since only the R state has a noninfinite substrate dissociation constant here, we shall use the symbol K_S in place of K_{SR} for this system.) The velocity equation for such a system, which can be derived from a rapid equilibrium set of assumptions, yields the following functional form:

$$v = \frac{V_{max} \dfrac{[S]}{K_S} \left(1 + \dfrac{[S]}{K_S}\right)}{L + \left(1 + \dfrac{[S]}{K_S}\right)^2} \tag{12.4}$$

For oligomers larger than dimers, a generalized form of Equation 12.4 can be derived (Segel, 1975):

$$v = \frac{V_{max} \dfrac{[S]}{K_S} \left(1 + \dfrac{[S]}{K_S}\right)^{h-1}}{L + \left(1 + \dfrac{[S]}{K_S}\right)^{h}} \tag{12.5}$$

where h is the total number of ligand binding sites on the oligomeric enzyme.

The MWC model provides a useful framework for understanding positive homotropic cooperativity, and it can be modified to account for heterotropic cooperativity as well (Segel, 1975; Perutz, 1990). This model cannot, however,

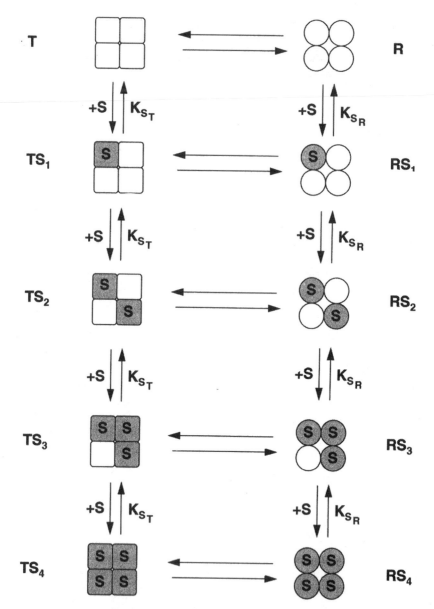

Figure 12.7 Schematic representation of the equilibria involved in the binding of substrates to a tetrameric enzyme according to the model of Monod, Wyman, and Changeux (1965). See text for further details.

account for the phenomenon of negative homotrophic cooperativity. When negative cooperativity is encountered, it is usually explained in terms of the Koshland sequential interaction model.

12.3 EFFECTS OF COOPERATIVITY ON VELOCITY CURVES

Referring back to the Koshland simple sequential interaction model, we can state that if the cooperativity is large, the concentrations of species with at least one substrate binding site unsaturated will be very small at any concentration of substrate greater than K_S. In the case of a tetrameric enzyme, for example, under these conditions Equation 12.2 reduces to the much simpler equation:

$$v = \frac{V_{max}[S]^4}{K' + [S]^4} \tag{12.6}$$

where $K' = a^3 b^2 c K_S^4$. Equation 12.6 is a specific case (i.e., for tetrameric enzymes) of the more general simple equation:

$$v = \frac{V_{max}[S]^h}{K' + [S]^h} \tag{12.7}$$

in which h is the total number of substrate binding sites on the oligomeric enzyme molecule and K' is a constant that relates to the individual interaction coefficients a through h, and the intrinsic dissociation constant K_S. Note that in the absence of cooperativity, and when $h = 1$, Equation 12.6 reduces to an equation reminiscent of the Henri–Michaelis–Menten equation from Chapter 5. When cooperativity occurs, however, the constant K' no longer relates to the concentration of substrate required for the attainment of half maximal velocity.

Equation 12.7 is known as the Hill equation, and the coefficient h is referred to as the Hill constant. This simple form of this equation can be readily used to fit experimental data to enzyme velocity curves, as introduced in Chapter 5 (see Figure 5.15). When the degree of cooperativity is moderate, however, contributions from intermediate occupancy species (i.e., number of bound substrate molecules $< h$) may contribute to the overall reaction. In these cases, the experimental data are often still well modeled by Equation 12.7, although the empirically determined value of h will no longer reflect the total number of binding sites on the enzyme, and may not in fact be an integer. In this situation the experimentally determined coefficient is referred to as an *apparent h* value (sometimes given the symbol h_H). The next integer value above the apparent h value is considered to represent the *minimum* number of binding sites on the oligomeric enzyme.

For example, suppose that the experimentally determined value of h is 1.65. This could be viewed as representing and enzyme with 1.65 highly cooperative substrate binding sites, but of course this makes no physical sense. Instead we might say that the minimum number of binding sites on this enzyme is 2 and that the sites display a more moderate level of cooperativity. However, there is no compelling evidence from this experiment that the enzyme has only two binding sites. It could have three or four or more binding sites with weaker intersite cooperativity. This is why the value of 2 in this example is said to represent the minimum number of possible binding sites.

As we saw in Chapter 5, the Hill equation can be linearized by taking the logarithm of both sides and rearranging to yield:

$$\log\left(\frac{v}{V_{max} - v}\right) = h \log[S] - \log(K') \tag{12.8}$$

This equation can be used to construct linearized plots from which the values of h and K' can be determined graphically. An example of a linearized Hill plot was given in Chapter 5 (Figure 5.16). Despite the form of Equation 12.8, the experimental graphs usually deviate from linearity in the low substrate region, where species with fewer than h substrate molecules bound can contribute to the overall velocity. Typically, the data conform well to a linear function between values of [S] yielding 10–90% saturation (i.e., V_{max}). The slope of the best fit line between these limits is commonly taken as the average value of h_H.

The degree of sigmoidicity of the direct velocity plot is a measure of the strength of cooperativity between sites in an oligomeric enzyme. This is best measured by taking the ratio of substrate concentrations required to reach two velocities representing different fractions of V_{max}. Most commonly this is done using the substrate concentrations for which $v = 0.9V_{max}$, known as $[S]_{0.9}$, and for which $v = 0.1V_{max}$, known as $[S]_{0.1}$. The ratio $[S]_{0.9}/[S]_{0.1}$, the *cooperativity index*, is an inverse measure of cooperatiave interactions; in other words, the larger the difference in substrate concentration required to span the range of $v = 0.1V_{max}$ to $v = 0.9V_{max}$, the larger the value of $[S]_{0.9}/[S]_{0.1}$ and the weaker the degree of cooperativity between sites. The value of the cooperativity index is related to the Hill coefficient h, and K' as follows:

When $v = 0.9V_{max}$

$$v = 0.9V_{max} = \frac{V_{max}[S]_{0.9}^h}{K' + [S]_{0.9}^h}$$

$$\therefore [S]_{0.9} = (9K')^{1/h} \tag{12.9}$$

and when $v = 0.1V_{max}$,

$$v = 0.1V_{max} = \frac{V_{max}[S]_{0.1}^h}{K' + [S]_{0.1}^h}$$

$$\therefore [S]_{0.1} = \left(\frac{K}{9}\right)^{1/h} \tag{12.10}$$

Therefore:

$$\frac{[S]_{0.9}}{[S]_{0.1}} = \frac{(9K')^{1/h}}{(K'/9)^{1/h}} = (81)^{1/h} \tag{12.11}$$

or

$$h = \frac{\log(81)}{\log\left(\dfrac{[S]_{0.9}}{[S]_{0.1}}\right)} \tag{12.12}$$

Thus the Hill coefficient and the cooperativity index for an oligomeric enzyme can be related to each other, and together they provide a measure of the degree of cooperativity between binding sites on the enzyme and the minimum number of these interacting sites.

While the Hill coefficient is a convenient and commonly used index of cooperativity, it is not a direct measure of the change in free energy of binding ($\Delta\Delta G$) that must exist in cooperative systems. A thermodynamic treatment of cooperativity for a two-site system presented by Forsen and Linse (1995) discusses the changes in binding affinities in terms of changes in binding free energies. This alternative treatment offers a straightforward means of describing the phenomenon of cooperativity in more familiar thermodynamic terms. It is well worth reading.

Another useful method for diagnosing the presence of cooperativity in enzyme kinetics is to plot the velocity curves in semilog form (velocity as a function of log[S]), as presented in Chapter 8 for dose–response plots of enzyme inhibitors. Such plots always yield a sigmoidal curve, regardless of whether cooperativity is involved. The steepness of the curve, however, is related to the degree of positive or negative cooperativity. When the enzyme displays positive cooperativity, the curves reach saturation with a much steeper slope than in the absence of cooperativity. Likewise, when negative cooperativity is in place, the saturation curve displays a much shallower slope (Neet, 1980). The data in these semilog plots is still well described by Equation 12.7, as illustrated in Figure 12.8 for examples of positive cooperativity. The steepness of the curves in these plots is directly related to the value of h that appears in Equation 12.7.

These plots are useful because the presence of cooperativity is very readily apparent in these plots. The effect of positive or negative cooperativity on the steepness of the curves is much more clearly observed in the semilog plot as opposed to the linear plot, especially in the case of small degrees of cooperativity. The steepness of the curves in such semilog plots is also diagnostic of cooperative effects in ligands other than substrate. Thus, for example, the IC_{50} equation introduced in Chapter 8 (Equation 8.20) can be modified to include a term to account for cooperative effects in inhibitor binding to enzymes as well.

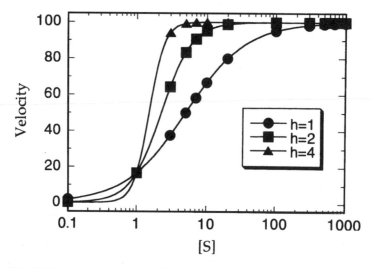

Figure 12.8 Velocity as a function of substrate concentration plotted in semilog fashion: data for a noncooperative enzyme; squares and triangles, data for enzymes displaying positive cooperativity. Each solid line through the data represents the best fit of an individual data set to Equation 12.7.

12.4   SIGMOIDAL KINETICS FOR NONALLOSTERIC ENZYMES

The appearance of sigmoidal kinetics in enzyme velocity curves for allosteric enzymes is a reflection of the cooperativity of the substrate binding events that precede the catalytic steps at the enzyme active sites. The same cooperativity should be realized in direct studies of ligand binding by the enzyme, which can be performed by equilibrium dialysis, certain spectroscopic methods, and so on (Chapter 4). If true allostery is involved, the cooperativity of ligand binding should be measurable in the enzyme velocity curves and in the separate binding experiments as well. In some cases, however, the direct ligand binding experiments fail to display the same cooperativity observed in the velocity measurements. One must assume that such ligand binding events are not cooperative, which means that some other explanation must be sought to account for the sigmoidal velocity curve.

One way of observing sigmoidal kinetics in the absence of true cooperativity entails an enzyme preparation containing a mixture of enzyme isoforms that have different K_m values for the substrate (Palmer, 1985). In such cases the velocity curve will be the superposition of the individual curves for the varied isoforms. If two or more isoforms differ significantly in K_m for the substrate, a nonhyperbolic curve, resembling the sigmoidal behavior of cooperative enzyme, may result.

Also, it has been noted that a two-substrate enzyme that follows a random ordered mechanism can display sigmoidal kinetics without true cooperativity.

This occurs when one of the two ordered reactions proceeds faster than the competing ordered reaction: when, for example, formation of $E \cdot AX$ then $E \cdot AX \cdot B$ and subsequent product release is faster than formation of $E \cdot B$ then $E \cdot B \cdot AX$ and product release. In the case of two ordered reactions of unequal speed, the affinity of the free enzyme for substrate B is less than the affinity of the $E \cdot AX$ complex for B. If $[E_t]$ and $[B]$ are held constant while $[AX]$ is varied at low concentrations of AX the enzyme will react mainly with substrate B first, and thus will proceed through the slower of the two pathways to product. As the concentration of AX increases, there will be a greater probability of the enzyme first binding this substrate and proceeding via the faster pathway. The observed result of this pathway "switching" with increasing substrate concentration is a sigmoidal plot of velocity as a function of $[AX]$.

Finally, sigmoidal kinetics can be observed even for a monomeric single binding site enzyme if substrate binding induces a catalytically required conformational transiton of the enzyme. If the isomerization step after substrate binding is rate limiting, the relative populations of the two isomers, E and E′, can influence the overall reaction velocity. If the equilibrium between E and E′ is perturbed by substrate, the relative populations of these two forms of the enzyme will vary with increasing substrate concentration. Again, the end result is the appearance of a sigmoidal curve when velocity is plotted as a function of substrate concentration.

12.5 SUMMARY

In this chapter we presented the concept of cooperative interactions between distal binding sites on oligomeric enzymes, which communicate through conformational transitions of the polypeptide chain. These allosteric enzymes display deviations from the normal Henri–Michaelis–Menten behavior that is seen with single substrate binding enzymes, as introduced in Chapter 5. Examples of allosteric proteins and enzymes were described that provide some structural rationale for allosteric interactions in specific cases, and two theoretical models of cooperativity were described. The classic signature of cooperativity in enzyme kinetics is a sigmoidal shape to the curve of velocity versus [S]. The appearance of such sigmoidicity in the enzyme kinetics is not sufficient, however, to permit us to conclude that the substrate binding sites interact cooperatively. Direct measurements of ligand binding must be used to confirm the cooperativity of ligand binding. We saw that in some cases sigmoidal enzyme kinetics exist in the absence of true cooperativity—when, for example, multisubstrate enzymes proceed by different rates depending on the order of substrate addition, and when rate-limiting enzyme isomerization occurs after substrate binding.

The understanding of allostery and cooperativity in structural terms is an active area of research today. This fascinating subject was reviewed by one of the leading experts in the field of allostery, Max Perutz, who spent most of his career studying the structural determinants of cooperativity in hemoglobin. The text by Perutz (1990) is highly recommended for those interested in delving deeper into this subject.

REFERENCES AND FURTHER READING

Abelson, P. H. (1954) *J. Biol. Chem.* **206**, 335.

Forsen, S., and Linse, S. (1995) *Trends Biochem. Sci.* **20**, 495.

Koshland, D. E., Nemethy, G., and Filmer, D. (1966) *Biochemistry*, **5**, 365.

Monod, J., Wyman, J., and Changeux, J. P. (1965) *J. Mol. Biol.* **12**, 88.

Neet, K. E. (1980) *Methods Enzymol.* **64**, 139.

Palmer, T. (1985) *Understanding Enzymes*, Wiley, New York, pp. 257–274.

Perutz, M. (1990) *Mechanisms of Cooperativity and Allosteric Regulation in Proteins*, Cambridge University Press, New York.

Segel, I. H. (1975) *Enzyme Kinetics*, Wiley, New York, pp. 346–464.

APPENDIX I

SUPPLIERS OF REAGENTS AND EQUIPMENT FOR ENZYME STUDIES

Some of the commercial suppliers of reagents and equipment that are useful for enzyme studies are given here. A more comprehensive listing can be found in the ACS Biotech Buyers Guide, which is published annually. The Buyers Guide can be obtained from the American Chemical Society, 1155 16th Street N.W., Washington, DC 20036. Telephone (202) 872-4600.

Aldrich Chemical Company, Inc.
940 West Saint Paul Avenue
Milwaukee, WI 53233
(800) 558-9160

Amersham Corporation
2636 South Clearbrook Drive
Arlington Heights, IL 60005
(800) 323-9750

Amicon
24 Cherry Hill Drive
Danvers, MA 01923
(800) 343-1397

Bachem Bioscience, Inc.
3700 Horizon Drive
King of Prussia, PA 19406
(800) 634-3183

Beckman Instruments, Inc.
P.O. Box 3100
Fullerton, CA 92634-3100
(800) 742-2345

BioRad Laboratories
1414 Harbour Way South
Richmond, CA 94804
(800) 426-6723

Biozymes Laboratories International
  Limited
9939 Hilbert Street, Suite 101
San Diego, CA 92131-1029
(800) 423-8199

Boehringer-Mannheim Corporation
Biochemical Products
9115 Hague Road
P.O. Box 50414
Indianapolis, IN 46250-0414
(800) 262-1640

Calbiochem
P.O. Box 12087
San Diego, CA 92112
(800) 854-9256

Eastman Kodak Company
343 State Street
Building 701
Rochester, NY 14650
(800) 225-5352

Enzyme Systems Products
486 Lindbergh Ave.
Livermore, CA 94550
(888) 449-2664

Hampton Research
27632 El Lazo Rd
Suite 100
Laguna Niguel, CA 92677-3913
(800) 452-3899

Hoefer Scientific Instruments
P.O. Box 77387
654 Minnesota Street
San Francisco, CA 94107
(800) 227-4750

Millipore Corporation
80 Ashby Road
Bedford, MA 01730
(800) 225-1380

Novex, Inc.
4202 Sorrento Valley Boulevard
San Diego, CA 92121
(800) 456-6839

Pharmacia LKB Biotechnology AB
800 Centennial Avenue
Piscataway, NJ 08854
(800) 526-3618

Pierce Chemical Company
P.O. Box 117
Rockford, IL 61105
(800) 874-3723

Schleicher & Schuell, Inc.
10 Optical Avenue
Keene, NH 03431
(800) 245-4024

Sigma Chemical Company
P.O. Box 14508
St. Louis, MO 63178
(800) 325-3010

Spectrum Medical Industries, Inc.
1100 Rankin Road
Houston, TX 77073-4716
(800) 634-3300

United States Biochemical
  Corporation
P.O. Box 22400
Cleveland, OH 44122
(800) 321-9322

Upstate Biotechnology, Inc.
199 Saranac Avenue
Lake Placid, NY 12946
(800) 233-3991

Worthington Biochemical
  Corporation
Halls Mill Road
Freehold, NJ 07728
(800) 445-9603

APPENDIX II

USEFUL COMPUTER SOFTWARE AND WEB SITES FOR ENZYME STUDIES

There is available a large and growing number of commercial software packages that are useful for enzyme kinetic data analysis. Also, several authors have published the source code for computer programs they have written specifically for enzyme kinetic analysis and other aspects of enzymology. I have listed some of the programs I have found useful in the analysis of enzyme data, together with the source of further information about them. This list is by no means comprehensive, but rather gives a sampling of what is available. The material is provided for the convenience of the reader; I make no claims as to the quality or accuracy of the programs.

COMPUTER SOFTWARE

Cleland's Package of Kinetic Analysis Programs. This is a suite of FOR-TRAN programs written and distributed by the famous enzymologist W. W. Cleland. The programs include methods for simultaneous analysis of multiple data for determination of inhibitor type and relevant kinetic constants, as well as statistical analyses of one's data. Reference: W. W. Cleland, *Methods Enzymol.* **63**, 103 (1979).

Enzfitter. A commercial package for data management and graphic displays of enzyme kinetic data. [See *Ultrafit* for similar version, compatible with Macintosh hardware.] Available from Biosoft, P.O. Box 10938, Ferguson, MO. Telephone: 314-524-8029. E-mail: *ab47@cityscape.co.uk.*

Enzyme Kinetics. A commercial package for data management and graphic displays of enzyme kinetic data. Distributed by ACS Software, Distribution Office, P.O. Box 57136, West End Station, Washington, DC 20037. Telephone: 800-227-5558.

EZ-FIT. A practical curve-fitting program for the analysis of enzyme kinetic data. Reference: F. W. Perrella, *Anal. Biochem.* **174**, 437 (1988).

Graphfit. A commercial package for data management and graphic display of enzyme kinetic data and other scientific data graphing. This program has extensive preprogrammed routines for enzyme kinetic analysis and allows global fitting of data of the form $y = f(x, z)$, which is very useful for analysis of inhibitor modality, and so on. Available from Erithacus Software Limited, P.O. Box 35, Staines, Middlesex, TW18 2TG, United Kingdom [*http://www.erithacus.com*]. Also distributed by Sigma Chemical Company.

Graphpad Prism. A general graphic package written for scientific applications. Contains specific equations and routines for enzyme kinetics and equilibrium ligand binding applications. The user's guide and related Web site are quite informative. Available from Intuitive Software for Science, 10855 Sorrento Valley Road, Suite 203, San Diego, CA 92121. Telephone: 858-457-3909. E-mail: *support@graphpad.com*.

Kaleidagraph. A commercial software package for general scientific graphing. Available from Synergy Software, 2457 Perkiomen Ave., Reading, PA 19606. Telephone: 610-779-0522. [*http://www.synergy.com*].

K·cat. A commercial package for data management and graphic displays of enzyme kinetic and receptor ligand binding data. Available from Bio-Metallics, Inc., P.O. Box 2251, Princeton, NJ 08543. Telephone: 800-999-1961.

Kinlsq. A program for fitting kinetic data with numerically integrated rate equations. Provides data analysis routines for tight binding inhibitors as well as classical inhibitors. Reference: W. G. Gutheil, C. A. Kettner, and W. W. Bachovchin, *Anal. Biochem.* **223**, 13 (1994).

Kinsim. A very useful program that allows the researcher to enter a chemical mechanism for a reaction in symbolic terms and have the computer translate this into a set of differential equations that can be solved to predict the concentrations of products and reactants as functions of time, based on the kinetic scheme and the values of the rate constants used. Reference: B. A. Barshop et al., *Anal. Biochem.* **130**, 134 (1983). This program has been greatly expanded and is available free from KinTek Corporation under the name KinTekSim. For instructions on downloading this freeware see the KinTeck Corporation Web site [*http://www.kintek-corp.com*].

MPA. A program for analyzing enzyme rate data obtained from a microplate reader. Provides a convenient means of transforming and analyzing data directly from 96-well formated data arrays. Reference: S. P. J. Brooks, *BioTechniques*, **17**, 1154 (1994).

Origin. A very robust commercial graphic program that allows fitting of data to equations of the form $y = f(x, z)$ and three-dimensional displays of the resulting fits. Available from Microcal Software, Inc., One Round-house Plaza, Northampton, MA 01060. Telephone: 800-969-7720. [*http:// www.microcal.com*].

Ultrafit. Similar to Enzfitter software, but designed for Apple Macintosh computers. Available from Biosoft, P.O. Box 10938, Ferguson, MO. Telephone: 314-524-8029. E-mail: *ab47@cityscape.co.uk.*

WEB SITES

Another useful source of programs and information about enzymes and protein biochemistry in general is the Internet. A recent search of the term "Enzyme" yielded over 600 Internet addresses with useful information and analysis packages that the reader can assess. It is well worth the reader's time to do a little "surfing" on this exciting medium. Rather than attempting a comprehensive listing of these many useful sites, I provide four particularly good starting points for exploring the Internet. Each of these sites contains a wealth of useful information to the biochemist and further provide links to additional Web sites that are of interest.

Data/Information for Enzymologists and Kineticists

http://www.med.umich.edu/biochem/enzresources/realenzymes.html

This site is part of an online contents page for an enzymology course taught at the University of Michigan. It contains a collection of links to useful Web sites of interest to enzymologists.

The Enzyme Data Bank

http://192.239.77.6/Dan/proteins/ec-enzymes.html

This site provides information on EC numbers, recommended and alternative names, catalytic activities, cofactor utilization, disease association, and other useful facts about enzymes. It is an excellent starting point for gaining information on a particular enzyme of interest.

ExPASy Molecular Biology Server

http://www.expasy.ch

This is the Web site for the molecular biology server of the Swiss Institute of Bioinformatics. It contains a number of useful databases and protein analysis tools. It also provides direct links to other molecular biology and biochemistry Web sites.

Pedro's BioMolecular Research Tools

http://www.public.iastate.edu/ ~ pedro/research_tools.htm

This is an excellent starting point for exploring useful Web sites. "Pedro" has compiled an extensive collection of Web site links to information and services for molecular biology and biochemistry. The site is subdivided into five major parts: Part 1, Molecular Biology Search and Analysis; Part 2, Bibliographic Text and WWW Searchers; Part 3, Guides, Tutorials, and Help Tools; Part 4, Journals and Newsletters; and Part 5, Extras (a list of biological resources).

INDEX